Pyrolysis Oils from Biomass

ACS SYMPOSIUM SERIES 376

Pyrolysis Oils from Biomass

Producing, Analyzing, and Upgrading

Ed J. Soltes, EDITOR
Texas A&M University

Thomas A. Milne, EDITOR
Solar Energy Research Institute

Developed from a symposium sponsored by
the Cellulose, Paper, and Textile Division
and the Division of Fuel Chemistry
at the 193rd Meeting
of the American Chemical Society,
Denver, Colorado,
April 5-10, 1987

American Chemical Society, Washington, DC 1988

Library of Congress Cataloging-in-Publication Data

Pyrolysis oils from biomass producing, analyzing, and upgrading.

(ACS Symposium Series; 376).

Developed from a symposium sponsored by the Cellulose, Paper, and Textile Division and the Division of Fuel Chemistry at the 193rd Meeting of the American Chemical Society, Denver, Colorado, April 5–10, 1987.

Includes index.

Bibliography: p.

1. Biomass energy—Congresses. 2. Pyrolysis—Congresses. 3. Hydrocarbons—Congresses.

I. Milne, Thomas A. II. American Chemical Society. Cellulose, Paper, and Textile Division. III. American Chemical Society. Division of Fuel Chemistry. IV. Title. V. Series.

TP360.P95 1988 662′.8 88–24172
ISBN 0–8412–1536–7

PRINTED IN THE UNITED STATES OF AMERICA

ACS Symposium Series

M. Joan Comstock, ***Series Editor***

1988 ACS Books Advisory Board

Foreword

The ACS SYMPOSIUM SERIES was founded in 1974 to provide a medium for publishing symposia quickly in book form. The format of the Series parallels that of the continuing ADVANCES IN CHEMISTRY SERIES except that, in order to save time, the papers are not typeset but are reproduced as they are submitted by the authors in camera-ready form. Papers are reviewed under the supervision of the Editors with the assistance of the Series Advisory Board and are selected to maintain the integrity of the symposia; however, verbatim reproductions of previously published papers are not accepted. Both reviews and reports of research are acceptable, because symposia may embrace both types of presentation.

Contents

INDEXES

Preface

PYROLYSIS OF BIOMASS HAS undergone a technological renaissance in the past decade, thanks to new concepts and processes for high oil yields, new analytical techniques to decipher the mechanisms of formation and constitution of the oil, and catalytic approaches to upgrading the oil fuel used for transportation. Governments have given priority to pyrolysis and direct liquefaction, following bench and pilot success in gasification and indirect liquefaction, to offer industry concepts for renewable liquid fuels, particularly premium fuels. Following the Workshop[1] on Fast Pyrolysis in 1980, and a series of liquefaction workshops in Canada[2] in 1980, 1982, and 1983, it seemed timely to convene scientists and engineers from North America and elsewhere to report on the present progress in integrated pyrolysis oil production, characterization, and upgrading studies.

As the book name implies, we attempted to bring together specialists to present the state of the science in the complete fuel cycle, from feedstock to upgraded liquid fuels suitable as replacements for petroleum-derived fuels. The introductory chapter contains a discussion of biomass pyrolysis and its place in a renewable fuel economy. Contributions to this book were received from five countries, a fact indicating the widespread interest in this conversion option for biomass, and permitting the enhancement and establishment of collaborative efforts.

At the time this preface was written, the U.S. newspapers reported a two-year low in crude oil prices ($15/Bbl), fleet auto efficiency standards were being relaxed, and oil imports were rising steadily. At the same time, through perhaps a short-term statistical fluctuation in rainfall, the public was finally becoming aware of the implication of the real, long-term rise in CO_2 levels from deforestation and the burning of fossil fuels. Perhaps the promise of a renewable source of liquid transportation fuels, coupled with extensive reforestation, as mitigators of the rise in CO_2 levels, will spur progress in the conversion of CO_2, through photosynthesis, to plant materials. Such plant materials could provide facile and economic routes to substitutes for fossil-based fuels, chemicals, and biobased materials.

---

1. Diebold, J. P., Editor. *Proceedings of the Specialists Workshop on the Fast Pyrolysis of Biomass.* Copper Mountain, Colorado, Oct. 20–22 SERI/CP-622-1096.

2. ENFOR Program and National Research Council of Canada. Biomass Liquefaction Specialists' Review Meetings in Toronto, Aug. 13, 1980; Saskatoon, Feb. 16–17, 1982; and Sherbrooke, Sept. 29–30, 1983.

We thank the sponsoring divisions and all contributors to this volume and hope that their results will foster both interest in, and support of, continuing work on the fascinating chemistry and engineering of biomass pyrolysis, oil characterization, and upgrading for many uses.

ED J. SOLTES
Texas A&M University
College Station, TX 77843

THOMAS A. MILNE
Solar Energy Research Institute
Golden, CO 80401

July 5, 1988

Chapter 1

Of Biomass, Pyrolysis, and Liquids Therefrom

Ed J. Soltes

Department of Forest Science, Texas Agricultural Experiment Station, Texas A&M University System, College Station, TX 77843–2135

Renewable biomass (harvest and process residues in forestry and agricultural operations, specific terrestrial or aquatic crops grown for fuel, often animal wastes, refuse derived fiber, etc.) represents an important energy resource in the United States, with future potentials especially important as fossil fuels are depleted (1). Biomass is however generally poorly suited for direct energy use, with pretreatments necessary for altering physical and chemical form. Moisture content can be in excess of 50%, so costs of collection, transportation, and energy conversion become relatively inefficient. Further, biomass availability (e.g., from row crops) is sometimes seasonal, so storage is necessary. As biomass can spoil, it has to be covered and processed soon after receipt. Fortunately, thermal conversion processes are somewhat insensitive to type, form and shape, and can convert biomass into stable, storable and transportable energy forms, and in physical or chemical forms that can be used in higher efficiency energy conversion processes developed for liquid petroleum, coal and natural gas. Under pyrolysis process conditions, a liquid tar (pyrolysis oil) can be produced which can be upgraded to liquid engine fuels (2).

Why liquid? It can be argued that the principal advantage of petroleum is that it is a liquid (3): liquid fuels, besides being energy dense, are especially easy to store, transport and meter, thus being really the only choice for transportation fuels. Biomass conversion processes (specifically pyrolysis) which have potential for producing liquid fuels, especially liquid fuels that can be direct replacements for gasoline or diesel engine fuels, are then of much interest.

The Nature of Biomass Pyrolysis

Biomass thermochemical processes have been studied for at least two reasons: (1) a better understanding of the combustion process to control biomass flammability; and, (2) research into improved processes for converting biomass into useful energy forms. The late Fred Shafizadeh (see, e.g., 4,5) laid the groundwork for all recent studies in both arenas. The work on combustion mechanisms continues at the Wood Chemistry Laboratory of the University of Montana in Missoula (see, e.g., 6). Antal has recently (7,8) reviewed all aspects of biomass pyrolysis, and the reader is directed to his reviews for detail study of the processes involved. Chatterjee has produced an excellent summary (9) of biomass pyrolysis

0097–6156/88/0376–0001$06.00/0

processes relative to application in lesser developed countries. This author has also written several reviews on biomass thermochemical processes, and specifically pyrolysis (2,3,13). Several volumes have been published recently on thermal conversion (10,11), with another on the way for the 1988 Phoenix conference (12).

The opening paragraph to this overview chapter relates the need for pyrolysis relative to liquid transportation fuels production. This is a rather specific energy need, and in fact, biomass can serve other energy needs as well. Charcoal from wood is the principal fuel of rural regions of most lesser developed countries. The gas from biomass gasification has found use in the retrofit of natural gas furnaces and engines, and highly efficient cogeneration plants. The energy content of residues from forestry and agricultural operations can often serve on-site process energy needs, but the biomass materials are generally poorly suited for direct use. Modern agricultural and process machinery seldom use solid fuels, therefore pretreatments are often required to change their chemical or physical nature to liquids or gases. Similar considerations apply to the conversion of biomass crops and to the use of biomass fuels off-farm. Thermochemical processes for biomass are basically pretreatment processes, processes that alter the chemical and physical nature of biomass to permit higher efficiency use.

Thermochemical processes are often characterized as combustion, gasification, pyrolysis, carbonization or tarification processes. Except for pyrolysis, these names reflect types of products produced. Thermochemical processing is variable and flexible. Depending on conditions used, (primarily the temperature, the oxygen-to-fuel ratio, and residence time at temperature), biomass can be altered very slightly, or be completely changed. These three variables define conditions for pyrolysis, gasification and combustion but there is often little distinction between these processes. In fact, there is a continuum of process conditions. Selection of treatment conditions permits a variety of outcomes of importance to the production of bioenergy products. The mechanisms for the formation of these products are indeed complex, and are still unfolding (see e.g., 7,8,14). Knowledge of these mechanisms has permitted the identification of conditions under which it is possible to produce tars in good yield from large, moist wood particles (15), or novel reactor design for the production and capture of desired tar products (16). Still, gases, liquids and solids are always produced, with relative yields and chemical or elemental compositions dependent on process variables.

Definition of Pyrolysis. The word pyrolysis has had some problems in definition, especially when applied to biomass. The older literature generally equates pyrolysis to carbonization, in which the principal product is a solid char. Today, the term pyrolysis is generally used to describe processes in which liquid oils are preferred products. This symposium was concerned with the latter pyrolysis - processes which offer enhanced yields of liquid oils, especially those with desirable chemical compositions and physical attributes for liquid fuels, fuel supplements and chemical feedstocks.

Definition of Pyrolysis Oil vs. Tar. Throughout this volume, "pyrolysis oil", "tar" and "pyrolytic tar" are used almost interchangeably. Tar or pyrolytic tar is a more generic term, now becoming used in reference to its formation in a secondary sense, e.g., as undesired in gasification, or the "creosote tar" of incomplete combustion or the usually wasted tar byproducts of charcoal manufacture. Oil or pyrolysis oil refers specifically to the liquid product of biomass pyrolysis when the oil is the principal product of interest. Thus, not only "pyrolysis" but also "tar" and "oil" definitions are being cast while research is being conducted in this area. To add to this confusion, this author more than once tried (in retrospect fortunately unsuccessful) to coin the term "tarification" as any pyrolytic process where the product of interest was the liquid tar product!

Other Feedstocks and Processes. Although most of the papers in the symposium deal with thermochemical conversion of wood in what may be called the conventional pyrolysis process mode (heating in the absence of air, or under air-starve conditions), other papers were invited which cover the conversion of other biomass materials, such as from municipal solid waste (17) or black liquor (18), or conversion under more novel processing conditions. Mention is made of liquefaction relative to pyrolysis. Although there has not been an attempt to define liquefaction in the literature, to the author's knowledge, it usually refers to the one-step high liquid yield process incorporating heat in combination with reducing gases and/or catalysts (19), but may now also include water-based processes (20). Liquids are also produced in acid solution (21), alkaline solution (22) and in solvolysis (23).

Characterization of Biomass Pyrolysis Oils

The chemical compositions of pyrolysis oils are very complex. High temperature reactions in the absence of oxygen or under air-starve conditions are not specific, and the availability of sufficient energy for alternative pathways results in a series of complex concurrent and consecutive reactions which provide a wide spectrum of pyrolytic products, usually in small yields (3,14).

Most of the literature in the past, including that by the author (13), reflected composition of tars produced as a byproduct of charcoal manufacture. These tars, from different biomass materials, exhibit like chemical compositions. The chars produced under similar carbonization conditions also exhibit similar physical and chemical properties (24). During the last decade or so, pyrolysis process research has confirmed that both the yield and chemical composition of pyrolysis oils are very dependent on reaction conditions (see e.g., 14-16,25,26). All biomass oils are not the same: oil formation conditions determine oil composition. Similarly, it has been reported that different biomass feedstocks pyrolyzed under similar process conditions can give oil products with similarities in chemical composition (e.g., 27-30).

Characterization Tools for Pyrolysis Oils. It wasn't too many years ago that the only tools available to the scientist interested in pyrolysis oil composition were gas chromatography and thermogravi-metric analysis. The complexity of the pyrolysis oils demands high performance equipment, and a list of such equipment mentioned during the symposium would include proton and carbon nuclear magnetic resonance spectroscopy, free-jet molecular beam/mass spectrometry (16,25), diffuse reflectance Fourier transform infrared spectrometry (31), photoelectron spectroscopy (31), as well as procedures such as computerized multivariate analysis methods (32) - truly a display of the some of the most sophisticated analytical tools known to man, and a reflection of the difficulty of the oil composition problem.

Not all scientists have access to this type of equipment, and gas chromatography (GC) is still used extensively, especially in capillary mode (see, e.g., 26,27). The higher resolution of capillary GC is preferred over column chromatography because of the many oxygenated species of similar polarity in pyrolysis oils. Much (usually most) of the pyrolysis oil is also non-volatile, and this causes additional problems for capillary columns: non-volatiles can cause column deterioration of the stationary phase. It is usually with much trepidation that the average scientist injects for the first time a black, viscous, mostly non-volatile pyrolysis oil into his pristine capillary column (especially when coupled to someone else's mass spectrometer), but be assured that it works, and that the column can live a long and useful life after many such experiences! Column deterioration can be minimized by periodic cleaning or rejuvenation of the column, as well as by occasionally removing the first centimeter or two of the column at the injection end containing the nonvolatiles.

Prior separations of complex oils into volatile or functional fractions such as with the use of High Performance Size Exclusion Chromatography (HPSEC) (27,33), or silica gel column chromatography (26) can help the researcher in obtaining more suitable fractions for capillary GC work. Although a number of phases are used in capillary GC for separating the volatiles of pyrolysis oils, the most popular appears to be a bonded silica of moderate polarity, such as the DB-5 column in 30 cm length. HPSEC by itself is becoming increasingly used as a tool for characterizing pyrolysis oils (see, e.g., 27,34).

Post-Pyrolysis Processing

The utility of biomass pyrolysis oils in any fuel or chemical sense must recognize this complexity in chemical composition, and it is commonly suggested that the tar either be fractionated into simpler mixtures, or be reprocessed or converted into more useful mixtures. Unfortunately, oils usually do not fractionate well (depends on storage conditions and type of oil), and will even undergo changes upon storage, resulting in higher concentrations of non-volatile high molecular weight materials. Extraction techniques work, but these processes are unwieldy. Further, functional fractions are generally poor feedstocks for conversion. Other than a limited amount of work conducted in using the phenolic fractions for adhesives (35-37), most product research has been concerned with reprocessing the oils into a more useful mixture of compounds, usually fuel hydrocarbons.

As detailed above, composition of the oils will depend on pyrolysis process conditions. If the oils are primarily phenolic, then hydrotreating (oxygen removal) is necessary to produce hydrocarbon fuels. Single ring phenolics and cyclic ketones present in the biomass pyrolytic oils can be upgraded through deoxygenation to hydrocarbons (a low oxygen content mixture of all types) in the gasoline (19,29) and diesel (29) boiling point ranges. Heavier, higher molecular weight products like the polycyclic aromatics need also be hydrocracked, with at least partial ring saturation prior to cracking. A number of catalysts have been tried, initially at high pressures with typical petroleum hydrotreating or hydrocracking catalysts (19,29), but more recently at lower pressures with acidic zeolites (e.g., 38,39). Kinetics of the hydrotreating/hydrocracking reactions are being studied (33). Other post-pyrolysis process variations include the treating of primary pyrolysis vapors to gasoline hydrocarbons (40), and pyrolysis oil cofed over zeolite catalyst with methanol derived from char gasification

(41). New characterization tools such as molecular beam mass spectrometry (42) help in understanding the mechanisms by which wood vapor and related compounds react over catalysts.

Prospects for Biomass Pyrolysis

It's easy to say that the key to commercial implementation of biomass pyrolysis for tar production will be the identification of economically competitive technology for the production of higher-valued products. As the primary virtues of pyrolysis oils are those attributable to petroleum (liquid fuels and, under some pyrolytic conditions, also olefins), it can be assumed that pyrolysis can become an avenue to petroleum-type products from renewable biomass. Is biomass pyrolysis, coupled with oil upgrading, the renewable route to petroleum? Pyrolysis, after all, allows for the production of biomass-derived fuels in efficient-to-use petroleum forms.

Fuels are however not the highest values obtainable from biomass (nor are they from petroleum, coal or natural gas), nor oil the only product of pyrolysis - nor is pyrolysis the only, or even most efficient, biomass conversion process. Perhaps biomass pyrolysis can provide the economic base that would permit the further exploitation of the chemical values of biomass materials to enhance overall process profitability (cf. history of olefins usage in petroleum). This author is bold in suggesting that, more appropriately, pyrolysis should be considered as an upgrading process for residues from chemical/ materials options for biomass, that efficient fuel/energy processes such as pyrolysis be considered in tandem with, if not after, efficient biochemical or chemical processing of biomass for polymer and oxychemical production. Biomass is composed of complex structures and polymers that have merit in their complexity, and usage of biomass should first consider chemicals/ materials production based on this complexity, geared to market demand, with only higher entropy states of biomass, process residues, relegated to fuel use status.

Literature Cited

1. Stevens, D.J. "An overview of biomass thermochemical liquefaction research sponsored by the U.S. Department of Energy." In *Production, Analysis and Upgrading of Oils from Biomass*, Vorres, K.S., Ed., American Chemical Society, Division of Fuel Chemistry Abstracts, 1987, 32(2), 223.
2. Soltes, E.J. "Thermochemical processes for bioenergy production." In *Biomass Energy Development*, Smith, W.H., Ed., Plenum Press: New York, 1986, 321.
3. Soltes, E.J. "Thermochemical routes to chemicals, fuels and energy from forestry and agricultural wastes." In *Biomass Utilization*, Cote, W.A., Ed., Plenum Press: New York, 1983, 537.
4. Shafizadeh, F. "Introduction to pyrolysis of biomass." In *Proc. Specialists' Workshop on Fast Pyrolysis of Biomass*, Copper Mountain, Diebold, J., Ed., SERI/CP-622-1096, Solar Energy Research Institute: Golden CO, 1980, 79.
5. Shafizadeh, F. "The Chemistry of Pyrolysis and Combustion." In *The Chemistry of Solid Wood*, Rowell, R.M., Ed. Advances in Chemistry Series 207, American Chemical Society: Washington DC, 1984.
6. DeGroot, W.F.; Pan, W-P.; Rahman, M.D.; Richards, G.N. "Early products of pyrolysis of wood." In *Production, Analysis and Upgrading of Oils from Biomass*, Vorres, K.S., Ed., American Chemical Society, Division of Fuel Chemistry Abstracts, 1987, 32(2), 36.

7. Antal, M.J., Jr. "Biomass pyrolysis: a review of the literature, Part I-carbohydrate pyrolysis." In Adv. in Solar Energy, Boer, K.W.; Duffie, J.W.; Eds., American Solar Energy Society: Boulder CO, 1984, 1, 61.
8. Antal, M.J., Jr. "Biomass pyrolysis: a review of the literature, Part II-lignocellulose pyrolysis." In Adv. in Solar Energy, Boer, K.W.; Duffie, J.W.; Eds., American Solar Energy Society: Boulder CO, 1985, 2, 175.
9. Chatterjee, A.K. State-of-the-Art Review on Pyrolysis of Wood and Agricultural Biomass. Final Report, Contract No. 53-319-R-0-206, AC Project P0380, USDA Forest Service: Washington DC, 1981.
10. Bridgwater, A.V. Thermochemical Processing of Biomass, Butterworths: London, 1984.
11. Overend, R.P.; Milne, T.A.; Mudge, L.K.; Eds. Fundamentals of Thermochemical Biomass Conversion, Elsevier: New York, 1985.
12. Proceedings of the International Thermochemical Biomass Conversion Conference, Phoenix, AR. Elsevier: New York, 1988, in press.
13. Soltes, E.J.; Elder, T.J. "Pyrolysis." In Organic Chemicals from Biomass, Goldstein, I.S., Ed., CRC Press: Boca Raton FL, 1981, 63.
14. Elliott, D.C. "Relation of reaction time/temperature to the chemical composition of pyrolysis oils." In This Volume.
15. Lede, J.; Li, H.Z.; Villermaux, J. "Pyrolysis of biomass: evidences for a fusion-like phenomena." In This Volume.
16. Diebold, J.P.; Scahill, J.W. "Production of primary pyrolysis oils in a vortex reactor." In This Volume.
17. Helt, J.E.; Agrawal, R.K. "Liquids from municipal solid waste." In This Volume.
18. McKeough, P.J.; Johansson, A.A. "Oil production by high-pressure thermal treatment of black liquors: aqueous-phase products." In This Volume.
19. Baker, E.G.; Elliott, D.C. "Catalytic hydrotreating of biomass-derived oils." In This Volume.
20. Boocock, D.G.B.; Allen, S.G.; Chowdhury, A.; Fruchtl, R. "The production, evaluation and upgrading of oils from the steam liquefaction of poplar chips." In This Volume.
21. Nelson, D.A.; Hallen, R.T.; Theander, O. "Formation of aromatic compounds from carbohydrates: X. Reaction of xylose, glucose and gluconic acid in acidic solution at 300°C." In This Volume.
22. Krochta, J.M.; Hudson, J.S.; Tillin, S.J. "Kinetics of alkaline thermochemical degradation of polysaccharides to organic acids." In This Volume.
23. Bouvier, J.M.; Gelus, M.; Maugendre. S. "Direct liquefaction of wood by solvolysis." In This Volume.
24. Wenzl, H.F.J. The Chemical Technology of Wood. Academic Press: New York, 1970, 267.
25. Piskorz, J.; Scott, J.S.; Radlein, D. "The composition of oils obtained by the fast pyrolysis of different woods." In This Volume.
26. Pakdel, H.; Roy, C. "Chemical characterization of wood oils obtained in a vacuum pyrolysis multiple hearth reactor." In This Volume.
27. Soltes, E.J.; Lin, S-C.K. "Chromatography of non-derivatized pyrolysis oils and upgraded products." In Production, Analysis and Upgrading of Oils from Biomass, Vorres, K.S., Ed., American Chemical Society, Division of Fuel Chemistry Abstracts, 1987, 32(2), 178.
28. Soltes, E.J.; Lin, S-C.K.; Sheu, Y-H.E. "Catalyst specificities in high pressure hydroprocessing of pyrolysis and gasification tars." In Production, Analysis and Upgrading of Oils from Biomass, Vorres,

K.S., Ed., American Chemical Society, Division of Fuel Chemistry Abstracts, 1987, 32(2), 229.

29. Soltes, E.J.; Lin, S-C.K. "Hydroprocessing of biomass tars for liquid engine fuels." In Progress in Biomass Conversion, Tillman, D.A.; Jahn,E.C., Eds., Academic Press: New York, 1984, 5, 1.
30. Elliott, D.C.; Sealock, J.J., Jr.; Butner, R.S. "Product analysis from direct liquefaction of several high-moisture biomass feedstocks." In This Volume.
31. Grandmaison, J.L.; Ahmed, A.; Kalaiguine, S. "Solid residues from supercritical extraction of wood: characterization of their constituents." In This Volume.
32. Hoesterey, B.L.; Windig, W.; Meuzelaar, H.L.C.; Eyring, E.M.; Grant, D.M.; Pugmire, R.J. "An integrated spectroscopic approach to the chemical characterization of pyrolysis oils." In This Volume.
33. Sheu, Y-H.E. Kinetic Studies of Upgrading Pine Pyrolytic Oil by Hydrotreatment, Ph.D. Dissertation, Texas A&M University, 1985.
34. Johnson, D.K.; Chum, H.L. "Some aspects of pyrolysis oils characterization by high performance size exclusion chromatography (HPSEC)." In This Volume.
35. Elder, T.J. The Characterization and Potential Utilization of the Phenolic Compounds Found in a Pyrolytic Oil, Ph.D. Dissertation, Texas A&M University, 1981.
36. Chum, H.; Diebold, J.; Scahill.; Johnson, D.K.; Black, S.; Schroeder, H.A.; Kreibich, R.E. "Biomass pyrolysis oil feedstocks for phenolic adhesives". In Adhesives from Renewable Resources, Conner, A.; Hemingway, R., Eds., American Chemical Society: Washington DC, 1988, in press.
37. Soltes, E.J.; Lin, S-C.K. "Adhesives from natural resources." In Progress in Biomass Conversion, Tillman, D.A.; Jahn,E.C., Eds., Academic Press: New York, 1983, 4, 79.
38. Renaud, M.; Grandmaison, J.L.; Roy, C.; Kaliaguine, S. "Low pressure upgrading vacuum pyrolysis oils from wood." In This Volume.
39. Dao, L.H.; Haniff, M.; Houle, A.; Lamothe, D. "Reactions of model compounds of biomass pyrolysis oils over ZSM-5 zeolite catalysts." In This Volume.
40. Diebold, J.P.; Scahill, J.W. "Biomass to gasoline (BTG): upgrading pyrolysis vapors to aromatic gasoline with zeolite catalysts at atmospheric pressure." In This Volume.
41. Chen, N.Y.; Walsh, D.E.; Koenig, L.R. "Fluidized bed upgrading of wood pyrolysis liquids and related compounds." In This Volume.
42. Evans, R.J.; Milne, T.A. "Molecular beam mass spectrometric studies of wood vapor and model compounds over HZSM-5 catalyst." In This Volume.

RECEIVED May 26, 1988

Chapter 2

Biomass Pyrolysis Technology and Products

A Canadian Viewpoint

R. D. Hayes

Renewable Energy Branch, Energy, Mines, and Resources, Ottawa, Canada K1A 0E4

By way of introduction, this paper presents four "views" on pyrolysis, including an international view, a technical review, a strategic point of view, and a general Canadian bioenergy overview. From a basic definition of pyrolysis as the thermal conversion of material in the absence of oxygen, the following discussion includes vacuum, atmospheric and pressurized pyrolysis technologies for the production of gas, liquid, or char products.

Canadian Bioenergy Overview

Canada is well endowed with energy alternatives to conventional petroleum. A number of these resources are vast, including tarsands, arctic and offshore oil, natural gas, coal, nuclear, bioenergy and other renewables. They are also, in general, expensive relative to current domestic and imported light crude oil prices. Bioenergy, however, has carried out an impressive impact on Canada's primary energy supply, by rising from 3% in 1978 to 6.8% in 1987. Its current contributions in the industrial and residential sectors are 17% and 14%, respectively.

Most of the oil substitution from biomass in Canada has been in light and heavy fuel oil markets. So far, very few of Canada's alternatives have made much impact in the transportation sector, nor are there any obvious expectations of near-term oil substitution in transportation other than a small niche of vehicles running on propane or natural gas.

Against this backdrop, Canadian bioenergy R&D is very broadly targetted to near-term opportunities for non-transportation uses and to medium and longer term opportunities for transportation uses. The term "broadly" is emphasized because there are exceptions to this generalization.

Canada's bioenergy R&D activities did not begin in earnest until 1978, lagging somewhat behind other parts of the world. In 1980, the federal National Energy Program greatly increased R&D funding in

0097–6156/88/0376–0008$06.00/0

all conservation and alternative energy areas. Bioenergy R&D effort grew to $22 million (Cdn) in 1983-84. Then as oil prices settled, Canada's energy R&D program experienced the "Crash of '84" the government decreased spending by 50%. Bioenergy R&D is currently centred in two federal departments. Energy, Mines and Resources Canada manages the Bioenergy Development Program at approximately $6 million (Cdn) for biomass conversion, and a level of approximately $1.5 million (Cdn) for biomass production is managed by the Canadian Forestry Service.

An approximate breakdown of 1987 Bioenergy Conversion R&D expenditures in general technology areas is as follows:

Biochemical conversion	39%
Thermochemical conversion	16%
Combustion	22%
Biomass handling and preparation	12%
Peat mining and processing	6%
Information transfer	5%

Of special interest to the audience of this paper is that the bulk of the thermochemical R&D is devoted to some form or other of pyrolysis.

History and Strategy, 1978-1987

Amidst the panic of petroleum shortages and rising prices of 1979, some individuals expected to produce oils from biomass, also known as bio-crude, proto-oil, bunker bio-oil... This expectation was fairly short-lived however. Although touted as something, someday destined to replace or reduce oil imports, pyrolysis oil from biomass had very little in common with petroleum. Its qualifications as "a crude" were very rudimentary. It looked black, smelled bad, and usually flowed if there was enough water in it; but that is where the resemblance ended.

Biomass, in those days, was modelled by some researchers as a young coal, and its pyrolysis oil as a young oil or a coal liquid. Neither assumption was very accurate. Those who followed coal liquefaction development had a head start in applying that enormous knowledge base to biomass. However, perhaps due to that knowledge base, early biomass process development did not fully take advantage of the unique opportunity for specialized processing that biomass could provide. Biomass is not a homogeneous chemical conglomerate, but rather a composite of three major polymers namely cellulose, hemicellulose and lignin.

Against this background, Canadian researchers joined the international flurry of research activity in this area. By 1980, there were twelve laboratories, mostly universities, involved in some variation of biomass pyrolysis research. The major focus of their work was to produce a fuel or heating oil. There was very little work done on upgrading bio-oil to a transportation fuel or ven a refinery feedstock. Also, there was little interest in the optimized production of gas or char in Canada since Canada is blessed with substantial reserves of natural gas and coal.

In 1980, the federal government embraced an "off-oil" policy under the National Energy Program. A number of programs provided assistance directly or indirectly to homeowners and industry for substituting heavy and light fuel oil, by other energy forms such as natural gas, electricity and biomass (via combustion). Bio-oil was not technically ready to take advantage of a market opportunity, and more importantly, it had no commercial or industrial infrastructure to build upon.

Therefore, in spite of world record petroleum prices, apparent shortfalls in imported and domestic oil supplies, and a policy environment that was actively encouraging energy alternatives in Canada, bio-oil became a long term curiosity. R&D was important, but it was supported under the objective of long term security of supply of liquid fuels.

Canada's R&D efforts continued to down-play upgrading over the next four years and concentrated almost exclusively on primary oil production. By 1984, oil prices appeared to be more settled and the time frame for commercializing transportation fuels from bio-oil was likely to be well beyond 2000. There appeared to be plenty of time to perfect oil upgrading at a later date, either through hydrotreating or using shape-selective zeolite catalysis.

Then came the "Crash of '84" referred to earlier. Bioenergy R&D funding resources were declining rapidly. Government seemed to want more near term results from an already shrinking energy funding base.

Between 1980 and 1984, biomass pyrolysis R&D funding had been slowly increasing while the number of Canadian laboratories working in this area had decreased from twelve to six. By 1984/85 thermochemical R&D and other long-term options within the Bioenergy Development Program of Energy, Mines and Resources had become vulnerable to funding reductions. It was important not to lose the advances already achieved and the world class expertise developed in Canada. The decision was taken therefore to wind up the research in Canada over the next two to three year period in order to leave a retrievable accounting of developments for future researchers. The second objective was to maintain (or salvage) a base core of excellence in Canada at a reduced funding levels.

This decision did not unfold as planned however. Three years later, the Canadian program in pyrolysis is stronger than ever. The 1987 R&D program of work includes work in five of the six laboratories, at funding levels slightly higher than in 1984. The following three explanations can help explain why the strategic plan changed so fundamentally.

1. Industry Interest. One of Canada's leading solar and biomass conversion equipment companies, Petro-Sun International Inc. became interested in scaling up the Université Laval/Université de Sherbrooke vacuum pyrolysis process on a cost-shared basis. Special Note. As of January 1988, Petro-Sun International Inc. was forced into receivership for reasons unrelated to their activities in biomass pyrolysis. The Université Laval is seeking alternative industrial partners. Another company, Ensyn Engineering Ltd., was established, and proposed a cost-shared, scaled-up development of the University of Western Ontario ultrapyrolysis. Similarly, there has been a number of industrial expressions of interest in the University of Waterloo flash pyrolysis.

2. Diversification. An approach has evolved in Canada whereby, unlike earlier efforts, a whole range of products and reactants are now considered. Products include oils of varying quality, sugar solutions in high yield, chemicals (olefins, phenolics, as well as high value specialty chemicals), gasoline or diesel fuel, and higher value carbon. Diversification of reactants include whole biomass (wood and straw), fractionated biomass components, peat, municipal solid waste, including used tires, etc. Another important area of research is the treatment, and especially, the conversion, of waste aqueous effluence from pyrolysis into value-added co-product credits. Left unprocessed, these effluence would otherwise incur a cost for waste treatment.
3. IEA Collaboration. By working together with three other member countries of the International Energy Agency's Bioenergy Agreement, Canada has been able to combine resources with USA, Sweden, and Finland to determine that Canadian biomass pyrolysis technologies appear to be as good as any in the world. Canada has also had its pyrolysis oils upgraded and assessed by collaborators in the USA. Preliminary results indicate that Canada should review its traditional strategy of not immediately pursuing upgrading R&D.

Technical Review

Over the past several years researchers W.J.M. Douglas and D.G. Cooper at McGill University have been studying an interesting thermochemical approach to wood liquefaction using aqueous hydrogen iodide in fairly mild conditions of pressure and temperature (125°C). Still far from certain is the exact nature of the liquid products and the techno-economic practicality of hydrogen iodide recovery and recycle. On the positive side, in addition to the low severity conditions of reaction, the process removes about 80% of the oxygen in the wood, and the char yield is very low.

D.G.B. Boocock and co-workers at the University of Toronto have undertaken the investigation of a steam pyrolysis or hydrothermolysis of wood. Based on their earlier work, they recently designed and constructed a laboratory scale cascade autoclave which can accommodate up to 100 g of wood chips or 170g of a single larger piece (3.8 cm reactor I.D., 600ml volume). It is rated at 24.1 MPa (3500 psi.) at 350°C allowing for 7.6 MPa (1100 psi) gas overpressure above the vapour pressure of water at that temperature. Results to date are preliminary since the unit was commissioned only in early 1987. Reproduced results indicate that oil yield increases with increased chip size. Product oil yields are very high (up to 50%) with no solids contamination, and the oil is easily separable from the aqueous phase. Coupled with an upgrading process, this technology may someday well lend itself well to commercialization. In addition to their process development work, Boocock's group has contributed greatly to the basic understanding of biomass liquefaction, especially through their scanning electron microscopic studies. Of potential interest too is their discovery in 1984 that a particular clone of hybrid poplar yielded 6% phenol.

E. Chornet, R.P. Overend and co-workers at the Université de Sherbrooke have been working on a liquefaction process in pressurized solvent for some years. Their approach involves an overall integration of biomass pretreatment, fractionation, acid processing, thermochemical and biochemical treatment.

D.S. Scott, J. Piskorz, D. Radlein and co-workers at the University of Waterloo are well known for their fluidized bed flash pyrolysis development, also known as the WFPP (Waterloo Fast Pyrolysis Process). The WFPP actually includes four process options as follows:

1. Direct thermal processing at 450-550°C, atmospheric pressure and about 500 ms vapour residence time. They report the highest liquid yields (80% including water, based on input wood) that is a suitable fuel for conventional boilers.
2. By varying the process conditions and adopting a mild sulphuric acid pretreatment followed by fluidized bed thermopyrolysis, the WFPP produces a high yield and concentration of anhydro-oligosaccharides, mostly levoglucosan rather than oil products. Their reproducible yields of sugars from pure cellulose are about 80% in a concentrated form. One can easily speculate whether this development could challenge some of the equally exciting bioconversion methods of converting lignocellulosics to fermentable sugars.
3. Waterloo's hydrogasification work has been technically highly successful, resulting in 75% conversion of carbon from wood to methane via pyrolysis over a nickel-alumina catalyst with hydrogen at about 550°C and 440 ms residence time.
4. Under current investigation is a fourth process options of producing unsaturated hydrocarbons in a catalysed reaction. Apart from the use of catalysts, the process equipment and operating conditions are very similar for all of the above process options.

M. Bergougnou, R. Graham and co-workers have developed an Ultra-Rapid Pyrolysis or Ultrapyrolysis process at the University of Western Ontario. Although there are similarities in this work and the research at the University of Waterloo, there are important differences. Whereas Waterloo utilizes fluidized bed heat transfer, Bergougnou employees a very rapid (30ms) mixing and heat transfer in a vortical contactor or vortactor followed by a plug-flow entrained bed down flow reactor (50-900ms) and quenching (30ms) with cryogenic nitrogen in a cryovortactor. Also dissimilar to the Waterloo process are the process conditions (650-1000°C, 50-900 ms residence time) and the main product at these temperatures is gas rather than liquid. Since pyrolytic fuel gas production has not been of high priority in Canada's bioenergy R&D strategy, the current objective, in collaboration with Ensyn Engineering, is chemical production, olefins in particular.

It is interesting to note here that the Universities of Waterloo and Western Ontario conducted an experiment of data comparison from each of their reactor systems. Using selected data from both groups at around 500 ms residence time, liquid and gas production data were plotted vs temperature. The temperature ranges were as follows: Waterloo at 400°-750°C, and Western Ontario at 650°-900°C. With combined data for each of the gas yield vs temperature and liquid yield vs temperature data, there was, as expected, considerable concurrence of overlapping data. In fact a simple first order kinetic model is able to describe the oil yield over the temperature range of both experiments.

Ensyn Engineering is a recently formed company whose principal investigator, R. Graham has scaled up the University of Western Ontario Ultrapyrolysis reactor by a factor of 20 to a 5-10 kg/hr capacity RTP (Rapid Thermal Processor). The reactor is designed to accept any carbonaceous feed (solid, liquid, or gas) by injecting into a turbulent cloud of hot solids. The mixed feed plus solids is carried through a tubular transport reactor to an inertial separator where vapour products are removed. Results to date are very preliminary. The concept formulated by Ensyn is that the RTP would be a central processor of primary pyrolysis oils from smaller, perhaps mobile, plants.

The multi-stage vacuum pyrolysis was developed by C. Roy and co-workers, initially at the Université de Sherbrooke and, currently, at the Université Laval. In collaboration with Petro-Sun International Inc. there is a pilot demonstration of a single-stage plant at St-Amable, Quebec. The unit has a capacity of 200 kg/hr and is designed for used tires. The technology is based on a multiple hearth vacuum pyrolysis process development unit located near the Université Laval and operating at a 30 kg/hr biomass capacity.

Although the multiple hearth concept suffers from low heat transfer relative to other pyrolysis processes, and, at first glance, is capital cost intensive, it has a number of features that show commercial promise as follows:

1. A reasonably high yield of pyrolysis oil (50% based on wood).
2. The production of co-product carboxylic acids and high value chemicals.
3. Reactive charcoal at 25% of input wood.
4. The aqueous phase is recovered separately as part of the process leaving a water-free pyrolysis oil ready for upgrading.
5. The multiple hearth performs a product fractionation function that could reduce extraction costs of high value chemicals.

Centralized Analysis

One final work of interest in Canada is the Centralized Analysis project at B.C. Research. In 1984, a trial project was set-up whereby different Canadian bio-oils could be compared in a consistent manner. The project embraced a three-pronged approach. Under the guidance of J. Howard and J. McKinley, B.C. Research performed and/or coordinated the centralized analyses of optimized oils produced by each researcher. If a particular analytical capability was not available at B.C. Research, another laboratory with that strength was contracted to do it. Individual researchers also did some of their own analyses to obtain immediate experimental feedback.

The second prong of the approach was that all researchers were provided with a standard wood sample *Populus deltoides* by Forintek Canada Corp. The idea was that when each process development became somewhat optimized, the researcher would submit oil from the standard wood sample to the centralized analysis team.

The third prong was a computer communications network link called CoSy, through the University of Guelph, to provide fast communication of analytical data. It was also used to encourage multilateral and bilateral collaboration and problem solving.

The centralized analysis project is entering Phase II. Learning from the successes and pitfalls of the first two and one half years, the scope of this project has changed somewhat. The consensus of program managers and researchers is that Phase II will have two major tasks.

Task 1 is a set of analytical techniques along the same lines as the initial project except the methods that were considered to be less interesting to the entire group are not included. The basic analyses include the following:

- Elemental Analysis
- Water Content
- Density
- Carboxylic Acids
- Gas Liquid Chromatography
- Carbon-13 NMR
- Gel Permeation Chromatography

Task 2 consists largely of special analytical projects to meet the needs of individuals or groups of thermochemical conversion researchers that may arise over the next two years.

International View - IEA

Under the International Energy Agency, eleven countries signed a three year Bioenergy Implementing Agreement, effective January 1, 1986 - December 31, 1988. Canada, USA, Sweden and Finland agreed to collaborate on a project entitled Direct Biomass Liquefaction (DBL). A Working Group of engineers and other specialists are preparing a detailed technical-economic assessment (TEA) for scale-up and operation at commercial size, of the most promising high and low pressure pyrolysis processes. Both primary oil production and upgrading are considered in the TEA. The upgrading work on Canadian atmospheric and vacuum pyrolysis oils has been conducted by D. Elliot at Batelle Pacific Northwest Laboratories. Yields of products in the gasoline boiling range have so far reached 35% of primary oil by hydrotreating. The Working Group is attempting two types of analyses, one based on current state of the art and the second based on projected improvements and developments in the technologies.

Canada has benefitted greatly from this project. In a funding environment that restricted major thrusts in upgrading R&D in Canada's national program, there is this element available in the international program. Another benefit to Canada is that the IEA Working Group, independent of self-interests or biases toward the various processes, concluded that Canadian primary pyrolysis oils have technical and economic advantages to other processes under development.

Acknowledgments

The author wishes to acknowledge the various researchers across Canada without whose ideas, open discussion, and dedication to research in thermochemical conversion of biomass, this paper would not have been possible. Each has contributed to this paper, either directly or indirectly. The following list is not complete, but includes the principal investigators in Canadian laboratories and other notable international collaborators. In alphabetical order, many thanks to Narendra Bakhshi, Dave Beckman, Dave Boocock, Jean Bouchard, Maurice Bergougnou, Esteban Chornet, Helena Chum, Jim Diebold, Murray Douglas, Dick Eager, Doug Elliot, Bob Graham, John Howard, Serge Kaliaguine, Bjorn Kjellstrom, Tom Milne, Hugh Menard, Jim McKinley, Ralph Overend, Hooshang Pakdel, Jim Pepper, Jan Piskorz, Desmond Radlein, Tom Reed, Christian Roy, Don Scott, Jacques Sicotte, and their many colleagues, staff, and students.

References

This paper contains technical contributions from all the Canadian and other work mentioned here. The reader is referred to these articles as primary references.

RECEIVED June 10, 1988

Chapter 3

Processing of Wood Chips in a Semicontinuous Multiple-Hearth Vacuum-Pyrolysis Reactor

Christian Roy, Richard Lemieux, Bruno de Caumia, and Daniel Blanchette

Department of Chemical Engineering, Université Laval, Sainte–Foy, Quebec F1K 7P4, Canada

A process for the pyrolysis under vacuum of biomass and waste materials has been under development since 1978 in Canada. Background data regarding the yields and qualities of the pyrolysis products derived from materials such as wood, bark, agricultural residues, peat, municipal solid wastes, activated sludges and scrap tires, have been obtained at the bench scale level. A Process Development Unit having a feedthrough capacity of up to 25 kg h^{-1} and using a multiple-hearth furnace reactor has been designed and built. The objectives of the P.D.U. was twofold. First, the unit was used to determine engineering data such as the overall thermal efficiency and the heat requirement for the reactions. Secondly, the configuration and mechanical operation of the ractor had to be tested before its scale-up for the pilot plant phase of the project. This paper reports on the role of several process parameters which are key factors to be considered for the design of a biomass vacuum pyrolysis unit.

The thermal decomposition of wood into charcoal and tar is an old process. One example of the existing technology for wood carbonization is the Lambiotte process. The ATOCHEM plant located in Premery, France, is based on the principle of external gas circulation and is a completely continuous process (1). The heating gas moves upward in the retort and constantly releases its heat into the wood, which is moving downward. The annual wood charcoal production of this plant is currently 20 000 t. The charcoal finds its use in barbecues and the iron industries. Another example is the Brazilian beehive kiln for charcoal production in a batch mode. In Brazil, charcoal is mainly sold to the iron, the cement and the barbecue industries. In 1985, the total annual production of wood charcoal in Brazil was 8,7 X 10^6 t (2).

0097–6156/88/0376–0016$06.00/0

The recovery of by-products is important for the economy of both processes. The French recover high-value chemicals such as food aromas from the pyroligneous liquors (3). The Brazilians market the wood tar by-product as a bunker fuel oil (2). However, further studies are still needed and are being conducted by the industry in order to make a better use of the tar and oil fractions.

Laboratory (4) and Process Development Unit (5,6) studies originally conducted at the Université de Sherbrooke, and now conducted jointly with the private industry at Université Laval, province of Quebec, have led to the conclusion that thermal decomposition under reduced pressure is an attractive approach for the conversion of biomass into chemicals and fuels products. The process uses a multiple-hearth furnace for wood pyrolysis. This approach is characterized by a low pressure and a short residence time of the vapor products in the reactor. When compared with conventional, atmospheric pressure carbonization, vacuum pyrolysis has the potential to significantly enhance the yields of organic liquid products with respect to solid and gaseous products. The pyrolysis oils (biooils) obtained from this process can be deoxygenated into transportation fuels upon further upgrading (7). Specialty as well as commodity (Pakdel, H.; Roy, C. Biomass, in press) chemicals can also be extracted from the pyrolysis oil product.

This paper discusses the preliminary engineering data leading to the construction of a vacuum pyrolysis pilot plant for the conversion of wood into oils, chemicals and charcoal.

Experimental

A schematic of the Process Development Unit (P.D.U.) used in this study is shown in Figure 1. The reactor is a multiple-hearth furnace 2 m high and 0.7 m diameter, with six hearths. Heat is provided from heating elements.

At the onset of an experiment wood chips are poured batchwise into a hopper that sits on top of the reactor. The hopper is equipped with a feeding device and is hermetically sealed. For the experiments reported, 6 to 16 kg of wood chips, with a granulometry of 1/4" to 1/2" Tyler Sieves, were fed at a constant rate of 0.8 to 4 kg h^{-1}.

A mechanical vacuum pump removed the organic vapor and gas products form the reactor through a series of outlet manifolds set along the reactor cylinder (H-I to H-VI). Each outlet was connected to a heat exchanger where the vapors were condensed and recovered as liquid in individual glass receivers. Cold tap water circulating on the shell side of the exchangers was used as cooling medium. The vapors from the heat exchangers were collected in a train of receivers that served as a secondary condensing unit (C-1 to C-4). The first reveiver was immersed in a bath of water-ethylene glycol mixture. Receivers 2 and 3 were immersed in baths of dry ice-acetone. Receiver 4 was filled with glass wool at room temperature.

Pressure in the system was lower than 11 kPa (absolute) under steady-state conditions. The noncondensable gas was continuously pumped into a 500 L vessel that was set under vacuum at the beginning of the run.

The solid residue was directed toward the bottom of the

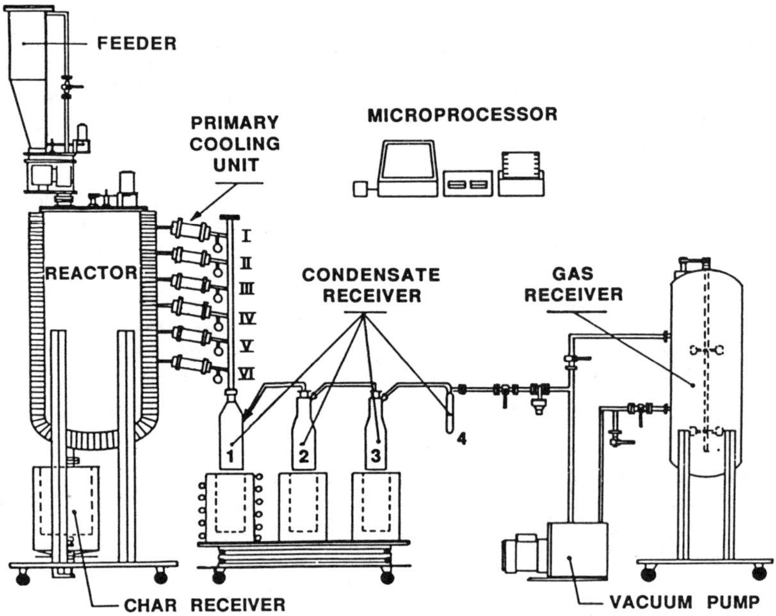

Figure 1. Schematic of the Vacuum Pyrolysis Process Development Unit

reactor. The residual charcoal was collected in a metallic vessel installed on a load cell.

At time zero of the run, wood chips were fed to the preheated reactor. The heating plate temperatures increased from top to bottom of the reactor. A typical temperature profile was 200°C to 450°C. The radial temperature gradient for any heating plate was lower than 5°C during a typical run.

The P.D.U. was attached to a central microprocessor which enabled simultaneous data acquisition and control of some 75 operating parameters (64 are recorded and 11 are controlled). Air leakage through the system was lower than 0.0013 atm L s^{-1}.

The experiments conducted with the P.D.U. were performed with *Populus deltoïdes*. The 8-year-old fast-growing poplar clone D-38 was grown in Brockville, Ontario. The sample was essentially all sapwood with no bark. It was supplied to our laboratories in the form of chips by Forintek Canada Corp., Ottawa. Its elemental composition was determined to be 48.2% C, 6.4% H, 45.3% O, 0.09% N, and 0.05% S. It gross heating values was 4660 kcal kg^{-1} with an average ash content of 0.6%. Moisture of the air dry feedstock was 5,9%.

Results and Discussion

Yield and Quality of Products. Results for the operation of the multiple-hearth furnace at varying final thermal decomposition temperatures and reactor pressures are presented in Table I of the paper. Biooil, pyrolytic water, charcoal and gas yields are presented along with the mass balance calculation for each run. Table I indicates that the largest amount of oil was produced at the lower pressure conditions. The oil yield drops rapidly with even a slight increase in pressure. Table I also indicates that a reactor temperature in the range of 425-450°C is optimum for the maximum oil production. The general characteristics of biooils, gas and charcoal produced under optimum conditions are presented in Table II of the paper. It can be observed that the gross calorific value of both charcoal and oil products increases compared with that of raw wood. The gas has a very low calorific value (1660 cal g^{-1}) but is burnable. The composition of the noncondensable gas phase is shown in Table III of the paper.

Separation of Water from the Organic Liquid Phase. One objective of the vacuum pyrolysis process is to produce large quantities of liquid fuels and chemicals from wood. However during the process organic substances are produced simultaneously with water (moisture and pyrolytic water). The aqueous phase is highly acidic in nature (Pakdel, H.; Roy, C. Biomass, in press). Its organic content ranges between 45 and 55% but further improvement of the process should reduce this concentration down to 10-20%. Since extraction of chemicals or further processing of pyrolytic oils mixed with water is difficult and expensive, it is highly desirable to recover in separate sections the bulk of the aqueous phase from the organic liquid phase.

The separation of water and the organics was achieved during

Table I. Yields and Mass Balances for Vacuum Pyrolysis of Wood

RUN	FINAL TEMPERATURE (°C)	REACTOR PRESSURE (kPa)	WOOD FEEDSTOCK (kg)	YIELDS (% wt., ash-free, dry wood basis)				MASS BALANCE CLOSURE (%)
				OIL	WATER	CHARCOAL	GAS	
C010	425	1.6	5.98	46.4	18.2	24.2	11.2	96.7
C012	363	2.4	5.99	41.6	14.9	33.0	10.5	97.5
C014	450	3.3	6.03	45.8	17.0	25.6	11.6	96.1
C015	425	1.9	6.00	50.1	15.2	25.0	9.7	96.3
C019	465	10.7	3.39	39.7	21.6	24.7	14.0	96.7
C023	450	1.6	15.43	50.9	16.5	21.3	11.3	98.3
C024	450	4.0	18.30	47.4	16.9	25.5	10.2	97.9
C025	450	1.3	15.99	50.0	15.6	23.0	11.4	98.6

Table II. General Characteristics of Biooil and Charcoal

	BIOOIL	CHARCOAL
Water content (%)	4.4	< 1
Gross calorific value ($kJ.kg^{-1}$, anhydrous basis)	22 500	32 000
Elemental analysis (%)		
C	55.6	84.4
H	5.8	3.5
N	0.7	0.2
S	0.6	0.8
O (by diff.)	37.3	11.1
Proximate analysis (%)		
Ash	0.13	2.8
Volatile matter		18.5
Fixed carbon		78.7
Specific surface (BET, m^2 g^{-1})		≈ 3

Table III. Gas Phase Composition During Vacuum Pyrolysis of Wood

RUN	FINAL TEMPERATURE (°C)	REACTOR PRESSURE (kPa)	TOTAL GAS YIELD (% wt., ash-free, dry wood basis	GAS COMPOSITION (% wt.)						CO/CO_2
				CO_2	CO	CH_4	H_2	C_2-C_4	OTHERS*	
C010	425	1.6	11.2	59.2	33.6	2.4	0.9	1.5	2.4	0.57
C012	363	2.4	10.5	60.4	34.9	0.9	0.1	0.9	2.8	0.57
C014	450	3.3	11.6	65.7	28.1	1.4	1.0	1.6	2.2	0.43
C015	425	1.9	9.7	63.8	30.5	1.5	0.8	1.3	2.1	0.48
C019	465	10.7	14.0	60.0	31.4	3.3	0.7	2.8	1.8	0.52
C023	450	1.6	11.3	60.7	31.6	2.7	0.0	1.4	3.6	0.52
C024	450	4.0	10.2	63.5	29.8	2.6	0.7	1.5	1.9	0.47
C025	450	1.3	11.4	64.6	30.3	1.9	0.4	1.0	1.8	0.47

* Others gases were among the followings: methanol, ethanol, acetone and acetaldehyde

this study by using a series of shell and tube heat exchangers with cool to warm water circulating in the shell side. This series of heat exchangers served as a primary condensing stage for the recovery of the liquid organic fraction. Water was primarily recovered in the series of traps that followed (see Figure 1). The relative proportion of water and oil in both condensing sections is shown in Table IV of the paper. Table IV indicates that the lower the pressure, the better the separation between oil and water (see runs C019 and C025 for example). On the other hand at similar operating pressure, the lower the cooling temperature, the more efficient the recovery of oil in the primary condensing section (see runs C023 and C025). Further studies are underway with the objective to find the optimum operating conditions for the maximum recovery of biooils with the least amount of water in it.

Oil Refining. Further fractionation of the wood oil product is necessary when the objective is to either recover pure chemical compounds, or upgrade or process specific chemical group components. Results which are reported elsewhere by Renaud et al (7) and Pakdel et al (8) show that the multiple-hearth reactor can be operated in a mode that enables the separation and recovery of selected fractions of liquid fuels and chemicals.

Mean Residence Time of the Vapor Products in the Reactor. One important feature of the vacuum pyrolysis process is the short residence time of the gases and vapors in the reactor. In a previous investigation (4), the residence time of the vapor products in the laboratory scale batch reactor used was estimated to be about 2 s. In the present investigation, a different pumping unit and a very large pyrolysis unit have been used, so that a new estimate of the mean residence time was necessary.

The mean residence time τ of the gas and vapor products in the reactor can be defined as the ratio between the reactor volume V and the volumetric feed rate v_o of the gases and vapours flowing through the reactor under steady state pressure conditions.

$$\tau = \frac{V}{v_o} \tag{1}$$

If the ideal gas law applies which is likely under the low pressure conditions used, we have:

$$PV = \left(\frac{m}{M}\right) RT \tag{2}$$

where P = gas pressure
V = gas volume
m = mass of gas
M = average molecular weight of the gas
T = gas temperature
R = gas constant

"Equation 2" can be expressed in a differential form with respect to time:

$$\frac{d(PV)}{dt} = \frac{RT}{M}\left(\frac{dm}{dt}\right) \tag{3}$$

where the mass of gas in the reactor is expected to vary with time while the average molecular weight of the gas should remain constant with time.

Development of "Equation 3" leads to "Equation 4"

$$P\left(\frac{dV}{dt}\right)_p + V\left(\frac{dP}{dt}\right)_v = \frac{RT}{M}\left(\frac{dm}{dt}\right) \tag{4}$$

Under study state operation of the P.D.U. at constant pressure, the term dP/dt in "Equation 4" is equal to 0 which reduces "Equation 4" to:

$$P\left(\frac{dV}{dt}\right)_p = \frac{RT}{M}\left(\frac{dm}{dt}\right) \tag{5}$$

In another special circumstance, the evacuation of gas from the reactor was interrupted during the experiment. As a result, the gas pressure built up in the reactor as depicted in Figure 2. In such case, V is constant and the term dV/dt in "Equation 4" is equal to 0 which reduces "Equation 4" to:

$$V\left(\frac{dP}{dt}\right)_v = \frac{RT}{M}\left(\frac{dm}{dt}\right) \tag{6}$$

"Equation 5" is identical to "Equation 6" if the term $\frac{RT}{M}\cdot\left|\frac{dm}{dt}\right|$ in "Equations 5 and 6" is the same. This is true if the yield and quality of the gas phase products which are present in the reactor remain constant with respect to time in both modes of operation of the reactor as shown in Figure 2. With this hypothesis, we have:

$$V\left(\frac{dP}{dt}\right)_v = P\left(\frac{dV}{dt}\right)_p \tag{7}$$

and

$$\frac{P}{\left(\frac{dP}{dt}\right)_v} = \frac{V}{\left(\frac{dV}{dt}\right)_p} \tag{8}$$

hence

$$\tau = \frac{P}{\left(\frac{dP}{dt}\right)_v} \tag{9}$$

"Equation 9" can be calculated using the data available in Figure 2. The average gas pressure in the reactor during steady state operation was 1.6 kPa. The rate of pressure built up in the closed reactor was determined to be approximately 150 kPa h^{-1}. As a result, the average residence time of the gas and vapour phase in the reactor was found to be about 40 s.

Heat Requirement for the Pyrolysis Reaction. Another engineering parameter to be considered when designing a full scale pyrolysis plant is the amount of energy required for the pyrolysis of each mass unit of wood fed to the reactor. Such value has been empirically determined using the P.D.U. described in this paper, and the detailed procedure has been published elsewhere (9). The determination was based on the difference of electric energy consumed before of after wood was fed to the reactor (heat loss to the atmosphere), and that of the electric energy required for maintaining the multiple-hearth furnace at predetermined set-point temperature with wood chips flowing through the reactor.

The amount of heat required to decompose wood (ΔH_r) will depend on the operating conditions used and the degree of conversion that has been reached. Under usual carbonization conditions, overall the reaction is known to be exothermic. Under vacuum pressure conditions, overall the reaction is slightly endothermic as illustrated in Table V of the paper. It can be concluded that some of the secondary reactions (e.g. thermal and catalytic cracking, repolymerization, recondensation, oxidation and reduction reactions) which occur under the higher pressure conditions are exothermic and hence contribute to the overall exothermicity of such reactions. In Table V, ΔH_r for run C019 at 11 kPa is less endothermic than ΔH_r for reactions C023, C024 and C025 which have been conducted at a similar temperature but under a significantly lower pressure.

Several D.T.A. studies suggest that the initial stage of thermal decomposition of wood is endothermic. As a consequence, it is expected that ΔH_r will vary with the degree of conversion of wood during pyrolysis. In Table V, the results for the ΔH_r calculations have been alternatively reported on the percentage of wood converted basis, as well as on an as-received anhydrous wood basis. Conversion is defined as:

$$\frac{\text{Amount of biooil + pyrolytic water + gas}}{\text{Amount of initial feedstock (org. basis)}}$$

In Table V, the maximum conversion which has been reached is similar for the four reported experiments. It can be concluded that the heat required for vacuum pyrolysis of aspen wood between 1.3 and 3.9 kPa is about 733 kJ kg^{-1} of organic wood converted. Overall, the reaction is slightly endothermic.

Table IV. Separation of Water and Biooils During Condensation

RUN	REACTOR PRESSURE (kPa)	FINAL TEMPERATURE (°C)	COOLING TEMPERATURE (°C)	TOTAL LIQUIDS (kg)	CONDENSING UNITS PRIMARY OIL (%)	CONDENSING UNITS PRIMARY WATER (%)	CONDENSING UNITS SECONDARY OIL (%)	CONDENSING UNITS SECONDARY WATER (%)
C019	10.7	465	11-28	2.06	52.2	19.2	7.4	21.2
C023	1.6	450	50-55	10.45	32.2	1.5	36.7	29.6
C024	4.0	450	30-35	11.85	39.8	1.5	27.2	31.4
C025	1.3	450	15-20	10.53	47.8	3.4	27.2	21.6

Table V. Variation of Heat of Pyrolysis With Pressure

RUN	FINAL TEMPERATURE (°C)	REACTOR PRESSURE (kPa)	CONVERSION (%)	ΔH_r ($kJ.kg^{-1}$ dry wood feedstock)	ΔH_r ($kJ.kg^{-1}$ of wood converted)
C019	465	10.7	75.3	450	600
C023	450	1.6	78.7	570	720
C024	450	4.0	74.5	550	730
C025	450	1.3	77.1	560	730

Heat Required for Cooling of the Vapors Products. The amount of heat removed when cooling the organic vapor products at the primary condensing stage was also experimentally determined (see Figure 1). The heat exchange (Q_w) between the cooling medium (water) and the hot vapor products was calculated using the "Equation 10".

$$Q_w = mC_p \quad (T_{w,in} - T_{w,out}) \tag{10}$$

where m is the water flowrate, C_p is the heat capacity for water and $T_{w,in}$ and $T_{w,out}$ is the water temperature at the inlet and the outlet of the heat exchanger, respectively. Taking into account the heat loss to the atmosphere, the overall heat exchanged for cooling of the gas and vapor product to 50°C was found to be 469 kJ kg^{-1} of air-dry wood.

The experimental set-up simultaneously enabled the calculation of the overall heat-transfer coefficient U for the heat exchangers. The data were used in "Equation 11".

$$Q_w = UA \ (\Delta T)_{in} \tag{11}$$

where A is the heat transfer area and $(\Delta T)_{in}$ is the temperature driving force. It was found that the U values varied betwen 8 and 15 $W.m^{-2}.°C^{-1}$ according to the position of the heat exchanger attached to the multiple-hearth furnace (9).

Determination of the Standard Heat of Reaction. The equipment used also enabled us to determine the standard heat of reaction for pyrolysis of air-dry wood chips. For matter of convenience the standard state for wood considered as "a pure substance" was 323 K (50°C) and 1.6 kPa. The standard heat of reaction was calculated using the "Equation 12":

$$\Delta H^o{}_{323} = \Delta H^o{}_t - \Delta H^o{}_r - \Delta H^o{}_p \tag{12}$$

where $\Delta H^o{}_r$ is the total enthalpy change for the reactants from temperature T to 323 K

$\Delta H^o{}_p$ is the total enthalpy change for the products from 323 K to temperature T

$\Delta H^o{}_t$ is the enthalpy change for the three-step process including the enthalpy change during the isothermal reaction at 323 K.

Figure 3 illustrates how each term in "Equation 12" was empirically determined during run C023. $\Delta H^o{}_r$ and $\Delta H^o{}_p$ in "Equation 12" were calculated using an average heat capacity of 2 $J.g^{-1}.K^{-1}$ for wood and 1 $J.g^{-1}.K^{-1}$ for wood charcoal. The value for $\Delta H^o{}_{323}$ in "Equation 12" was found to be 92 kJ kg^{-1} of air-dry wood which confirms that the reaction is slightly endothermic. Although this value has limiting practical use, it should be viewed as an attempt to improve our theoretical knowledge of thermodynamics of wood pyrolysis.

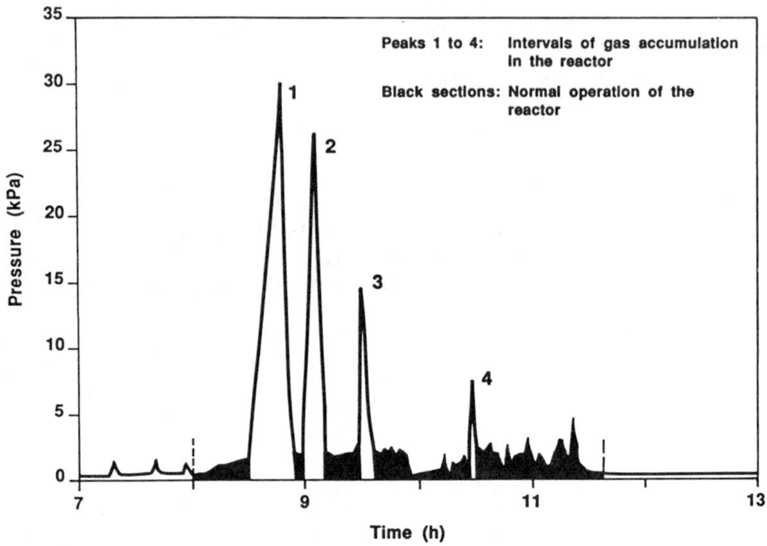

Figure 2. Variation of Reactor Pressure as a function of time

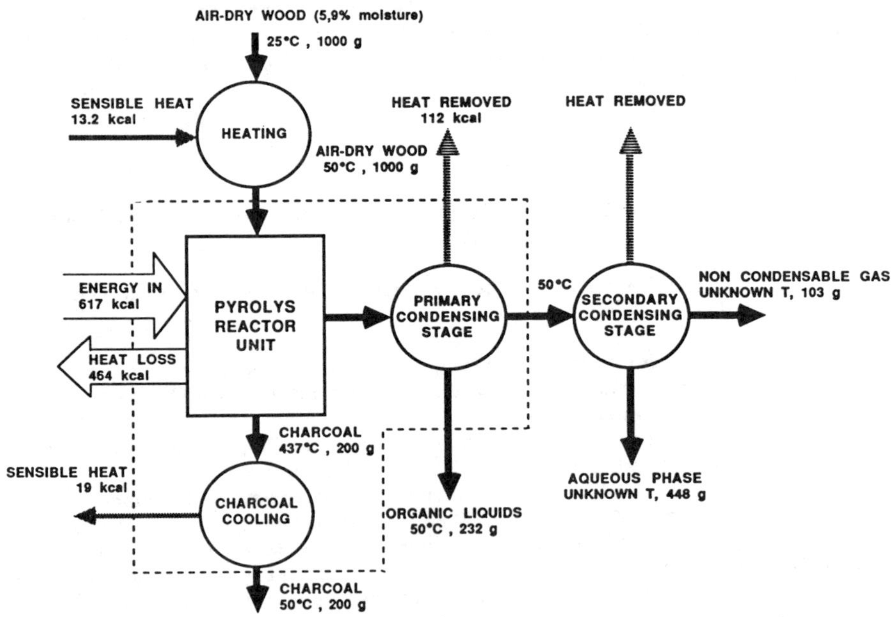

Figure 3. Mass and Energy balance around the Reactor

Calculation of the Thermal Efficiency of the Multiple-Hearth Reactor. The thermal efficiency of the system defined as the ratio of useful energy provided by the vacuum pyrolysis reactor, to the energy supplied to it during a specific run, has been calculated for experiment C023. The calorific values of the end-products represented the following percentages of the heating value of the initial feedstock: wood charcoal 36%, wood oil 41% and gas, 5%. Using these data together with values from Figure 3 enabled us to determine that the thermal efficiency was 82%. For this calculation it was assumed that the electric energy required for the mechanical pump was negligible and that the reactor was perfectly insulated.

Heat Transfer Phenomena in the Reactor. In the pyrolysis reactor, heat can be transferred by conduction, by radiation and by convection. Heat transfer by convection is negligible due to the low pressure conditions in the reactor. In principle, both conduction and radiation can play a significant role during pyrolysis. An overall heat-transfer coefficient of 25 $W.m^{-2}.^{o}C^{-1}$ was found in a previous study where the multiple hearth furnace was viewed as a heat exchanger. One weakness of this heat transfer model by conduction is that phase changes occur during the thermal decomposition of wood and this might lead to inaccurate conclusions.

A more recent study of the heat fluxes inside the reactor as well as a comparison of the theoretical data predicted by a radiation heat transfer model showed that radiation is the main mode of heat transfer in the multiple hearth reactor configuration used (10).

Conclusions

A multiple-hearth reactor has been successfully tested for the production of high yields of liquid fuels and chemicals. The reactor enabled the separation and recovery of water on the one hand and oil fractions or the other hand. Biooil yields reached about 50% by weight of the air-dry feedstock. Both the oil and charcoal have a higher calorific value than the raw wood feedstock.

Heat required for the pyrolysis of wood is slightly endothermic and was determined to be 733 kJ kg^{-1} of organic wood converted. The standard heat of reaction at 323 K and 1.6 kPa was found to be approximately 92 kJ kg^{-1} of air-dry wood. The thermal efficiency of the process is high at 82%. The mean residence time of the gas and vapour phase product in the reactor was empirically found to be about 40 s. Heat is mainly transferred by radiation in the type of reactor used.

Acknowledgments

This study was supported by the Bioenergy Development Program, Energy, Mines and Resources Canada and Petro-Sun R and D, Boucherville, Québec.

Literature cited

1. Wenzl, H.F.J. The Chemical Technology of Wood; Academic Press: New York, 1970; p 285.

2. Rodrigues de Almeida, M. International Workshop on Pyrolysis as a Basic Technology for Large Agro-Energy Projects, L'Aquila, Italy. 15-16 October, 1987.
3. Perdrieux, S.; Lemaire, A. Le bois National. 2 juillet 1983, 23-26.
4. Roy, C.; de Caumia, B.; Brouillard, D.; Ménard, H. In Fundamentals of Thermochemical Biomass Conversion: An International Conference. Overend, R.P., Milne, T.A. and Mudge, L.K., Eds. Elsevier Applied Science Publishers: New York, 1984; p. 237-256.
5. Roy, C.; de Caumia, B.; Blanchette, D.; Lemieux, R.; Kaliaguine, S. Energy from Biomass and Wastes IX, IGT Symposium. Lake Buena Vista, Florida. January 28-February 1, 1985, p. 1085-1106
6. Roy, C.; Lemieux, R.; de Caumia, B. and Pakdel, H. Biotechnology and Bioengineering. Symp. No. 15., 1985, p. 107-113.
7. Renaud, M.; Grandmaison, J.L.; Roy, C.; Kaliaguine, S. "Production, Analysis and Upgrading of Pyrolysis Oils from Biomass". ACS Division of Fuel Chemistry Preprints, Denver, Co. April 5-10, 1987; 32, (2), p. 276-286.
8. Pakdel, H.; Roy, C. Production, Analysis and Upgrading of Pyrolysis Oils from Biomass". ACS Division of Fuel Chemistry Preprints, Denver, Co. April 5-10, 1987; 32, (2), p. 203-214.
9. Lemieux, R. M.Sc.A. Thesis. Université de Sherbrooke, Québec, 1988. (In French).
10. Labrecque, B. M.Sc.A. Thesis, Université de Sherbrooke, Québec, 1988. (In French).

RECEIVED March 31, 1988

Chapter 4

Production of Primary Pyrolysis Oils in a Vortex Reactor

James Diebold and John Scahill

Chemical Conversion Research Branch, Solar Energy Research Institute, 1617 Cole Boulevard, Golden, CO 80401

A vortex tube has certain advantages as a chemical reactor, especially if the reactions are endothermic, the reaction pathways are temperature dependent, and the products are temperature sensitive. With low temperature differences, the vortex reactor can transmit enormous heat fluxes to a process stream containing entrained solids. This reactor is ideally suited for the production of pyrolysis oils from biomass at low pressures and residence times to produce about 10 wt % char, 13% water, 7% gas, and 70% oxygenated primary oil vapors based on mass balances. This product distribution was verified by carbon, hydrogen, and oxygen elemental balances. The oil production appears to form by fragmenting all of the major constituents of the biomass.

The pyrolysis of biomass follows a complex set of different chemical pathways, which have thus far not been well established. However, several global pathways have been established, which explain most of the observed phenomena. As shown in Figure 1, the first reaction in fast pyrolysis of biomass is the depolymerization of the lignocellulose macropolymers to form viscous primary oil precursors. These precursors are formed with almost no by-products, and consequently their elemental composition is very similar to the original biomass. With low heating rates, much of the primary oil precursors can repolymerize to thermally stable polymers through the elimination of mostly water to eventually form the material known as char. Evidence for a liquid or plastic phase intermediate in the formation of char is the physical shrinkage of the macrodimensions of wood, which takes place during charring (1) in a manner analogous to heat shrinkable polyethylene tubing. If the heating of the biomass proceeds very quickly to temperatures above 450°C, most of the primary oil

0097–6156/88/0376–0031$06.00/0

precursors can crack and vaporize before they form char. In the vapor state, the primary oil molecules are quite dilute, which slows possible second-order polymerization reactions. This dilution allows any unstable primary oil vapors to be converted by first-order reactions to more stable compounds, which can be collected from a reactor designed to have a short gaseous residence time followed by rapid quenching. Thermal stability is relative, however, and these stabilized primary oil vapors readily crack to gases following a global first-order reaction (2). The cracking of the primary oil vapors proceeds with a 10% loss in 36 ms at 700°C and extrapolated 10% losses in 6 ms at 900°C and 591 ms at 500°C.

Pyrolysis Reactor Design Considerations

Obviously, the lower the temperature of the primary vapors in the reactor, the greater the yield of primary vapors which can survive passing through the reactor to the quench zone. Minimizing the time required to travel from the vapor formation zone in the reactor to a lower temperature quench zone also helps to maximize the primary oil vapor yields. The ideal reactor would thus provide large heat fluxes preferentially to the pyrolyzing biomass particle, while not overheating the surface of the particle to cause cracking of the primary vapors to gases as the vapors escape the surface of the particle. The ideal reactor would allow the vapors to be immediately swept away by a cooler carrier gas stream out of the reactor to a cold quench zone in order to preserve as much of the vapors as possible. The residence time of the biomass particles in the ideal reactor must be long enough to ensure complete pyrolysis, but the accumulation of dead char in the reactor is undesirable. It would also be advantageous if the reactor could selectively remove dead char and recycle partially pyrolyzed particles.

The use of thermal radiation for fast pyrolysis has been explored, as this approach preferentially heats the solid with potentially high heat fluxes. However, heating the particle with a high temperature heat source can drive the surface temperature of the particle too high and some cracking of the vapors to gases would be expected. The use of hot flue gases or hot solids as a heat transfer medium usually requires that they be at very high temperatures to lessen the amount of the medium which must be generated or recycled; flue gases or hot sand at 900° to 1000°C have been used for fast pyrolysis, but tend to produce higher yields of noncondensible gases from cracking the primary pyrolysis oil vapors to gases as described above. Conversely, the use of gases or solids at lower temperatures for the heat source requires that much more of them be used, which increases the process energy required. Large amounts of gases dilute the process stream and promote the loss of volatile organics in the gas stream from the condensation part of the system. The ideal reactor for the fast pyrolysis of biomass to primary oils would achieve high heat transfer rates through the use of a mechanism which has an inherently high heat-transfer coefficient, rather than through the use of a high-temperature source. Such a heat transfer mechanism is attained by the conduction of heat from a moderately hot reactor wall directly to the biomass particle.

It can be readily demonstrated that when a stainless steel wire at 500° to $900^{o}C$ is contacted with a monolithic piece of biomass, the biomass surface is ablatively pyrolyzed and converted to liquids and vapors, which allow passage of the wire through the biomass. If the stored energy in the wire is transferred to the biomass by sliding the wire across the biomass, pyrolysis rates over 3 cm/sec are observed (3). This method of heat transfer has been studied by pushing a wooden rod into a heated, stainless steel disk, and the pyrolysis rate has been found to be proportional to the pressure exerted and to the temperature difference, where the biomass surface was calculated to be pyrolyzing at $466^{o}C$. Heat transfer coefficients as high as 8 W/cm^2 K were reported, which is over 300 times higher than thermal radiation from a wall at $900^{o}C$ having an emissivity of one (4).

Although a reactor can be designed to push wooden rods into a hot surface for research purposes (4-5), most practical biomass feedstocks are expected to be in the form of sawdust or chips. A modified entrained-bed reactor was selected in which the entrained particles enter the reactor tangentially so that centrifugal forces push the feedstock particles onto the externally heated cylindrical wall. Drag forces induced on the particle by the entraining gas stream serve to keep the particles moving on the wall. Since the particles are on or very nearly on the wall, they tend to intercept preferentially the heat, which is conducted through the reactor wall. With nonreacting solid particles in a heat exchanger made from a cyclone separator, the total heat transferred to the process stream was relatively independent of the solids' content at carrier-to-solids (C/S) mass ratios as low as one, whereas with more solids, the heat transferred increased dramatically. The temperature rise in the gas stream was as little as half of that seen in the solids at these low C/S ratios. The heat transfer coefficient from the wall to a solids-free gas was found to follow traditional convective heat transfer relationships, but to be 1.8 times higher in the cyclone than in a straight tube for the same entering tube diameter and entering gas velocities (6). A reported property of a cyclone is that above an entering Reynold's number of 3000, the cyclone has plug flow (7). The cyclone is an interesting reactor concept for the pyrolysis of biomass, as reported in the literature (7-8). However, the reactor of interest in this paper is a vortex tube, which has many similarities to a cyclone separator.

Vortex Reactor Considerations

The vortex tube of interest has a tangential entrance into one end of a cyclindrical tube and an axial exit at the other end of the tube. Research into the aerodynamics of the vortex tube revealed that the pitch of the vortex near the cylindrical wall was about 1.2 times the diameter of the vortex tube. This results in a coarse helical path of the gases near the wall, as measured by Pitot tubes (9-10). This coarse helical path on the wall also exists for the cylindrical part of the more conventional cyclone separator (11). The effect of the coarse, gaseous path is that entrained solids, which are centrifuged to the wall, follow the same coarse path through the reactor. This has two deleterious effects: only a narrow path of the cylindrical wall would be used for heat transfer; and the residence time of the solid particles is only a fraction of what it would be with a tighter helical path.

The vortex tube reactor which we developed is shown in Figure 2 and is about 14 cm in diameter and 69 cm in length. It has some unique features, which were found necessary to achieve the desired reactor performance in the fast pyrolysis of biomass. The carrier gas is usually steam between 500 and 1000 kPa, depending upon the desired flow rate, and passes through a supersonic nozzle. Biomass in the form of minus 3-mm sawdust is metered into, and is entrained by, the supersonic carrier gas stream. Cold-flow studies in a clear plastic model with 4000-frames-per-second movie coverage established that this entrainment method results in rapid acceleration of the sawdust particles to velocities over 125 m/s. The cold-flow studies also verified that the entrained particles were following the reported coarse path of the gas flow near the wall. This coarse helical path appeared to be independent of the entrance angle, the entrance duct shape, and the flow rate of the carrier gas (13). The pitch angle of the solids flow in the conical section of a conventional cyclone was observed to be about one-fifth that in the cyclindrical section (11), but the cylindrical vortex tube has more heat-transfer surface area per unit length. To force the entrained particles into a tight helical path, the 316 SS cylindrical vortex tube wall was machined to leave a 3-mm high and 3-mm wide raised helical rib. High-speed movies taken through a clear plastic bulkhead with cold-flow verified that the raised rib forced the solids to take the desired tight helical path (14). A tracer gas experiment, following the progress of propane pyrolysis, verified that this reactor design was essentially plug flow, with the inner vortices contributing a very small amount of internal recycling (15).

Vortex Reactor Operation

Initial operation with this vortex tube as a reactor for the fast pyrolysis of biomass was with heating the reactor wall to relatively high temperatures of 800°C or so. At that time, the goal was to crack the primary vapors to gases, rather than the preservation of the primary oils. Under these conditions, the sawdust had ample time to pyrolyze, as well as the char having time to partially gasify to produce char yields of only about 5%. However, as the vortex reactor wall became hotter, the tendency increased to accumulate a layer of secondary tar and char on the wall. By reducing the wall temperature to 625°C, the buildup of an insulating char-tar layer became negligible, but the rate of pyrolysis of the sawdust particles was so low that about 30% of the feed could be recovered in the char cyclone as scorched feed. A tangential exit was then added to the vortex tube reactor

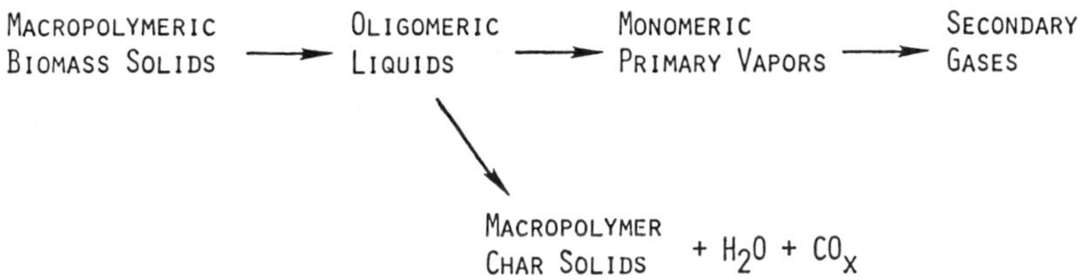

Figure 1. Global Reactions in Fast Pyrolysis

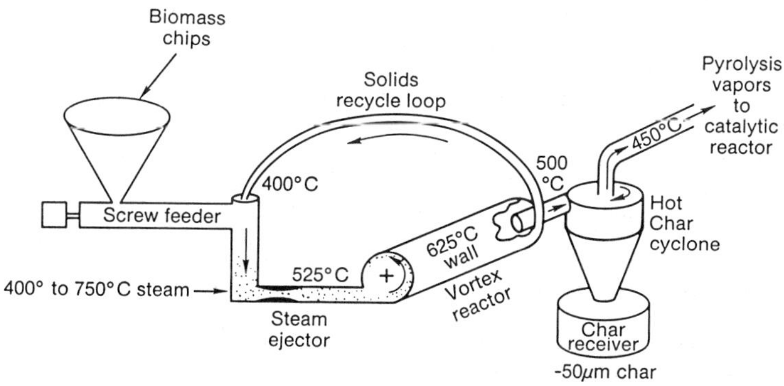

Figure 2. Vortex Reactor Schematic. (Reproduced with permission from ref. 19. Copyright 1985 Solar Energy Research Institute.)

to allow the unreacted feed and large char particles to be recycled to the entrance of the reactor as shown in Fig 2. The carrier gas nozzle acts as an ejector to create the pressure differential to drive the recycle loop. The recycle stream blows the sawdust off the feeder screws to positively entrain the feed to the carrier gas ejector.

The temperature at the exit of the insulated, but unheated, recycle loop is typically 400° to $450^{o}C$. The carrier gas is preheated to between 600° and $700^{o}C$ prior to expansion through the ejector nozzle. With these conditions, the temperature of the pyrolysis stream is 480° to $520^{o}C$, as it exits the vortex reactor system. About 10% of the feed is converted to char, which is recycled with the scorched feed until it is attrited to less than 50 micrometers in size. The vortex reactor system acts as a particle size classifier, with the char fines being entrained out of the vortex system to be removed by a cyclone separator having a higher collection efficiency. The fine char has a volatile content of 15% to 20% and burns readily, especially when hot. The bulk density of the fine char is between 0.18 and 0.24 g/mL, depending upon whether it was freshly poured or has been allowed to settle (the bulk density of the sawdust feedstock was 0.24 g/mL). The empirical formula for this volatile char is $CH_{0.53}O_{0.12}$, and it has a heating value (HHV) of 33 kJ/g (14,000 Btu/lb). A microscopic examination of the char fines shows that the particles have the appearance of broken thin-walled tubes; i.e., charred and broken cell walls. Although the reactor has not been optimized to maximum throughout, sawdust feeding rate of 22 kg/hr have resulted in good operation with a steam-to-biomass ratio of 1.0 (Run 93).

As noted above, the primary vapors are cracking significantly even at $500^{o}C$ and a residence time of half a second. If the recycle loop of the vortex reactor is removed, the yield of permanent gases is about 4%, based on the reacted feed. The initial gases, which are formed under these conditions, are extraordinarily rich in carbon dioxide and are associated with the formation of char. With the recycle loop open, some of the primary pyrolysis vapors are recycled along with the carrier gas, unreacted solids, and large char. The additional time, which the recycled primary vapors spend in the vortex reactor leads to a small loss in the yield of primary vapors and a higher yield of noncondensible gases of about 7% to 12%. The composition of the gases shifts considerably from the initial gases formed to that associated with a small loss of primary vapors. An even greater shift in the gas composition occurs with more extensive cracking of the primary vapors to produce an assymptotic gas composition as the primary vapors near extinction, which is low in carbon dioxide, as shown in Table I.

The experimental determination of the feed consumed, the char yield, and the noncondensible gas yields are relatively straightforward. However, the primary vapor and water yields have proven difficult to measure directly due to the formation of aerosols. These aerosols escape high-pressure sprays, cyclonic separators, and impingement or inertial collection techniques. The use of condensible steam as the carrier gas makes the water yield very sensitive to small measurement errors in the steam carrier gas flow. The use of noncondensible gases as the carrier tends to strip the volatile organics and the water of pyrolysis from the condensate. These considerations have led to the use of a noncondensible carrier gas, nitrogen, and to the determination of the water formed during pyrolysis and the primary oil yield by difference. By analyzing the recovered condensate for water, the yield of water may be determined. These techniques led to the conclusion that yields of about 70% primary vapors were achieved, based on taking the difference between the sawdust fed and the measured gas flow and char collected, correcting for the water content of the condensates. After elemental compositions were obtained for the feed and the collected products, an elemental balance was computed which verified the high primary vapor yields of 69 to 77 wt %, as shown in Table II, based on the yield of recovered char (2). However, our best recovery of dry oils has been 55 wt % of the dry feed with a 94% mass closure, when operating the system with 9 kg dry feed per hour and a nitrogen carrier gas flow rate of 20 kg per hour.

The primary pyrolysis oils, which have been recovered from the vortex reactor, are highly oxygenated and have nearly the same elemental composition as the biomass feedstock. The oils have a dark brown color and are acidic with a pH between two and three. The heating value (HHV) of the dry oils is 20 to 22 kJ/g (8700 to 9500 Btu/lb). The oils can absorb up to about 25% water before forming two phases. The viscosity of the oils was 1300 cp at 30°C, and the density was 1.3 g/mL (16). Although the primary vapors have a low molecular weight as determined by the FJMBMS (17), they rapidly polymerize upon physical condensation to form high-molecular-weight compounds in the oils (18). Attempts to slowly distill the oils led to the rapid polymerization of the oils boiling above 100°C (16). The oils have significant chemical activity, which suggests their potential use in low-cost adhesives, coatings, and plastics.

Scale-up Potential

The concept of supplying heat through the wall of a vortex reactor to drive endothermic processes is in its early development. The scale-up potential of this concept depends upon the angular momentum of the swirling carrier gases to keep the entrained feed particles moving on the wall. The heat flux delivered to tubular pyrolysis reactors typically ranges between 5 and 15 W/cm^2 (20,000 to 50,000 Btu/hr-ft^2). Reported data for vortex tubes indicates that with diameters larger than 2.5 cm most of the angular momentum is retained even after traveling a tube length equivalent to 20 tube diameters. The major momentum losses are due to the frictional contact of the solids and the gases with the vortex-tube wall. With larger vortex tubes needed for scale-up, the angular momentum of the process stream will increase more than the frictional losses. The heat transferred to the reactor will scale by the product of the diameter and the length. These considerations have led to calculations which suggest that a vortex reactor with a 250 TPD capacity would have a diameter of only

Table I. Pyrolysis Gas Composition at Various Cracking Severities (mol %)

	Vortex Exit Gases w/o Recycle (Run 34)	Vortex Exit Gases w/Recycle (Run 58)	Severely Cracked Vapor Gases (Run 58)
H_2	3.4	8.3	17.8
CO	46.2	49.2	52.2
CO_2	43.1	27.6	7.5
CH_4	4.6	8.9	12.0
C_2H_2	--	0.1	1.1
C_2H_4	1.3	2.4	5.9
C_2H_6	0.3	0.7	0.6
C_3H_8	0.1	0.1	--
C_3H_6	0.4	0.8	0.8
C_4H_8	0.3	0.3	0.6
C_5+	0.5	1.4	1.4
wt % yield of gases	f4	f6	65%

Source: Reprinted with permission from ref. 12. Copyright 1983 Solar Energy Research Institute.

Table II. Elemental Balance for Fast Pyrolysis to Primary Vapors

Feed T Primary Vapors + Water + Char + Gas

$$CH_{1.4}O_{0.62} \text{ T } W\ CH_{1.2}O_{0.49} + X\ H_2O + Y\ CH_{0.53}O_{0.12} + Z\ CH_{0.38}O_{1.33}$$

Exp. Char Yield	Calculated Product Values, Wt %		
	Primary Vapors	Water	Prompt Gas
7.5	76.8	11.7	4.0
10.5	73.1	12.8	4.1
12.7	69.0	14.0	4.3

Source: Reprinted with permission from ref. 12. Copyright 1983 Solar Energy Research Institute.

about 0.5 m and a length of 9 to 12 m. The fabrication technique would most likely be by the welding up of a spirally wrapped tube to form the raised, helical rib.

Conclusions

For the fast pyrolysis of biomass, a vortex reactor has significant advantages for the production of primary pyrolysis vapors, including: high heat transfer coefficients which allow the use of moderately low temperatures of the vortex reactor walls to supply the endothermic heat of pyrolysis; separation of the partially pyrolyzed feed particles from the char; the ability to recycle the partially pyrolyzed feed; the ability to accept a wide spectrum of particle sizes in the feed; short gaseous residence times; and preferential heating of the solid feed over the vapor stream to preserve the primary vapors. Primary pyrolysis vapor yields in the 70% range have been calculated by mass balances and verified by elemental balances, although physical collection of these vapors has proven to be elusive due to the formation of persistent aerosols and due to the volatility of the vapors in the carrier gas (methods to recover these vapors more completely with practical means are under development).

Acknowledgments

The financial support for the development of the vortex reactor and the catalytic conversion of the primary vapors to high-octane gasoline has been provided by the Biofuels and Municipal Waste Technology Division of DOE (FTP 646) with Mr. Simon Friedrich as the DOE program manager and Dr. Don J. Stevens of PNL as the technical monitor. The physical recovery of the primary oils for use as chemicals and adhesives is being developed through the financial support of the Office of Industrial Programs of DOE with Mr. Al Schroeder as the DOE program manager (FTP 587).

Literature Cited

1. McGuinnes, E. A. Jr.; Harlow, C. A.; Beall, F. C. Proc. Plant Sci. Appl. SEM, 1976, p. 543.
2. Diebold, J. P., MS Thesis T-3007, Colorado School of Mines Golden, CO, 1985.
3. Diebold, J. P., In Proceedings of Specialists Workshop on Fast Pyrolysis of Biomass, Diebold, J., ed.; Solar Energy Research Institute, Golden, CO 1980; SERI/CP-622-1096; p. 237.
4. Lede, J.; Panagopoulos, J.; Li, H. Z.; Vettermaux, J. Fuel 1985, 64, 1514-1520.
5. Reed, T. B., ACS Symposium on Production, Analysis, and Upgrading of Pyrolysis Oils from Biomass, Denver, CO, April 5-10, 1987.
6. Szekely, J.; and Carr, R. Chem. Eng. Sci., 1966, 21, 1119-1132.

7. Lede, J.; Verzaro, F.; Antoine, B.; Villermaux, J. In Proceedings of Specialists Workshop on Fast Pyrolysis of Biomass, Diebold, J., ed.; Solar Energy Research Institute, Golden, CO, 1980, SERI/CP-622-1096, p. 327.
8. Diebold, J. P.; Benham, C. B.; Smith, G. D. Monograph on Alternate Fuel Resources, Hendel, F., ed.; AIAA Monograph Series 1976, 20, 322-325.
9. Hartnett, J. P.; and Eckert, E. R. G., Trans. ASME 1957, May, 751-758.
10. Scheller, W.; Brown, G., Ind. Eng. Chem. 1957, 49, No. 6, 1013-1016.
11. ter Linden, A., Inst. Mech. Eng. J. 1949, 160, 233-251.
12. Diebold, J. P.; "Entrained-Flow, Fast Pyrolysis of Biomass. Annual Report. 1, December 1982 - 30, September 1983", 1984, Solar Energy Research Institute, Golden, CO, SERI/PR-234-2144
13. Diebold, J. P.; Scahill, J. W. In Proceedings of the 13th Biomass Thermocontractor's Meeting, Pacific Northwest Laboratory, Richland, WA, 1981, CONF-8110115. PNL-SA-10093; p. 332.
14. Diebold, J. P.; Scahill, J. W. Proceedings of the 15th Biomass Thermocontractor's Meeting, Pacific Northwest Laboratory, Richland, WA, 1983, CONF-830323, PNL-SA-11306, p. 300.
15. Diebold, J. P.; Scahill, J. W. AIChE Winter National Meeting, Atlanta, GA, 1984, paper 31d.
16. Elliott, D. C.; "Analysis and Comparison of Biomass Pyrolysis/Gasification Condensates - Final Report" Pacific Northwest Laboratory, Richland, WA, 1986. PNL-5943/UC - 61D.
17. Evans, R. J., Appendix C in "Entrained-Flow, Fast Pyrolysis of Biomass, Annual Report, 1 October 1983 - 30 November 1984," by J. P. Diebold and J. W. Scahill, Solar Energy Research Institute, Golden, CO, 1985. SERI/PR-234-2665.
18. Johnson, D. K.; Chum, H. L., ACS Symposium on Production, Analysis, and Upgrading of Pyrolysis Oils from Biomass, 1987, Denver, CO. pp. 167-177.
19. Diebold, J. P.; Scahill, J. W.; Evans, R. J. Entrained-Flow, Fast Ablative Pyrolysis of Biomass. Annual Report 1 December 1984-31 December 1985, 1986, Solar Energy Research Institute, Golden, CO. SERI/PR-234-3012.

RECEIVED April 14, 1988

Chapter 5

Conditions That Favor Tar Production from Pyrolysis of Large, Moist Wood Particles

Marcia Kelbon, Scott Bousman, and Barbara Krieger-Brockett

Department of Chemical Engineering, BF–10, University of Washington, Seattle, WA 98195

The experiments described in this paper will aid in identifying favorable feedstock properties and process settings for tar production from large (about 1 cm) wood particles. Conversion of biomass to pyrolytic oils will be greatly facilitated if large, undried particles can be used directly as a feedstock. The experiments have been conducted using well-defined, reactor-independent conditions and the results indicate that an optimum moisture content can facilitate tar production. The reacted fraction that is tar is greater for intermediate moisture contents, particle sizes, and heating rates, and the optimum tar production conditions for moist feeds change with both heating intensity and particle size.

The economics of making pyrolytic oils from biomass will be improved if large, and perhaps moist, particles can be used directly as a feedstock. This is a result of the high cost of size reduction and drying of fibrous and often green biomass. Both present and planned industrial biomass converters employ large particles, usually the readily available wood chips, sometimes using them in fluidized beds which are less sensitive to feed size distributions (1). However, the pyrolysis of realistically sized particles (about 0.01 x 0.01 x 0.005 m or greater) is complicated by non-uniform temperature profiles within the particle, and the synergistic effects that heating intensity, particle moisture, and size have in altering the intraparticle temperature histories.

Selectively producing from wood a particular pyrolysis fraction such as tar, or specific components in any fraction, is difficult. To enhance selectivity, the majority of studies to date have used finely ground (< 100 mm diameter) wood samples in which heat transfer rates are rapid enough to cause a uniform particle temperature, and mass transfer is fast enough to minimize secondary reactions. For these powders, pyrolysis with a high heating rate to moderately high temperatures generally favors tar formation, if the

0097–6156/88/0376–0041$06.00/0

tar can be quickly removed from the reaction zone. However, in wood chip pyrolysis, the particle interior heats slowly owing to its low conductivity, and secondary reactions of tar are significant. As yet, the best conditions for tar formation from chips have not been identified.

Since realistic wood feedstocks are heterogeneous, the uniform entity appropriate for fundamental studies to identify improved conditions for tar formation is the single particle. In it, the intraparticle conditions can be measured and related to process conditions that can be manipulated. In addition, the effect of heat and mass transfer rates on reaction products can be determined. The findings can be rationalized to all reactors in which the studied experimental conditions prevail. The investigation of reacting single particles has proven extremely successful in the development of catalytic reactors (2) and coal pyrolysis (3,4,5).

One goal of this study was to investigate the synergistic influence of particle size, heating intensity, and moisture on particle temperature, a dependent variable for large pyrolyzing particles. The effect of thermal or heat transfer properties was studied by manipulating the density (6). These manipulated variables, by determining the temperature profile, also determine the tar yield and composition. The 0.005 to 0.015 m thick particle size range was investigated because of its practical importance. In addition, Kung (7) demonstrated that this size range spans the transition from flat temperature profiles to steep temperature gradients for the heating intensities normally found in industrial converters. One dimensional (1-D) heating has been used experimentally (8,9) and typifies the type of heat transfer experienced by most wood chips owing to the large aspect ratios they usually have (10). One dimensional heating intensity is made precise experimentally by controlling the applied heat flux at one face of a wood cylinder with insulated sides. The lowest heating intensity studied, 8.4×10^{-4} W/m^2 (2 cal/cm^2-s), barely chars thick samples (or causes smoldering combustion if oxygen is present). The highest, 2.5×10^{-3} W/m^2 (6 cal/cm^2-s), is typical in furnaces or high temperature reactors designed for maximum heating of particles.

Biomass moisture varies with species and age as well as with growing region. Typical moisture contents (dry basis) found for feed piles in the Pacific Northwest range from 10% to 110%, hence the choices in our experiments (11). Simultaneous variations in two or three of the above experimental conditions were systematically investigated using a Box-Behnken experimental design (12).

Experimental Details

Apparatus. A description and diagram of the single particle pyrolysis reactor appears in Chan, et al. (13) but a brief presentation is given for completeness. A wood cylinder was placed snugly in a glass sleeve and reactor assembly. One face of the cylinder was heated by an arc lamp. It provided radiative, spatially uniform, 1-D axial heating as verified by absolute calibration of the heat flux (6,14, 15). The heating period was the same for all experiments, 12 min, chosen specifically to be sufficiently short to enable the study of active devolatilization, not char gasification, in the thinnest wood cylinders. Thermocouples in the wood at 2, 4, and 6 mm from the heated face measured the temperature. An infrared pyrometer, mounted off-axis from the arc lamp beam, approximately measured the surface temperature of the pellet. The design of the glass reactor and baffle allowed the uniform irradiation but prevented volatiles from condensing on the window. A large helium carrier gas flowrate was directed past the heated surface of the particle. The carrier quenched the volatiles and swept them without significant backmixing or reaction (6) to the analysis system. The helium pressure on the unheated face was slightly elevated to ensure volatiles flow in the particle toward the heated face, which in turn ensured maximum volatiles recovery. It has been verified (8) that during devolatilization of a large authentic particles, nearly all of the volatiles flow toward the heated surface owing to the decreased porosity behind the reaction front. A cold trap (packed with glass wool and at -40 C) immediately downstream from the reactor condensed tars and water from the volatiles. Permanent gases flowed through this trap without condensation and were sampled near its exit at preselected times during the experiment using two automated gas sampling valves. This provided information on evolution rates of gaseous products. All volatiles were later analyzed by gas chromatography.

<u>Sample Preparation</u>. The wood pellets were all cut and sanded, with great attention to grain direction, from uniform sapwood sections of the same lodge pole pine *(Pinus contorta)* tree provided by Weyerhauser Co. (Corvallis Mill). The cylinders were oven-dried at 90 C for at least three weeks. Dry particles used in some of the pyrolyses attained an equilibrium moisture content of about 5% during handling at the normal laboratory conditions. Moisture was quantitatively added (for some pyrolyses) using a microsyringe and balance, and allowed to come to a uniform distribution as described by Kelbon (14).

<u>Gas Analysis</u>. Permanent gases, operationally defined as all components passing through the -40°C cold trap, were collected in 30 stainless steel sample loops of known volume. Immediately after the experiment, samples were automatically injected into a Perkin-Elmer Sigma 2 Gas Chromatograph using a Supelco 100/120 mesh Carbosieve S column 0.5 mm diameter and 1.5 m long. The peaks were integrated and quantified by a Perkin-Elmer Sigma 15 Chromatography Data Station. Both a thermal conductivity detector and a flame ionization detector were used as described in Bousman (15).

<u>Tar Analysis</u>. The tar is washed from the trap and a higher molecular weight, but small amount, was washed from the reactor. An aliquot of the former was analyzed for water and low molecular weight compounds by gas chromatography using a Supelco 80/100 mesh Porapak Q column 0.5 mm diameter and 0.5 m long. The remainder from the tar trap was treated with approximately 2 g of $MgSO_4$ to remove water and then filtered. A known quantity of p-bromophenol was

added to both the trapped sample and reactor washings as an internal standard. Each was concentrated using a Kaderna Danish evaporator in order to remove THF while minimizing the loss of the other compounds. The resulting high molecular weight tar fraction was analyzed by gas chromatography using either a J and W Scientific DBWAX or Carbowax 20M fused silica capillary column 0.25 mm ID, 30 m long with a Perkin-Elmer splitting injection port. Only major peaks which appeared in all samples are presented in this paper as composite, chromatogram-like figures. The component identification number versus absolute yield is plotted for several conditions.

Char Analysis. The fragile char was carefully separated with a scalpel from the light color, unreacted portion of the wood pellet, and each fraction weighed. A qualitative chemical analysis of some char samples was performed using a Fourier Transform Infrared Spectrometer (FTIRS) with preparation as described in Bousman (15). The surface areas of some chars were also characterized using CO_2 adsorption.

Experimental Design and Data Analysis. With such lengthy experiments and chemical analyses, only a fraction of the independent variable combinations could be studied. These were picked according to a type of incomplete factorial experimental design which facilitates identification of optimum conditions (12). The resulting measurements (Table 1) were fit to empirical models of the class given in Equation 1 (12,16) to identify and quantify major effects:

$$y_k = b_0 + \Sigma\, b_i x_i + \Sigma\Sigma\, b_{ij} x_i x_j, \quad \text{for } i< j \qquad (1)$$

where

y_k = k-th product yield of interest, (or other dependent variable)

x_i, x_j = independent variables (process controllables) and

b_0, b_i and b_{ij} = least square parameter estimates obtained from multiple regression.

The independent variable settings were all equally spaced, and were scaled to a range between -1 and +1 making them dimensionless before fitting the data as shown in Table 1. In this way, the units or range of a studied variable do not contribute to the magnitude of the coefficients, the b's in Equation 1 above. The intrinsic importance of a squared term can also be determined by this scaling since the second order terms are of the same magnitude as the linear terms. The units of the coefficients b are the same as the dependent variable, y, and represent changes in it for a now "unit change" in the independent variable from the mean value studied. When process controllables such as moisture and particle size combine to affect the pyrolysis in a non-additive (synergistic) way, as when the last term of Equation 1 is large, non-linear dependence of yields on the process controllables is demonstrated. Trend lines in subsequent figures are plots of Equation 1 for particular values of the other variables.

Results and Discussion

The range of intraparticle temperatures measured for the pyrolysis conditions studied are shown in Figures 1-3. The temperatures achieved are consistent with those found in other studies for both small (17-19) and large (20) wood particles. Comparing Figure 1 to Figure 2 illustrates the effect of initial moisture at 110% and 10%, respectively, on temperature history. When moisture is present during pyrolysis, the temperature rises to a plateau at 100°C until the water has locally evaporated, and then more slowly attains a somewhat lower ultimate temperature than found in drier wood particles at the same conditions (see Figure 2). The duration of the 100°C plateau and time to reach ultimate temperature are prolonged the greater the initial moisture of the particle. The surface temperature measurement in Figs 1 and 2 is of some qualitative value but once volatiles are produced, the infrared pyrometer cannot accurately "see" the surface and the measured temperature is artificially low. In Figures 1 and 2, examine the relatively greater temperature difference (~200°C) between the traces at 2 and 4 mm depth compared to that (~50°C) between 4 and 6 mm. It is indicative of the steep temperature gradient experienced by the 15 mm thick particle. This is in contrast to the shallow temperature gradients in Figure 3. Presented are the temperatures in the very moist, 10 mm thick particle for the highest and lowest heating intensities studied (8.4 and 25.1 x 10^{-4} W/m^2), respectively. For low heating intensities, the temperature gradient is slight as suggested by Kung (7), but all zones of the particle pyrolyze at low temperature. At 25.1 x 10^{-4} W/m^2, the temperature of pyrolysis is high and in the range that favors tar formation, but the temperature drops off rapidly with distance from the surface. Note that reaction zones near the particle surface experience quite high heating rates even for moist particles, which should favor tar production early in the devolatilization of a large particle.

Fractional yields are an important measure of selectivity and are useful for downstream separation process considerations. When interpreting the fractional yield data from large particle pyrolysis, several points should be kept in mind. The individual product yields are expressed as weight fractions of *that which reacted* in the 12 min pyrolysis duration. The fractional yields presented here are directly comparable to data from other investigators if in those studies, particles had sufficient time to react completely, but not undergo extensive char gasification after all the volatiles were gone. The fraction of the particle which reacts in the fixed pyrolysis time is given in Table 1. It can be seen that the 12 min duration successfully studies the active devolatilization period because all but 1 of the % reacted values are under 100%. A second point is that the reported yields reflect the non-uniform temperature histories presented in Figures 1-3.

The appropriate measure of reproducibility of the experiments is the standard deviation for tar yield calculated from 3 replicate runs, and these appear at the bottom of Table 1. Their standard deviation is 2%. However, because of the inherent difficulty in recovering all products, especially tars, the mass balances do not always add to 100%. Thus, another measure of uncertainty is the

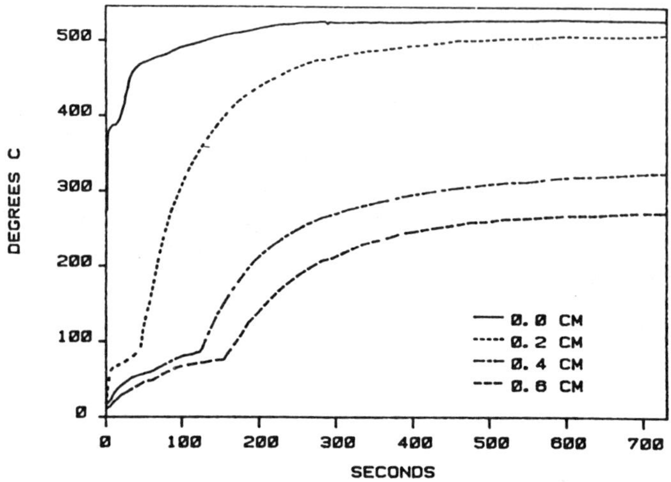

Figure 1. Temperature histories at the indicated locations in a 1.5 cm thick wood particle pyrolyzed with a constant heat flux of 1.6E-4 W/m^2. Initial particle moisture was 110% (dry basis).

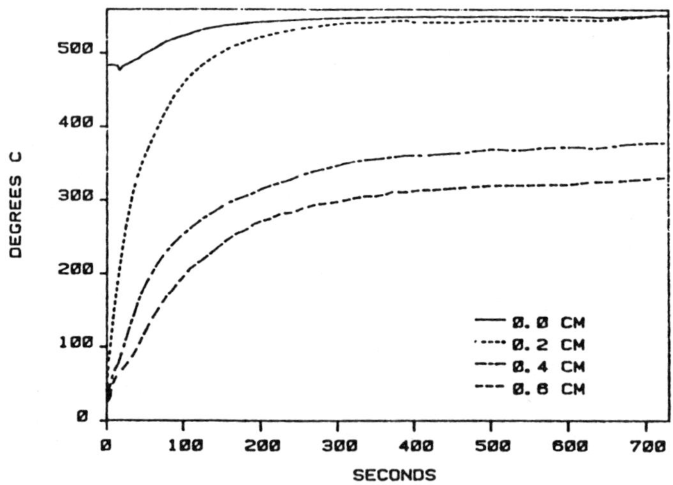

Figure 2. Temperature histories at the indicated locations in a 1.5 cm thick wood particle pyrolyzed with a constant heat flux of 1.6E-4 W/m^2. Initial particle moisture was 10% (dry basis).

Table 1 - Experimental Conditions and Pyrolysis Product Yields

Actual values of independent variables:			Scaled values of independent variables:					
moisture content %	heat flux W/m^2	particle thickness cm	M.C.	flux	thickness	percent reacted	% tar	% mass balance
110	2.5E-3	1	1	1	0	73.3	41.8	84.2
110	8.4E-4	1	1	-1	0	12.0	53.4	71.5
10	2.5E-3	1	-1.	1	0	95.6	51.4	83.0
10	8.4E-4	1	-1	-1	0	7.1	40.6	98.8
110	1.6E-3	1.5	1	0	1	33.5	60.8	85.0
110	1.6E-3	0.5	1	0	-1	78.5	63.8	61.6
10	1.6E-3	1.5	-1	0	1	35.6	62.6	83.7
10	1.6E-3	0.5	-1	0	-1	97.9	64.4	74.4
60	2.5E-3	1.5	0	1	1	67.8	58.6	79.5
60	2.5E-3	0.5	0	1	-1	100.0	60.8	66.5
60	8.4E-4	1.5	0	-1	1	8.8	66.4	86.2
60	8.4E-4	0.5	0	-1	-1	16.2	60.9	73.8
60	1.6E-3	1	0	0	0	73.3	59.0	67.9
60	1.6E-3	1	0	0	0	59.7	58.4	83.7
60	1.6E-3	1	0	0	0	59.3	61.0	57.9
mean of the replicates						64.1	59.5	69.8
standard deviation of the replicates						8.0	1.4	13.0

discrepancy in the mass balance. For all the experiments with moisture added, not just the replicated, the average mass balance closure was 80%, with a standard deviation of 11%. For another set of experiments not presented in Table 1 but shown in Figure 4 for dry particles, the mass balance closes to within 6%. The discrepancy was usually assumed to be tar that condensed on the system surfaces.

The synergy, or interaction of particle size and heat flux to alter the tar yield can be clearly seen in Figure 4. Presented is the weight fraction water-free tar produced from the fixed duration pyrolysis of dry wood particles of several thicknesses. The trend lines result from regression of all the data to Equation 1. Recall that the experimental uncertainty as evidenced by the replicate runs is about 2%. The data points show runs in which the density was changed by a factor of three using compressed sawdust pellets (6). It can be seen to have a small effect. If there were no "combined" effect of heating intensity and particle thickness, the tar trend lines for different particle sizes would be essentially parallel. However, the reduction in tar for an increase in heating rate, as characterized by the slope of the trend lines at any point, increases as the particle thickness increases. The thinnest particle, 5 mm, produces nearly a constant tar yield between 30 and 40% when heated over this wide, practical range of intensities. The thickest particle pyrolyzed at the lowest heat flux produces over 65% tar, and it decreases to 20% at the maximum heating intensity investigated. This is consistent with the extensive tar cracking that likely occurs in the particle surface char where high temperatures prevail for severe heating. Note that when heating at the intermediate heat flux studied, dry particles of all thicknesses yield approximately the same 30%-35% tar. This condition represents a tar yield that is insensitive to particle size over this range.

The temperature profile in a dry particle is affected principally by only two process variables, the particle size and applied heat flux. Thus, it can be plotted in a graph such as Figure 4. However, when initial moisture of the particle is also varied in combination with these variables, the presentation of the data must be divided into several graphs, one for each particle size, for example. The yield of tar produced from pyrolysis of moist wood is presented in Figure 5 as a function of both initial moisture content (abcissa) and heating intensity for the thickest particle size used in this study, 15 mm. The uncertainty in the tar from wet particle pyrolysis is about 5%. Note that moisture causes the optimum tar conditions to occur at 16.7 x 10^{-4} W/m^2, rather than the 8.4 x 10^{-4} W/m^2 for dry particles (Figure 4), consistent with the higher heating intensity required to vaporize the water and cause pyrolysis. Overall, the level of tar produced is higher, 50-70%, than for dry wood particles. It can be speculated that the reason for the higher tar yield is a combination of a lower average pyrolysis temperature, as well as a reduced extent of cracking reactions as the volatiles leave the cooler particle. Hydrogen is measurable in the gases, as well.

The data in Figure 5 are difficult to visualize since tar yields depend on more variables than can clearly be presented in a

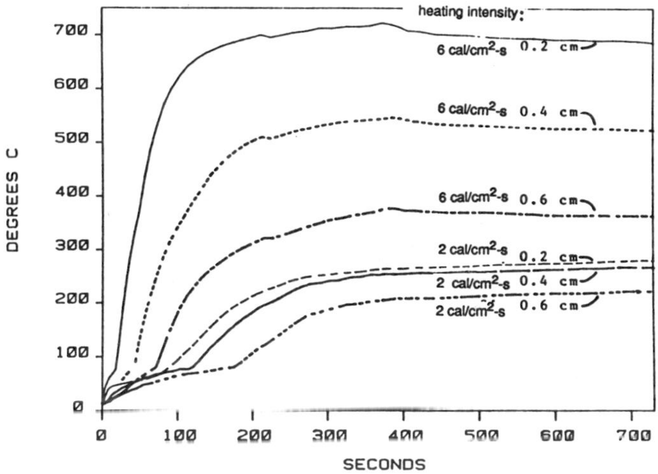

Figure 3. Temperature histories at the indicated locations in a 1.0 cm thick wood paricle pyrolyzed with two different heat fluxes. Initial particle moisture was 110% (dry basis).

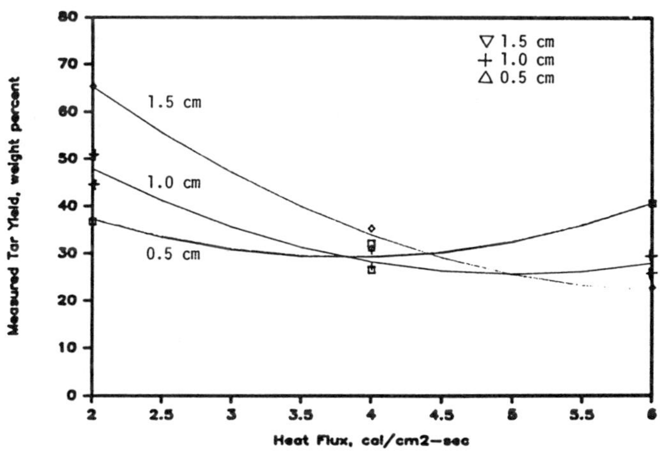

Figure 4. Weight fraction tar yield for the pyrolysis of *dry* wood particles at varying surface heat fluxes for three thicknesses; trend lines are least-square fits and symbols are some of the experimental data.

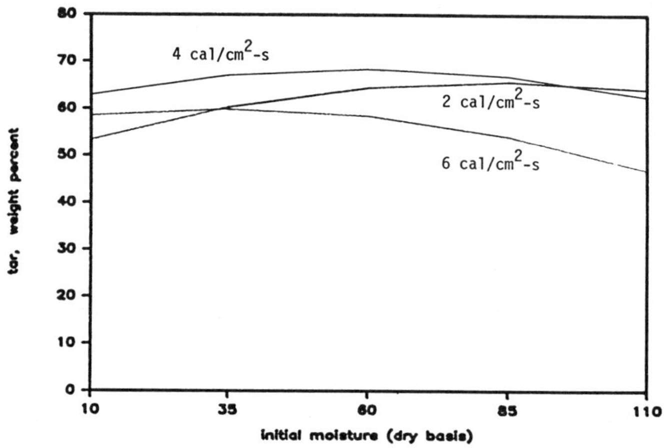

Figure 5. Measured weight fraction tar produced from a 1.5 cm thick wood particle as a function of moisture content for 3 different heating rates.

two dimensional plot. Another way to visualize this dependence uses contours of constant tar production predicted by the regression of all the tar yield data from the experiments where initial moisture was varied. Figure 6(a-c) shows how particle size and heating rate affect tar yield at each of the three initial moisture levels that were studied. Each contour is a 1% change in tar yield, and recall that the experimental uncertainty is estimated at 2% from replicates. Thus, the nature of the contours are not precisely known, but the graph serves to illustrate the complicated behavior of optimum yields. Note by the magnitudes on each contour line, that the tar yield increases from a "saddle point" for both larger and smaller particle thicknesses, but decreases from the "saddle" for large and smaller heating rates. Note the position of the saddle moves to lower heat fluxes for higher initial moisture of the particles being pyrolyzed. The highest tar yields observed approach 70% for the smallest and largest particles studied.

In Figure 7, a composition profile of some components of both the tar and gas fraction is presented for pyrolysis of a 1 cm thick, moist particle. Changes in this profile simply highlight changes in product composition as process conditions change, which is important information for the design of downstream separation operations. For different conditions that manipulate the particle temperature history after the fashion shown in Figures 1 3, different amounts of the particle react as can be verified in Table 1. If the amount reacted is normalized to represent 100%, the ordinate is the weight fraction of each component in the offgas. Thus all components plotted are each less than 2% of the moist wood particle pyrolysis products. The error as estimated from replicate determinations is about ±0.5% absolute, or 25% relative precision. The high molecular weight species composition appears to vary little as particle temperature is manipulated by changing both moisture and heating intensity of the pyrolysis. However, the low molecular weight tar compounds are in greater concentration for mild intensity heating (8.4×10^{-4} W/m^2) than for severe heating which appears to favor the hydrocarbon gases and hydrogen. Although the composition differences are nearly within the experimental error, moisture appears to slightly enhance the production of methanol and acetic acid for these experiments (1 cm particle).

Conclusions

Data has been presented which suggests that moisture can enhance the production of tar from the pyrolysis of large wood particles using conditions that occur in a large scale reactor where the heat flux a particle experiences is quite constant. The most favorable conditions result in about 70% of the reacted biomass becoming tar. If one assumes that the mass balance discrepancy results from tar condensing on reactor surfaces, this is a conservative estimate.

Acknowledgments

This work was supported by the National Science Foundation, Grant No. CPE-8400655. Marcia Kelbon gratefully acknowledges the partial support of a Weyerhauser Fellowship and Scott Bousman acknowledges the partial support of a fellowship from Chevron Corp.

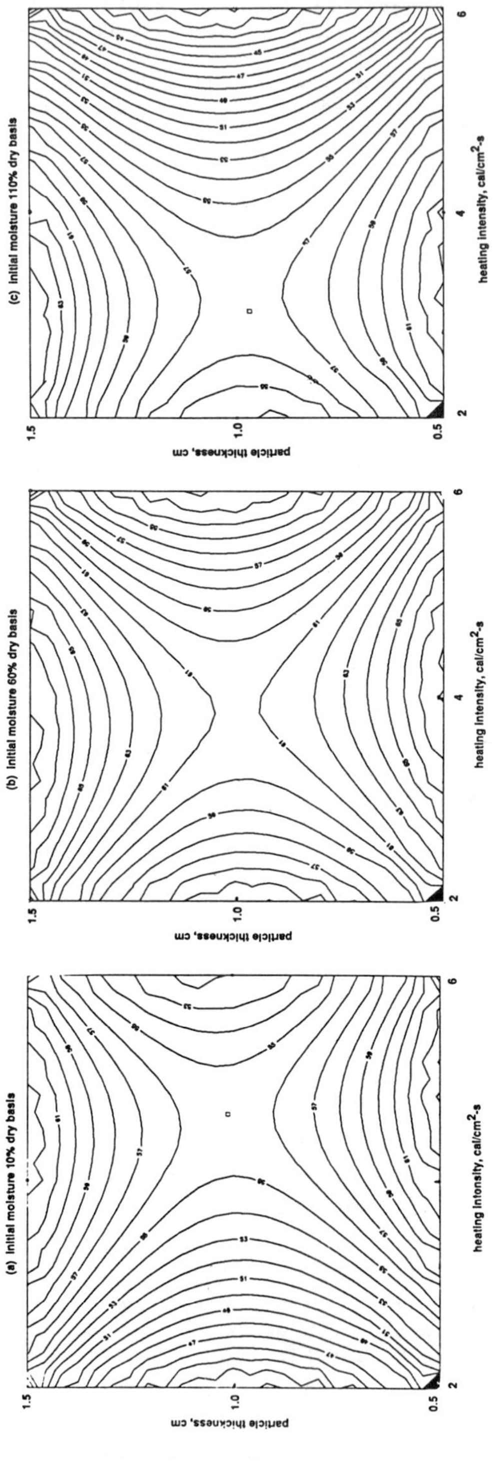

Figure 6. Contour lines of constant tar yield plotted against both particle thickness (vertical scale) and heating intensity (horizontal scale) for (a) 10% initial moisture; (b) 60% initial moisture; and (c) 110% initial moisture.

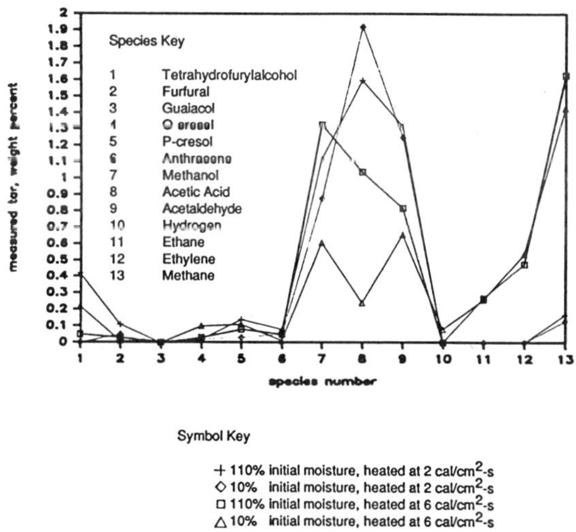

Figure 7. Tar product composition profile for 1. cm thick particles with two different initial moisture contents, heated at two different heating rates.

Literature Cited

1. Reuther, J.J.; Golike, G.P.; Taylor, P.H.; Karsner, G.G. In Applications of Chemical Engineering Principles in the Forest Products and Related Industries; American Inst. of Chemical Engineers, Forest Products Div.; Seattle, WA, 1986; p 160.
2. Petersen, E.E. In Chemical Reaction Analysis, Prentice Hall, N.Y., 1965.
3. Russel, W.A.; Saville, D.A.; Greene, M.I. AIChE J 1979, 25, 65.
4. Srinivas, B.; Amundson, N.R. AIChE J 1980, 26, 487.
5. Massaquoi, J.G.M.; Riggs, J.B. AICHE J 1983, 29, 975.
6. Chan, W.C.R. Ph.D. Thesis, Dept. Chem. Engrg., U. Washington, Seattle, WA, 1983.
7. Kung, H.C. Combustion and Flame 1972, 18, 185-192.
8. Lee, C.K.; R.F. Chaiken; Singer, J.M. 16th International Symposium on Combustion, Combustion Institute, Pittsburgh, PA, 1976, p 1459.
9. Ohlemiller, T.; Kashiwagi, T.; Werner, K. NBSIR 85-3127, Dept. of Commerce, National Bureau of Standards, Center for Fire Research, Gaithersburg, MD, April 1985.
10. Miles, T.R. In Thermochemical Processing of Biomass, Ed. by A.V. Bridgewater, Butterworths, London, 1984, p 69.
11. Chan, W.C.R.; Krieger, B.B., submitted to *I.E.C. Proc. Des. Dev.*, 1986.
12. Box, G.E.P.; Behnken, D.W. Technometrics 1960, 2, p 455.
13. Chan, W.C.R.; Kelbon, M.; Krieger, B.B. Fuel 1985, 64, 1505.
14. Kelbon, M., M.S. Thesis, Dept. Chem. Engrg., U. Washington, Seattle, WA, 1983.
15. Bousman, W.S. M.S. Thesis, Dept. Chem. Engrg., U. Washington, Seattle, WA, 1986.
16. Box, G.E.P.; Hunter, J.S.; Hunter, W.G. Statistics for Experimenters, Wiley, N.Y. 1978.
17. Scott, D.S.; Piskorz, J.; Radlein, D. I.E.C. Proc. Des. Dev. 1985, 24, 581.
18. Nunn, T.R.; Howard, J.B.; Longwell, J.B.; Peters, W.A. I.E.C. Proc. Des. Dev. 1985, 24, 836.
19. Bradbury, A.G.W.; Sakai, F.; Shafizadeh, F. J. Appl. Poly. Sci. 1979, 23, 3271.
20. Kanury, A.M. Combustion and Flame, 1972, 18, 75.

RECEIVED March 31, 1988

Chapter 6

Relation of Reaction Time and Temperature to Chemical Composition of Pyrolysis Oils

Douglas C. Elliott

Pacific Northwest Laboratory[1], P. O. Box 999, Richland, WA 99352

Biomass pyrolysis tars have widely varying chemical compositions. As a result of this variation, the physical properties of the tars also vary over a wide range. For short-residence-time processing there is a direct correlation between chemical composition and operating temperature. As the temperature increases the oxygen content decreases and the hydrogen to carbon ratio decreases. Biological activity, as measured by mutagenic and tumor-initiating activity, also correlates with temperature. The activity appears only in the high-temperature tars which contain high levels of polycyclic aromatic hydrocarbons. The effect of pressure and residence time on pyrolysis tar yield and quality is also shown. The addition of reducing gas and catalysts to the pressurized system results in product improvements which make comparisons to pyrolysis less straightforward.

The recognition of the production of tars by pyrolysis of wood dates to ancient times (1). The composition of these complex mixtures was studied over the years and many components were identified (2). Until the last several decades the results reported in the literature accurately portrayed the composition of wood pyrolyzate tar since essentially all of such tars were byproducts of charcoal production, a slow process operated at temperatures limited to less than 500°C. As the processing technology has developed during this century, the processing environment has moved to higher temperatures and shorter residence times in attempts to optimize product distribution to either liquid or gas products and minimize solid char production. As processes developed to minimize tar production (as for gasification) or maximize tar production (as for liquefaction), the variation in the tar composition has not always been recognized.

As a result of the study of biomass pyrolysis tar composition which has occurred over the past decade, it is now obvious that all biomass tars are not the same (3). When considering the tar product

[1]Operated by Battelle Memorial Institute for the U.S. Department of Energy under Contract DE–AC06–76RLO–1830.

0097–6156/88/0376–0055$06.00/0

of a given biomass conversion process it is not enough to deal only with the quantity of tar but the quality of the tar must also be considered. It is apparent that further processing, either chemically or biologically, of the tar as a byproduct or as the major product will be entirely dependent on the tar formation conditions which determine the tar composition.

Early recognition of the changes in tar composition as a result of pyrolytic processes was reported by Diebold in 1980 (4). The pyrolysis pathways reported (see Figure 1) depict the conversion of primary pyrolysis tars to two types of secondary tar. The "secondary tar" from Figure 1 is the high-pressure liquefaction product which has been widely studied. The other secondary tar is the "Vapor Phase Derived Tar." This material is also suggested as being favored at high pressure with further reaction at high pressure leading to carbon black. However, neither of these secondary tar formation pathways addresses the change in tar composition as it is held at high temperatures without high pressure. Only transient oxygenated fragments and gases are produced from primary tar at low pressure and high temperature according to these reaction pathways.

A somewhat more complete series of pyrolysis pathways was presented recently by Evans & Milne (5). In this series of reactions (see Figure 2) biomass is pyrolyzed at low pressure to primary oil which is further converted sequentially to oxygenates, aromatics, polycyclic aromatics and, eventually, soot. This series correlates well with the progression of chemicals earlier identified by this author (6) which is repeated below.

Mixed Oxygenates	→	Phenolic Ethers	→	Alkyl Phenolics	→	Heterocyclic Ethers	→	PAH	→	Larger PAH
400°C		500°C		600°C		700°C		800°C		900°C

These chemical groups are representative of the type of components found in condensate produced by short-residence-time processing of wood at the given temperature. As such these chemical groups suggest the types of chemical mechanisms occurring in high-temperature pyrolysis of biomass, i.e., dehydrogenation and condensation (aromatization), dealkylation and deoxygenation.

Experimental Systems

The data in this paper are drawn from a number of studies performed at Pacific Northwest Laboratory (3,7-9). We have examined biomass pyrolysis tar samples from many types of pyrolytic gasification and liquefaction systems through the U.S. Department of Energy support of domestic biomass thermochemical conversion research and international cooperative efforts. Many of these can be compared directly as a function of temperature since they are all produced at short residence time, approximately one second. Others produced at longer residence time or pressure require discussion separately.

The bulk of this chapter will deal with tar produced in short-residence-time reactors. There are four convenient groups in which to categorize these processes: conventional flash pyrolysis (10-13), high-temperature flash pyrolysis,(14) conventional steam gasification (15-17) and high-temperature steam gasification (18). The effects of these four process groups and their operating temperature regimes on product distribution is presented in Figure 3.

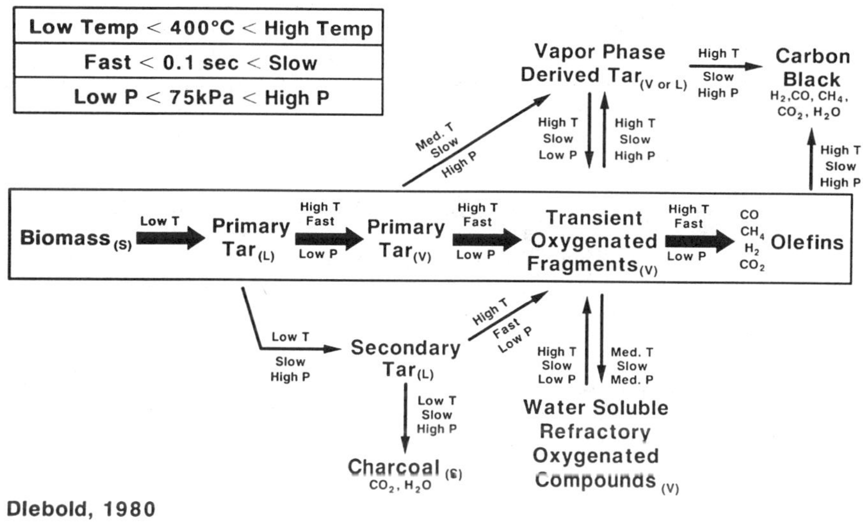

Figure 1. Biomass pyrolysis global mechanism, circa 1980

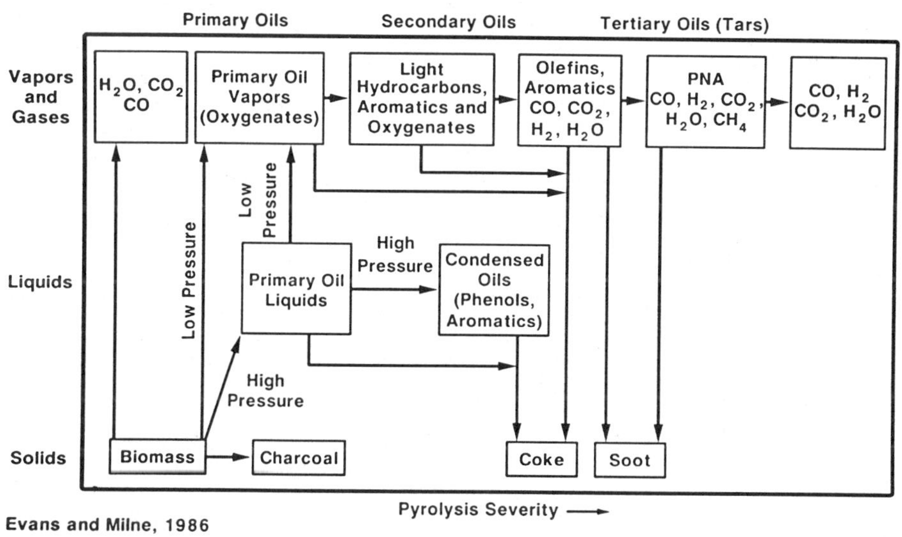

Figure 2. Biomass pyrolysis global mechanism, circa 1986

As has been reported numerous times previously in the literature the amount of tar product decreases steadily above 500°C while the amount of gas product increases. The amount of char is low even at 500°C and also decreases over the temperature range. The quality of both the tar and the char as indicated by the oxygen content changes dramatically.

Results and Discussion

Chemical Properties of Tars. In addition to the change in quantitative yield of wood pyrolysis tar there is also a substantial change in chemical composition of the tar as a function of temperature. As shown in Figure 4 the oxygen content as well as the hydrogen to carbon atomic ratio can be plotted vs. temperature to show a strong relationship. These data are for tar elemental compositions calculated to a water-free basis from analyses of crude tars. As stated earlier these data are from a number of different reactor systems yet the strong correlation is evident. With increasing processing temperature the oxygen content declines as does the hydrogen to carbon ratio. These changes represent the conversion of the highly oxygenated pyrolyzates to more thermally stable less oxygenated species which are eventually deoxygenated leaving highly aromatic structures.

The relationship of tar composition to processing temperature is further substantiated by the listing of chemical components of the tars as found in Table I. These chemicals were identified through the use of gas chromatography-mass spectrometry (for procedures see ref. 3). Since gas chromatography was used for separating these components, certain limitations of the data can be assumed. For example, low volatility of a compound will interfere with the identification of a compound in a gas chromatography system. Therefore, high-molecular-weight components would not be identified in this system nor would highly polar compounds. High-molecular-weight hydrocarbons, up to 276 molecular weight, have been identified in our analyses. Single-hydroxy, double-ring aromatic compounds

Table I. Chemical Components in Biomass Tars

Conventional Flash Pyrolysis (450-500°C)	Hi-Temperature Flash Pyrolysis (600-650°C)	Conventional Steam Gasification (700-800°C)	Hi-Temperature Steam Gasification (900-1000°C)
Acids	Benzenes	Naphthalenes	Naphthalene
Aldehydes	Phenols	Acenaphthylenes	Acenaphthylene
Ketones	Catechols	Fluorenes	Phenanthrene
Furans	Naphthalenes	Phenanthrenes	Fluoranthene
Alcohols	Biphenyls	Benzaldehydes	Pyrene
Complex Oxygenates	Phenanthrenes	Phenols	Acephenanthrylene
Phenols	Benzofurans	Naphthofurans	Benzanthracenes
Guaiacols	Benzaldehydes	Benzanthracenes	Benzopyrenes
Syringols			226 MW PAHs
Complex Phenolics			276 MW PAHs

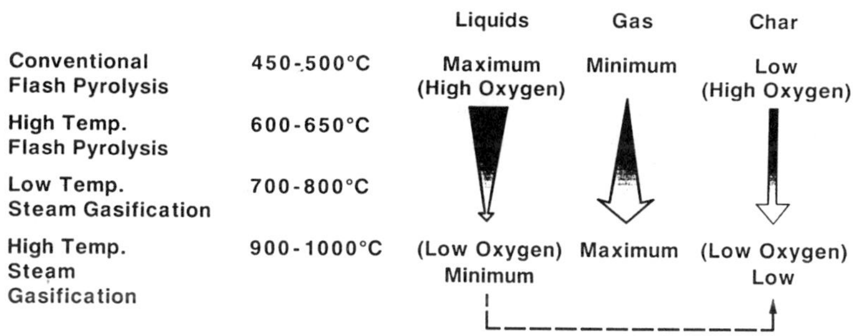

Figure 3. Biomass pyrolysis tar producing systems

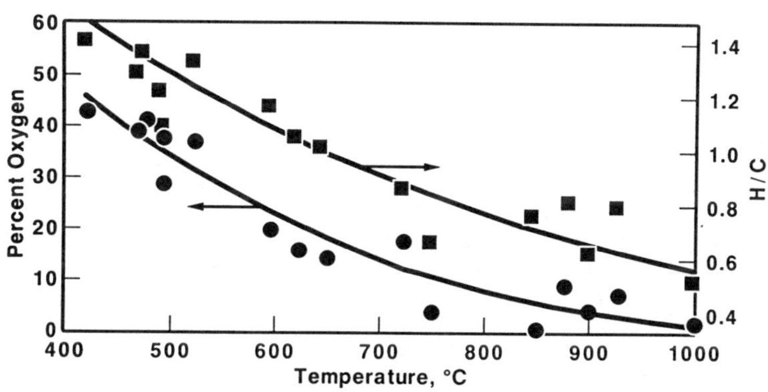

Figure 4. Elemental composition of dry tars produced over a range of temperatures

(phenylphenols and naphthols) seem to be the upper limit for high-molecular-weight polar compounds. Specific isomer identification is difficult with our type of analysis. Often no reference listing can be found for many of the alkylated aromatics and phenolics found in these tars. Therefore, identification is usually limited to functional type and molecular weight.

The listings in Table I further confirm the multistep transformation as a function of temperature of the highly oxygenated pyrolyzates to the condensed aromatics. The scission of labile bonds around aromatic rings to eliminate side chains containing carbonyl and methoxyl groups leads to catechols. For example, the pyrolysis of guaiacol to catechol is well documented in the literature (19). Other condensation reactions to produce benzofurans, biphenyls and naphthalenes from phenolics have also been recently demonstrated in these high-temperature environments (20). At higher temperatures the phenolics and furans disappear while the molecular weight of the more stable aromatics continue to increase. In addition to elimination of oxygen, alkyl side chains are no longer observed at the highest temperatures tested.

This same series of reaction pathways can also be found in the coal pyrolysis literature. The relationship of phenol and naphthalene content in coal tar is the same as that found for wood tar. As graphically presented in Lowry (21) naphthalene increases while phenol decreases as the coal tar processing temperature increases. Similarly, Figure 5 shows data for aqueous condensate components from wood gasification systems. However, in the case of Figure 5 the weight percent of all phenolics and C_9 to C_{12} aromatics are summed for the data presentation. The thermal environment at temperatures above 800°C leaves little difference between the coal tar composition and that produced from wood.

Physical Properties of Tars. Due to the differences in chemical composition of the tars, the physical properties also vary significantly. The pyrolytic oxygenates exhibit considerable water solubility. In addition to water, other lightweight polar compounds such as acetic acid, hydroxyacetaldehyde, methanol, and hydroxypropanone are present which also act as solvents. As a result, the conventional flash pyrolysis oil can be thought of as a solution of polar organics in a water, acid, ketone solvent. This solution has a high density (>1.2 g/ml) and a relatively low viscosity (60 cps at 40°C).

Higher temperature products show a trend to lower density indicative of the transformation of oxygenates to hydrocarbons. Viscosity can vary widely depending on the degree of water incorporation into the tar and the condensation temperature which determines the amount of light hydrocarbons collected in the tar. In the higher temperature ranges the high molecular weight hydrocarbons can be collected from the tar as toluene insoluble char. The amount of this material increases from 19.0 weight percent to 41.2 weight percent in tar samples recovered over the operating temperature range from 880°C to 1000°C (7).

Biological Properties of Tars. Just as physical properties of tars can be related to their chemical composition, so can biological activities of the tars. Biological activity can be categorized as either short-term, as in acute toxicity, or long-term, as in

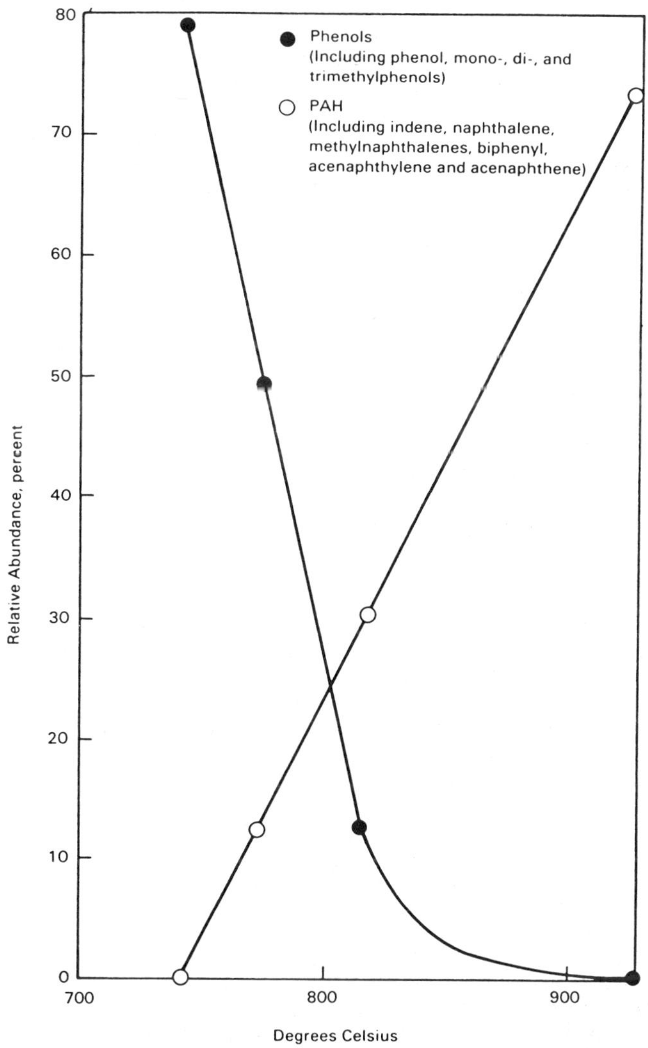

Figure 5. Shift in chemical composition of aqueous phase condensate from fluidized bed gasifiers

carcinogenicity which is often measured as mutagenicity or tumor-initiating activity. The short-term acute toxicity of these tars is a valid concern. However, one would expect a wide range of effects based on the variation in chemical composition. The lack of such toxicological tests for biomass tars is a major shortcoming of the current state of knowledge. Long-term toxicity as measured by Ames Assay for mutagenicity and mammalian cell tumor-initiating activity was measured in this study and was found to vary widely depending on the biomass processing temperature. Ames Assay was used as an initial survey for all the tars and aqueous condensates. Activity was limited to high-temperature process tars (> 800°C). Results of more lengthy tests of mouse skin painting and tumor counting (30 mice per treatment group) confirm the Ames Assay results. The results depicted in Figure 6 and 7 show that activity in the high temperatures was equal to or greater than that of the benzo(a)pyrene activity standard. The activity of the tar produced at 750°C was much less, yet still statistically significant. The low-temperature tar (485°C) showed no activity above that of the control. The activity in these tars is attributed to polycyclic aromatic hydrocarbons. For example, the concentration of benzo(a)pyrene in the high-temperature tars was so great that the amount added to the animal skin was actually greater than the amount of pure compound used as a standard. There were essentially no polycyclic aromatic hydrocarbons in the low-temperature pyrolysis tar.

<u>The Effect of Time and Pressure on Tar Properties</u>. An inadequate data base is currently available to truly evaluate the effect of time and pressure on biomass pyrolysis tar composition. Most pressurized pyrolysis tests with biomass have been performed at lower temperatures and longer residence times than flash pyrolysis. In addition, a reducing gas and catalyst are sometimes added in pressurized experiments. Considering the elemental composition of high-pressure processed biomass oils (<u>22</u>) a trade-off in temperature and residence time can be envisioned in order to make the data comparable to flash pyrolysis tars as in Figure 4. Typical catalyzed, high-pressure oil composition (12% oxygen, 1.23 H/C) (<u>9</u>) can be compared to flash pyrolysis oils produced at between 550°C (based on the H/C ratio) and 700°C (based on the oxygen content). Apparently due to the reducing gas and catalysis, the oxygen content is reduced while the hydrogen to carbon ratio remains higher in the high-pressure oils than is typical for flash pyrolyzates. However, the chemical components identified in the high-temperature flash pyrolyzates representative of this temperature (see Table 1) are not obviously different from those found in the high-pressure oils (<u>9</u>).

In the case of uncatalyzed, high-pressure processing, as reported by Boocock (<u>22</u>), the oil has a composition (24% oxygen, 1.10 H/C) which falls at the 600°C level for a flash pyrolyzate. In extending this comparison to its logical conclusion, one finds that a similar product can be produced at lower temperature (300°C less) with an increase in residence time (100 times) and increased pressure (100 times). Equivalent comparisons with the catalyzed case are not so straightforward, apparently due to different reaction mechanisms favoring an improved product composition, i.e., lower oxygen and higher hydrogen to carbon.

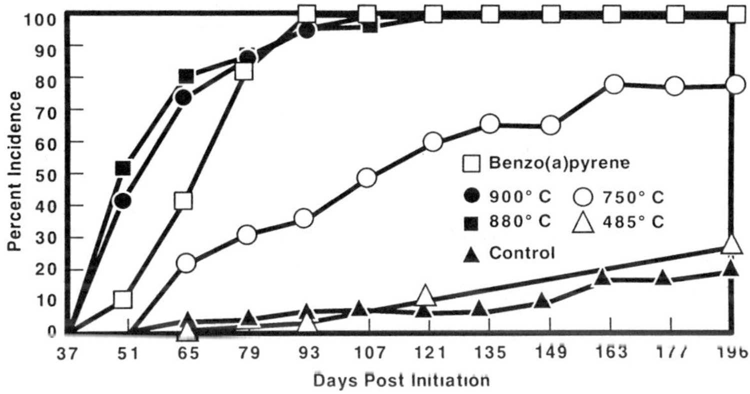

Figure 6. Tumor incidence as a percentage of animals affected

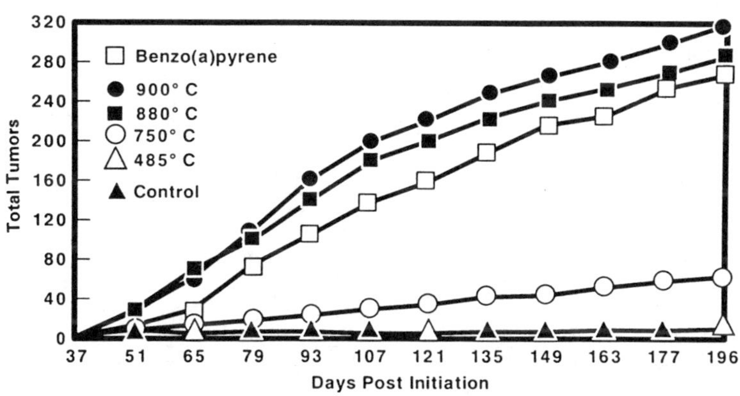

Figure 7. Total number of tumors resulting from skin application

Acknowledgments

The following researchers provided biomass pyrolysis tar samples, the analyses of which produced the data that are the basis for this paper:

M. J. Porter	Biomass Liquefaction Experimental Facility	Albany, OR
D. S. Scott	University of Waterloo	Waterloo, Ontario Canada
J. A. Knight	Georgia Tech Research Institute	Atlanta, GA
J. P. Diebold	Solar Energy Research Institute	Golden, CO
M. D. Brown	Pacific Northwest Laboratory	Richland, WA
V. J. Flanigan	University of Missouri-Rolla	Rolla, MO
L. K. Mudge	Pacific Northwest Laboratory	Richland, WA
S. P. Babu	Institute of Gas Technology	Chicago, IL
H. F. Feldman	Battelle Memorial Institute	Columbus, OH

Financial support for this research was provided by the Biofuels and Municipal Waste Technology Division of the U.S. Department of Energy, through the Biomass Thermochemical Conversion Technical Field Office at Pacific Northwest Laboratory. The author appreciates the support of Mr. Simon Friedrich, manager of the Biomass Thermochemical Conversion Program at DOE Headquarters and Mr. Gary F. Schiefelbein, Mr. Mark A. Gerber and Dr. Don J. Stevens of the Technical Field Office.

Literature Cited

1. Emrich, W. Handbook of Charcoal Making: The Traditional and Industrial Methods; D. Reidel Publishing Co.: Dordecht, Holland.
2. Soltes, E. J.; Elder, T. J. In Organic Chemicals from Biomass, Goldstein, I. S., Ed. CRC Press Inc.: Boca Raton, FL, 1981; Chapter 5, 74-81.
3. Elliott, D. C. Analysis and Comparison of Biomass Pyrolysis/Gasification Condensates - Final Report, PNL-5943, Pacific Northwest Laboratory: Richland, WA, 1986.
4. Diebold, J. In Proc. Specialists' Workshop on Fast Pyrolysis of Biomass, SERI-CP-622-1096, Solar Energy Research Institute, Golden, CO, 1980, p.3.
5. Evans, R. J.; Milne, T. A. Fundamental Pyrolysis Studies: Final Report 1 October 1980 - 30 December 1985, SERI/PR-234-3026, Solar Energy Research Institute: Golden, CO, 1986.
6. Elliott, D. C. Analysis and Comparison of Biomass Pyrolysis/Gasification Condensates - An Interim Report, PNL-5555, Pacific Northwest Laboratory: Richland, WA, 1985.
7. Elliott, D. C. Analysis of Medium-BTU Gasification Condensates: June 1985 - June 1986, PNL-5979, Pacific Northwest Laboratory: Richland, WA, 1987.
8. Elliott, D. C. Analysis and Upgrading of Biomass Liquefaction Products, Final Report, Volume 4 of International Energy Agency Cooperative Project D-1, Biomass Liquefaction Test Facility. National Energy Administration: Stockholm, 1985.

9. Elliott, D. C. In Fundamentals of Thermochemical Biomass Conversion; Overend, R. P., Milne, T. A., and Mudge, L. K., Eds.; Elsevier Applied Science Publishers: London, 1985; Chapter 55.
10. Diebold, J. P.; Scahill, J. "Progress in the Entrained Flow Pyrolysis of Biomass," presented at the 12th Biomass Thermochemical Conversion Contractors' Meeting, Washington, D.C., March 1981. SERI/PR-622-1151, Solar Energy Research Institute: Golden, CO, 1981.
11. Knight, J. A., Gorton, C. W.; Kovac, R. J. In Proceedings of the 16th Biomass Thermochemical Conversion Contractors' Meeting, CONF-8405157, National Technical Information Service: Springfield, VA, 1984, pp. 287-297.
12. Scott, D. S.; Piskorz, J. Can. J. Chem. Eng. 1984, 62, 404-412.
13. Roy, C.; deCaumia, B.; Brouilland, D.; Ménard, H. In Fundamentals of Thermochemical Biomass Conversion, Overend, R. P., Milne, T. A., and Mudge, L. K., Eds.; Elsevier Applier Science Publishers: London, 1985; Chapter 13.
14. Elliott, D. C. "A Process for Direct Production of Hydrocarbon Fuels from Biomass" Invention Report E-639 Battelle-Northwest: Richland, WA, 1985.
15. Mudge, L. K.; Baker, E. G.; Brown, M. D.; and Wilcox, W. A. In Proc. of the 1985 Biomass Thermochemical Conversion Contractor's Meeting, CONF-8510167, National Technical Information Service: Springfield, VA, 1986, pp. 235-256.
16. Flanigan, V. J.; Punyakumleard, A.; Sitton, O. C. Biotech. and Bioeng. Symp. 14, 1985, pp. 3-14.
17. Kosowski, G. M.; Onischak, M.; Babu, S. P. In Proc. of the 16th Biomass Thermochemical Conversion Contractors' Meeting, CONF-8405157, National Technical Information Service: Springfield, VA, 1984, pp. 39-60.
18. Feldman, H. F.; Paisley, M. A.; Appelbaum, H. R. In Proc. of the 16th Biomass Thermochemical Conversion Contractors' Meeting, CONF-8405157, National Technical Information Service: Springfield, VA, 1984, pp. 13-38.
19. Vuori, A. "Thermal and Catalytic Reactions of the C-O Bond in Lignin and Coal Related Aromatic Methyl Ethers." Acta Polytechnica Scandinavica, Chem. Tech. and Met. Series No. 176, 1986.
20. Cypres, R. Fuel Proc Tech. 1987, 15, 1-15.
21. Weiler, J. F. In Chemistry of Coal Utilization, Supplementary Volume, Lowry, H. H., Ed.; John Wiley & Sons: New York, 1973; Chapter 14, 582.
22. Boocock, D. G. B.; Chowdhury, A.; Allen, S. A.; Hayes, R. D. Amer. Chem. Soc., Div. Fuel Chem. Prepts, 1987, 32(2), 90-97.

RECEIVED March 31, 1988

Chapter 7

Pyrolysis of Biomass

Evidence for a Fusionlike Phenomenon

Jacques Lede, Huai Zhi Li, and Jacques Villermaux

Laboratoire des Sciences du Génie Chimique, Centre National de la Recherche Scientifique–ENSIC, rue Grandville 54042 Nancy, France

This paper presents an outline of the main ideas and results illustrating the fusion like behaviour of biomass pyrolysis. The ablative pyrolysis rate of wood rods applied on a hot spinning disk or on a stationnary heated surface is studied. The observations are quite similar to those made in the same conditions with rods of true melting solids (ice, paraffin, ...) in agreement with a theoretical approach. The "fusion" temperature of wood is found close to 739 K in agreement with a limited number of experiments of wood sawdust fast pyrolysis in a hot cyclone reactor. The results show also that in most of the experimental devices, the direct measurement of wood pyrolysis reaction rate constant is impossible above about 800 K.

Considering a thermal reaction of a Solid → Fluid type, the apparent rate of reaction can be controlled by chemistry, thermal and mass transfer resistances. If the chemical processes are very fast, and if the fluid products are easily eliminated from the medium, the overall rate of reaction is controlled by heat transfer. This is the case of the ablation regime (1), characterized by a steep temperature gradient at the wood surface and consequently by a thin superficial layer e of reacting solid moving at a constant velocity v towards the cold unreacted parts of the solid (Figure 1).

Suppose now that heat is provided by a surface at T_w. A theoretical increase of the surface temperature T_d of the solid (by increasing T_w) would lead to a subsequent increase of the heat flux demand. Such a demand would be satisfied by an equal external heat flux supply, a condition fullfilled only if large temperature gradients exist (large T_w - T_d differences). The consequence is that T_d seems to reach a stable value, leading to a fusion like behaviour of the reaction.

Wood pyrolysis carried out in conditions of high available heat flux and efficient elimination of products has been found to occur in ablation regime with production of very low fractions of char (2,3,4) and could therefore be interpreted as a simple fusion. This

0097–6156/88/0376–0066$06.00/0

paper presents a brief outline of the main ideas and results illustrating this effect observed under different experimental conditions. More details can be found in related papers (5,6,7,8).

The reaction has been carried out in three different conditions: heating against a hot spinning disk ; against a fixed heated surface; in a continuous cyclone reactor. In the first two cases, the behaviour of the reaction is compared to that of solids undergoing simple fusion in the same conditions.

Spinning Disk Experiments

The melting of ice, paraffin and "rilsan" (polyamide 11) and the pyrolysis of wood have been carried out by applying under known pressures p, rods of the corresponding solids against a hot spinning stainless steel disk (temperature T_w)(5,6). In wood experiments, the reaction produces almost exclusively gases and liquids, the solids being mainly ashes deposited on the disk. The liquids produced are rapidly extracted from the wood surface and eliminated by the fast moving disk on which they undergo further decomposition to gases at a rate depending on T_w. The presence of the thin liquid layer acts as a kind of lubricant.

Figure 2 reveals that under comparable values of p, the behaviour of v as a function of v_R is similar, the orders of magnitude of v being the same for the four types of solids.

For $v > 2$ m s^{-1}, v increases with p following :

$$v = a\ p^F \tag{1}$$

a depends on T_w and F on the material. The mean values of F (ice : 0.035 ; paraffin : 0.29 ; "rilsan" : 0.83 ; wood : 1) can be fairly well represented by :

$$F(\text{melting solid}) = \frac{Cp_s(T_f-T_o)}{Cp_s(T_f-T_o)+L} \quad \text{or} \quad F(\text{wood}) = \frac{Cp_s(T_d-T_o)}{Cp_s(T_d-T_o)+\Delta H} \tag{2}$$

F being close to 1 for wood shows that it is probable that ΔH, the enthalpy of pyrolysis, is small with respect to sensible heat in agreement with the litterature.

The equations of heat flux density balances between the disk and the rod are :

$$\begin{aligned} \text{melting solid} &: h(T_w - T_f) = v\rho_s Cp_s(T_f - T_o) + v\rho_s L \\ \text{wood} &: h(T_w - T_d) = v\rho_s Cp_s(T_d - T_o) + v\rho_s \Delta H \end{aligned} \tag{3}$$

Assuming that the heat transfer coefficient h is the only parameter depending on the pressure ($h = Kp^F$) it can be deduced :

$$\frac{v}{p^F}(\text{melting solid}) = \frac{K}{\rho_s}\frac{T_w - T_f}{Cp_s(T_f-T_o)+L} \ ; \ \frac{v}{p^F}(\text{wood}) = \frac{K}{\rho_s}\frac{T_w - T_d}{Cp_s(T_d-T_o)+\Delta H} \tag{4}$$

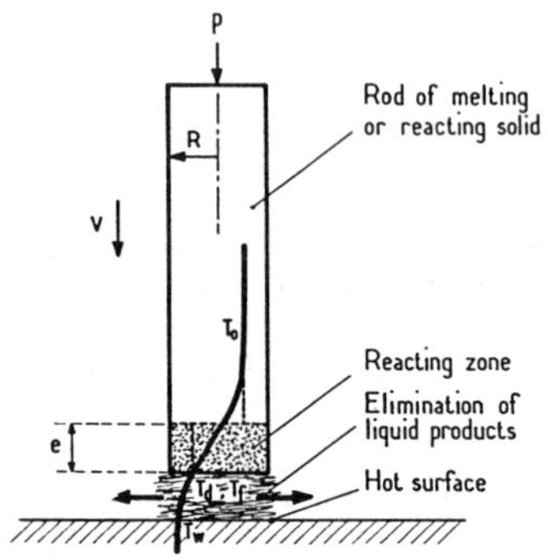

Figure 1. Ablative pyrolysis (melting) of a solid cylinder pressed against a heated surface.

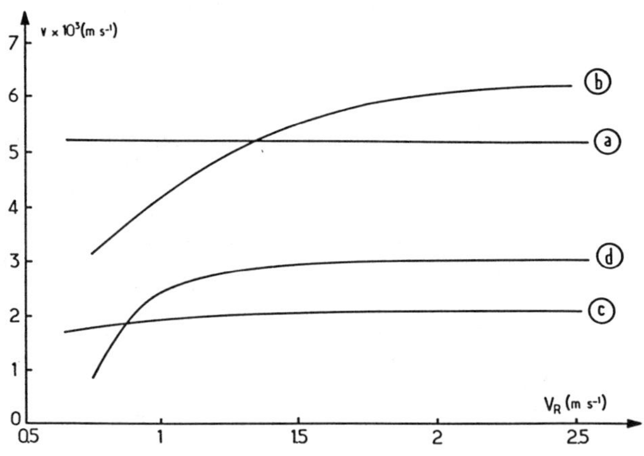

Figure 2. Experimental variations of ablation velocities v with disk velocity V_R for four kinds of solids : a (ice, T_w = 348 K, p = 2 x 10^5 Pa), b ("rilsan", T_w = 723 K, p = 3,45 x 10^5 Pa), c (paraffin, T_w = 373 K, p = 3.45 x 10^5 Pa), and d (wood, T_w = 1073 K, p = 3.7 x 10^5 Pa).

In agreement with Equation 4, Figures 3(a,b,c,d) show that the variations of v/p^F with T_w are linear for the three melting solids and also for wood. The values of T_f calculated from the extrapolation of the straight lines to v = 0 are in very good agreement with the known values of melting points (better than 2 % accuracy). The corresponding "fusion" temperature of wood is then calculated close to 739 K. The apparent large magnitude of errors observed in figure 3d for wood (the error bars correspond to extreme measured values of v/p^F for a given T_w) is a consequence of the slight theoretical dependance of T_d with T_w and h as mentioned in the final discussion. This "fusion" temperature of 739 K must then be considered as a mean value (figure 8).

The values of heat transfer coefficients obtained from the slopes of the straight lines in figures 3(a,b,c,d) are of the same order of magnitude (around 10^4 W m^{-2} K^{-1}) whatever the solid showing that the mechanisms of heat transfer are probably similar for wood and melting solids (6). For wood, h varies as h = 0.017 p(W m^{-2} K^{-1}) (5). These values reveal very efficient transfers.

Fixed Heated Wall

The same experiments as before have been made with rods of ice, paraffin and wood pressed against a stationnary piece of brass heated at T_w.

An analytic solution has been found for representing the rate of ablative melting of a solid cylinder pressed against a horizontal wall maintened at T_w (7). In steady state, a liquid layer of constant thickness is formed between the hot surface and the rod, with a radial flow of liquid. The resolution of the equation of liquid flow associated with that of energy balance between the two surfaces allows to derive the following relationship :

$$v = \frac{\rho_\ell}{\rho_s}\left[\frac{2}{3}\,\frac{\alpha_\ell^3}{\mu_\ell R^2}\,Pe^3\,(Ph)\,p\right]^{1/4} \qquad (5)$$

where Pe is a Peclet number, a function of a phase change number Ph :

$$Ph = \frac{Cp_\ell(T_w - T_f)}{L} \quad \text{with } Pe = \frac{Ph}{1 + 0{,}33\,Ph^{5/6}}$$

The Equation 5 shows that v varies as $p^{0.25}$ whatever the type of solid. In reduced form, the Equation 5 can be written as follows :

$$V = \frac{2}{3}\left[Pe^3\,P\right]^{1/4} \qquad (6)$$

$$\text{with } V = \frac{\rho_s}{\rho_\ell}\frac{R}{\alpha_\ell}\,v \text{ and } P = \frac{R^2}{\mu_\ell\,\alpha_\ell}\,p$$

By plotting V against $\left[\frac{2}{3}\,Pe^3\,P\right]^{1/4}$, one should obtain a straight line of slope one whatever the nature of melting solid and wall temperature.

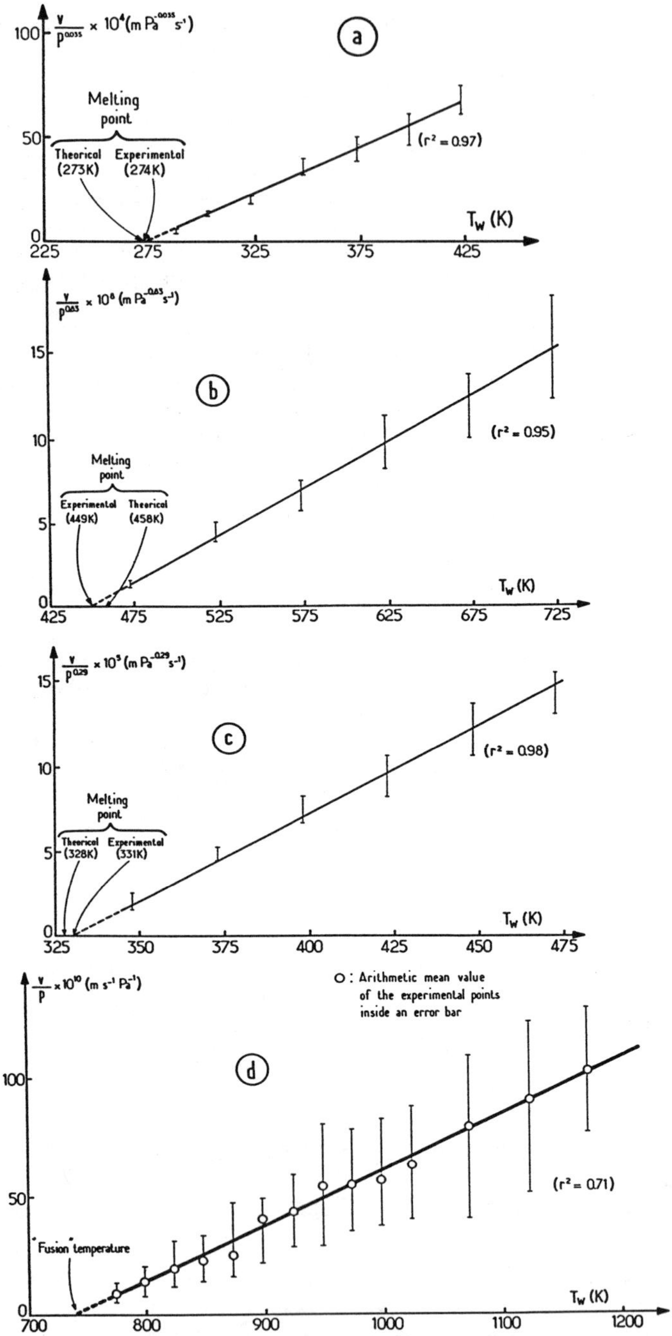

Figure 3. Experimental variations of v/p^F with disk temperature T_w. The straight lines have been determined from the least square method. Fig. 3a : ice ; Fig. 3b : "rilsan" ; Fig. 3c : paraffin ; Fig. 3d : wood.

As in the case of ice and paraffin (7), ablation rate v for wood pyrolysis varies as p^F with a mean value of F close to the theoretical one : 0.25 (figure 4). The figure 5 gathers all the experimental points. It shows a similar behaviour for wood, ice and paraffin in agreement with the theoretical straightline. The physical properties used for the calculation of V and P for ice and paraffin are reported in ref. (7,8). In the case of wood, the factor containing these properties in Equation (5) has been fitted to the experimental results (figure 4) leading to $v = 1.55 \times 10^{-3}\ (Pe^3\ p/p_o)^{1/4}\ (m.s^{-1})$. The fitted constant associated with estimated values for ρ_ℓ (500 kg. m^{-3}), Cp_ℓ (3.65 $kJ.kg^{-1}K^{-1}$) and α_ℓ ($0.3 \times 10^{-7}\ m^2\ s^{-1}$) allows us to calculate $\mu_\ell = 72.5 \times 10^{-3}$ Pa.s, a reasonable value for the viscosity of a liquid at a melting point of 739 K (7,8).

Life Time of Liquids Produced

The same pyrolysis conditions can be achieved with a moving piece of wood pressed upon a fixed heated surface. In that case, it is easy to measure the necessary time t_∞ of decomposition of the liquids left behind the wood on the surface. Figure 6 reports the linear variations of the experimental values of $1/t_\infty$ (pseudo first order kinetic constant) as a function of $1/T_w$. Assuming that the liquids are rapidly heated to surface temperature before decomposition it is possible to estimate the kinetic parameters of the reaction of liquids decomposition : $A = 2.7 \times 10^7\ s^{-1}$ and E = 116 kJ. Compared to the parameters used in Diebold kinetic model (14) the experimental points could represent the two possible processes : "Active" → primary vapors or "Active" → char.

Cyclone Reactor Experiments

The continuous fast pyrolysis of wood sawdust has been studied in a Lapple type (2.8×10^{-2} m diameter) cyclone reactor heated between 893 and 1330 K (2). The wood particles carried away by a flow of steam enter tangentially into the cyclone on the inner hot walls of which they move and undergo decomposition. Mass balances show in all the cases, a very low fraction of char (< 4 %) while the gasification yield X ($X = \frac{\text{weight of dry gas evolved}}{\text{weight of dry wood pyrolysed}}$) increases with wall temperature T_w. It appears from figure 7 that the reaction seems to occur only for wall temperatures greater than about 800 K in good agrement with "fusion" temperature of 739 K. The variation of X with T_w is approximately linear up to about 1300 K. For higher temperatures, the gasification yield must tend to the theoretical constant value of 1. In such a model, the decomposition temperature of particles being roughly constant, the gasification yield increase with T_w would then result from further vaporization and/or decomposition of primary products (mainly liquids) at the wall and/or in the gas phase with an efficiency depending on T_w.

Discussion

All these results obtained in different experimental conditions, show striking similarities between ablative wood pyrolysis and melting of

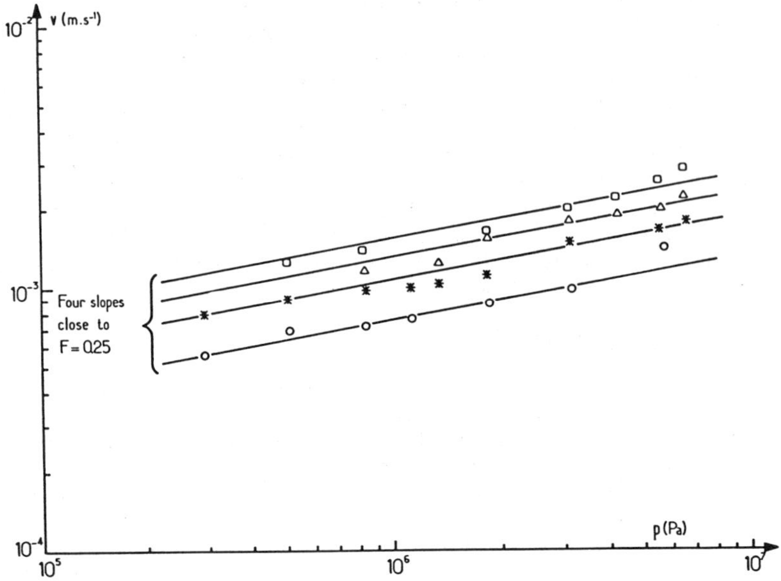

Figure 4. Ablative pyrolysis rate of wood v as a function of applied pressure p for different wall temperatures. O : 823 K ; ✱: 873 K ; △ 923 K ; □ : 973 K.

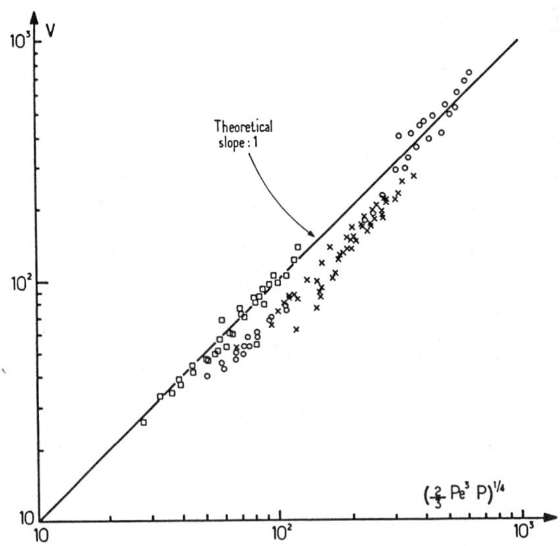

Figure 5. Dimensionless representation of reduced velocity V as a function of reduced pressure P for three kinds of solids. O: paraffin ; X : ice ; □ : wood.

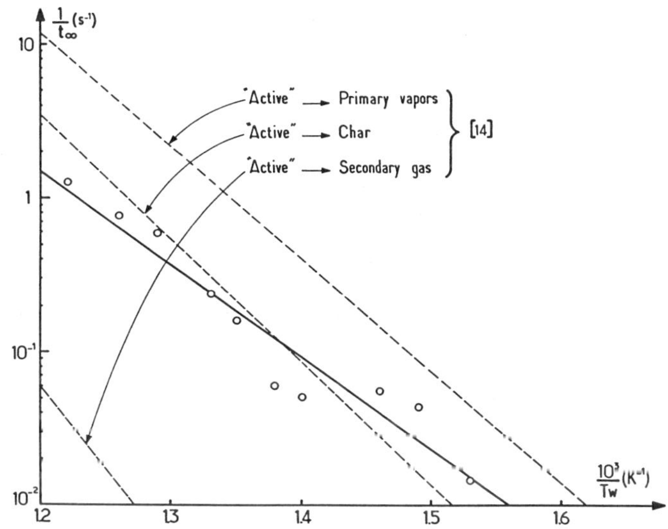

Figure 6. Variations of the observed kinetic constant of liquid decomposition with wall temperature in an Arrhenius representation. A comparison with the scheme of Diebold (<u>14</u>).

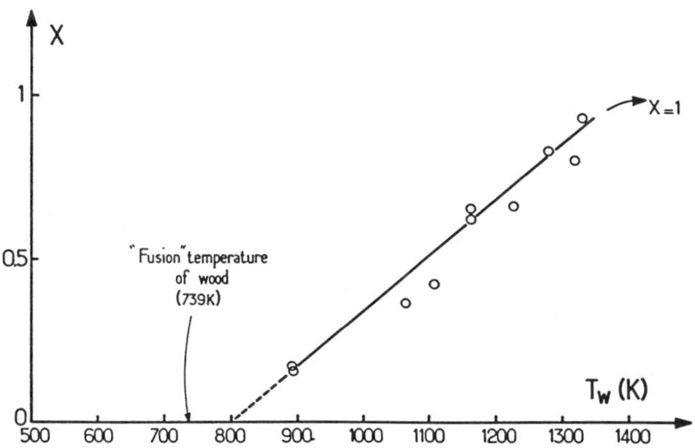

Figure 7. Variation of the gasification yield X as a function of wall temperature T_W in the fast pyrolysis of wood sawdust in a cyclone reactor : comparison with the "fusion" temperature of 739 K.

solids. Nevertheless the equivalent fusion temperature of 739 K has been calculated from Equation 4 based on the assumption that T_d is constant. Let suppose now that T_d depends on external physical conditions (p, T_w) under the assumption that $\Delta H << Cp_s\ (T_d - T_o)$. The heat flux density balance equation is :

$$h(T_w - T_d) = v\ \rho_s\ Cp_s(T_d - T_o) \quad (7)$$

The reaction occuring in ablation condition concerns only a superficial wood layer of thickness e moving at a constant velocity v towards the unreacted parts of the rods as the reaction proceeds. Assuming, in a first approximation, that the temperature is constant (= T_d) inside this reacting volume eS (S being the cross section of the rod) the equation of mass balance inside this volume is $k\ e(S\rho_s) = v(S\rho_s)$ where k is the chemical first order kinetic constant of the reaction. This equation associated with the relation between v and e ($ev = \alpha_s$ (5)) leads to $k = v^2/\alpha_s$ and finally to :

$$T_d = \frac{h\ T_w + T_o \sqrt{k\ \lambda_s \rho_s\ Cp_s}}{h + \sqrt{k\ \lambda_s \rho_s\ Cp_s}} \quad (8)$$

Figure 8 shows the variations of T_d with T_w ($\lambda_s = 0.2\ W\ m^{-1}K^{-1}$; $Cp_s = 2800\ J\ kg^{-1}\ K^{-1}$; $\rho_s = 700\ kg\ m^{-3}$) with h as a parameter. The first order rate constant for the formation of "active cellulose" (9) ($k(s^{-1}) = 2.83 \times 10^{19} \exp(-29000/T_d)$) has been postulated to fit the present case of wood primary decomposition.

It can be observed that the smaller the value of h, the shortest the domain of wall temperature where $T_d = T_w$ (for the lowest values of h, T_d/T_w becomes less than one as wood begins to decompose). In most of usual experimental devices wood temperature is then very different from source temperature. Consequently, the direct determination of pyrolysis rate laws would make sense only for low wall temperatures (< 800 K).

Figure 8 shows that T_d varies with T_w and h indicating that strictly speaking, the fusion model is not appropriate. But it can be observed (specially for low h) that as soon as $T_d/T_w < 1$, T_d increases more and more slowly with T_w and rapidly reaches a roughly constant value. The fusion model seems then to be an excellent first approximation.

The hatched zone reported in figure 8 is bounded by the extreme values of h determined in ref. (5) and by the extreme values of T_w explored. The "fusion" temperature of 739 K appears to be centrally situated inside the hatched surface. Such a fair agreement shows that the chosen kinetic law is a good approximation for wood decomposition. The "fusion" temperature must then be considered to have a mean value lying roughly between 660 and 725 K for T_w = 773 K and between 700 and 800 K for T_w = 1173 K.

Conclusion

The behaviour of wood rods undergoing ablative pyrolysis by more or

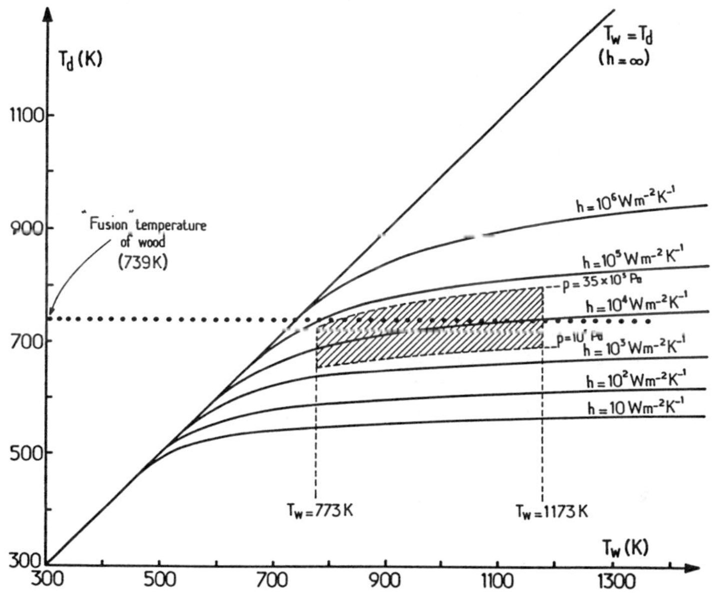

Figure 8. Theoretical variations of wood surface temperature T_d as a function of heat source temperature T_w for different values of the external heat transfer coefficient. The hatched surface corresponds to the experimental domain ($776 < T_w(K) < 1176$ and $10^5 < p(Pa) < 35 \times 10^5$ corresponding to $2 \times 10^3 < h(W\ m^{-2}\ K^{-1}) < 6 \times 10^4$).

less intimate contact with a hot surface has revealed strong similarities with a phase change phenomenon. The principal reasons developed are the following : similar behaviour of wood rods to those of true melting solids when applied on moving or fixed surfaces ; similar orders of magnitude of v and h ; same dependence of ablation rate on an exponent (F) of the applied pressure showing probable low values of the enthalpy of reaction ; the same $p^{0.25}$ dependence of v in the case of a fixed surface ; similar law of variations of v with wall temperature. Ablative pyrolysis carried out with sawdust in a cyclone reactor has shown that no fast reaction occurs for wall temperatures lower than $\sim$ 800 K in good agreement with the fusion temperature of 739 K.

A consequence of these conclusions is that the accurate direct determination of kinetic rate constant of wood decomposition is a difficult task, in fact impossible over wide ranges of temperatures in most of experimental devices (upper limit around about 800 K). Even if such high temperatures could be reached, the system would have to be designed in such a way that the products of the reaction be removed from the reacting surface with high efficiencies. For example, figure 8 shows that $T_d = T_w$ = 800 K would be observed for h = 10^6 W m^{-2} K^{-1}. Assuming that the available heat flux is controlled by conduction through the oil layer, such a heat transfer coefficient would be effective for an equivalent layer thickness of 0.1 μm ! (calculation made with a thermal conductivity of 0.1 W m^{-1} K^{-1} for oil). Of course, an efficient removal of these liquids would prevent also the extent of their subsequent decomposition to secondary products and then reduce the formation of new thermal isolating layers.

All these conclusions are in agreement with the analysis of other authors. With his demonstration of sawing biomass with a hot wire, Diebold illustrated in 1980 (4) the efficiency of "solid convection" for carrying out the reaction of ablative pyrolysis of biomass. The same author stated also recently that cellulose passes probably through a liquid or plastic unstable state("active cellulose") during pyrolysis before further decomposition (14). Considering the simplified cellulose pyrolysis reaction scheme proposed by this author, figure 6 shows that the values of the kinetic constants measured in the present work are close both to the process "active" → char and "active" → primary vapors with similar activation energies. In the same way, Antal pointed out the strong analogies observed between cellulose pyrolysis at high heating rates and phase change phenomena (10,11) with an upper limit at which pyrolysis occurs of 773 K. Evidence of such an upper limit is explained by a competition between heat demand from biomass and available external heat flux (12). The same author also noted the difficulty and indeed impossibility of achieving conditions whereby pyrolysis kinetics could be studied at very high temperatures. Finally, it must be recalled that in 1980, Reed (13) proposed a model for estimating the enthalpy of flash pyrolysis of wood based on several steps : heating of biomass up to a reaction temperature of 773 K, followed by a depolymerisation to form a solid which subsequently melts, melted matter being afterwards able to vaporize, following the temperature.

Legend of Symbols

a	Constant ($m\ s^{-1}\ Pa^{-F}$)
A,E	Kinetic parameters of liquids decomposition (s^{-1}, kJ)
Cp	Specific heat capacity ($J\ kg^{-1}\ K^{-1}$)
e	Thickness of reacting wood layer (m)
F	Exponent
h	Heat transfer coefficient ($W\ m^{-2}\ K^{-1}$)
H	Specific enthalpy ($J\ kg^{-1}$)
K	Constant ($W\ m^{-2}\ K^{-1}\ Pa^{-F}$)
k	First order kinetic constant (s^{-1})
L	Heat of fusion ($J\ kg^{-1}$)
p	Pressure (Pa)
p_o	Atmospheric pressure (Pa)
P	Reduced pressure
Pe	Peclet number
Ph	Phase change number
r^2	Correlation coefficient
R	Radius of the solid cylinder (m)
S	Cross section of the rod (m^2)
t_∞	Life time of liquids (s)
T_d	Wood surface temperature (K)
T_f	Fusion temperature of a melting solid (K)
T_o	Ambient temperature (K)
T_w	Wall temperature (K)
v	Ablation velocity of the solid cylinder ($m\ s^{-1}$)
V	Reduced velocity
V_R	Relative velocity between the disk and the rod ($m\ s^{-1}$)
X	Gasification yield
α	Thermal diffusivity ($m^2\ s^{-1}$)
λ	Thermal conductivity ($W\ m^{-1}\ s^{-1}$)
μ	Viscosity (Pa s)
ρ	Density ($kg\ m^{-3}$)

Subscripts :

s	solid cylinder
ℓ	liquid layer

Literature cited

1. Villermaux, J.; Antoine, B.; Lédé, J.; Soulignac, F. Chem. Eng. Sci. 1986, 41, 151.
2. Lédé, J.; Verzaro, F.; Antoine, B.; Villermaux, J. Chem. Eng. Process. 1986, 20, 309.
3. Lédé, J.; Verzaro, F.; Antoine, B.; Villermaux, J. Proc. of Specialists Workshop on Fast Pyrolysis of Biomass. Copper Mountain Co, October 1980, 327.
4. Diebold, J.P. As Réf. 3, p. 237
5. Lédé, J.; Panagopoulos, J.; Li, H.Z.; Villermaux, J. Fuel. 1985, 64, 1514
6. Lédé, J.; Li, H.Z.; Villermaux, J.; Martin, H. J. Anal. Appl. Pyrolysis. 1987, 10, 291.
7. Martin, H.; Lédé, J.; Li, H.Z.; Villermaux, J.; Moyne, C.; Degiovanni, A. Int. J. Heat Mass Transfer. 1986, 29, 1407.

8. Li, H.Z. "Pyrolyse éclair de baguettes de bois. Modèle de fusion". DEA, (INPL, Nancy), 1984.
9. Bradbury, A.G.W.; Sakai, Y.; Shafizadeh, F. J. Appl. Polym. Sci. 1979, 23, 3271.
10. Kothari, V.; Antal, M.J. Fuel. 1985, 64, 1487.
11. Antal, M.J. Fuel. 1985, 64, 1483.
12. Antal, M.J. Biomass Pyrolysis. A review of the literature. Part 2 : Lignocellulose Pyrolysis. In Advances in Solar Energy. Plenum Press, New-York, 1985, vol. 2
13. Reed, T.B.; Diebold, J.P.; Desrosiers, R. As Ref. 3, p. 7
14. Diebold, J.P. Thesis, T-3007, Colorado School of Mines, Golden, CO, USA 80401, 1985.

RECEIVED March 31, 1988

Chapter 8

Liquids from Municipal Solid Waste

James E. Helt and Ravindra K. Agrawal[1]

Argonne National Laboratory, 9700 South Cass Avenue, Argonne, IL 60439

The pyrolysis of municipal solid waste (MSW) is a promising method of producing useful fuels. However, the reaction mechanisms are not well understood. Current DOE/ANL research is directed at gaining a better understanding of the basic thermokinetic mechanisms associated with the pyrolytic conversion of MSW.

Initial results are given for determinations of pyrolytic reactivity for MSW components in a nonisothermal bench-scale reactor. The effects of heating rate and sample weight on overall degradation and product formation over the temperature range 300-475°C were determined. Gas chromatographic and gas chromatographic/mass spectrometric analyses of both the gaseous and liquid products are being performed. Tentative identifications were made of some liquid components from pyrolysis of MSW.

Municipal solid waste (MSW) is a highly variable "raw material," by both season and location. However, it is generally accepted to have a composition within the ranges shown in Table I (1). Cellulosic materials, including paper, newsprint, packaging materials, wood wastes, and yard clippings, constitute over 50% of MSW.

A basic understanding of the pyrolytic reactions is important and relevant to both combustion and conversion of MSW. As MSW is heated the different components react differently at different temperatures. The volatile species can evaporate without major change, while the rest of the cellulosic components partially break down to volatile components leaving a carbonaceous char

[1]Current address: KRW Energy Systems, Inc., P.O. Box 334, Madison, PA 15663–0334

0097–6156/88/0376–0079$06.00/0

Table I. Composition Ranges for Several MSW Samples

Component	Composition Range, wt %
Paper	30-50
Glass	8-10
Metals	7-10
Plastics	1-5
Rubber-Leather	1-3
Wood	1-4
Textiles	1-5
Food Wastes	10-20
Yard Wastes	5-20
Other	1-4

Source: Reprinted with permission from ref. 1. Copyright 1983 Argonne National Laboratory.

containing in addition the ash and noncombustibles. One fraction of the volatile pyrolysis products is a noncondensable gaseous fraction comprising CO, CO_2, some hydrocarbons, and H_2. Another fraction is a light condensable fraction containing H_2O, volatile hydrocarbons and low molecular weight degradation products such as aldehydes, acids, ketones, and alcohols. Finally, there is a tar fraction containing higher molecular weight sugar residues, furan derivatives, and phenolic compounds. The proportion and composition of these products are highly dependent on the cellulosic composition of the MSW, the pyrolysis temperature and the presence of inorganic compounds that could influence (catalyze) the pyrolysis reactions.

Pyrolysis of cellulose at temperatures below 300°C results mainly in char formation. Any lignin present in the MSW (Kraft paper, cardboard, and wood waste contain significant proportions) tends to char, even at higher temperatures. On the other hand, the cellulose and hemicelluloses readily decompose to volatile products at temperatures above 300°C. Most of the plastics present thermally degrade at a significantly higher temperature (400-450°C) (2).

BASIC MECHANISMS RESEARCH

The ANL/DOE program on pyrolysis of municipal solid waste (MSW) has two overall objectives: (1) to understand the basic thermokinetic mechanisms associated with the pyrolytic conversion of MSW and (2) to seek new processing schemes or methods of producing a liquid or gaseous fuel from MSW feedstock. To meet these objectives, we are performing laboratory experiments with the aim of determining the effects of different operating parameters on the pyrolysis-product compositions and deriving an analytical model of the pyrolytic process that describes the chemical kinetics.

This DOE-sponsored research involves both ANL activities and subcontracted work. Argonne is performing closely controlled laboratory-scale parametric tests. The work is being performed on two experimental facilities: (1) a TGA to study the thermal degradation versus temperature, and (2) a bench-scale reactor to produce significant quantities of products to permit characterization. The goal is to determine how different operating parameters influence the product compositions.

Subcontracted activities are being performed at the Solar Energy Research Institute (SERI) and the Chemical Engineering Department at the University of Arizona. SERI has used their direct high-pressure molecular beam mass spectrometric sampling system to collect qualitative "fingerprints" and experimental data on the pyrolysis products generated from components of MSW and various refuse derived fuel (RDF) samples (3). They are also using the same experimental apparatus to distinguish between the primary and secondary reactions leading to the formation of the pyrolysis products (4). This data will be part of an overall data base to describe the influence of sample properties and reaction conditions on the solid phase and gas phase processes of low-temperature (<500°C) MSW pyrolysis to oils. Additionally, SERI has recently started a new task on the development of a rapid method of characterizing the liquid products from pyrolysis based on mass spectrometric data. This new task is composed of three parts: 1) compound class analysis by advanced pattern recognition techniques, 2) liquid product analysis via compound class analysis, and 3) correlation of chemical composition to fuel properties.

The University of Arizona has completed a small research effort on the fundamentals of direct liquefaction of MSW (5). They modified an existing autoclave and a real-time digital microprocessor control system so that it could be operated in a semi-continuous mode. Various components of MSW were studied in order to obtain meaningful data, not confused by the different thermokinetics of different MSW components. Feedstocks included wood flour, cardboard, newsprint and rice (starch), as well as the important model compounds alpha cellulose and lignin. It was found that these MSW components could be converted to liquid oils and a high-heating value residual solid at temperatures of 325°C to 400°C and pressures of 1000 psi to 3000 psi.

A task which is related to the basic mechanisms work is also being performed by ANL (2). This task explores the possibility of using catalytic hydrotreating to upgrade the liquid products produced during conventional pyrolysis of MSW. The liquid products obtained from MSW pyrolysis processes are generally unsuitable for use as liquid fuels. Heating values are low and the liquids are very corrosive, viscous and unstable during storage. A major reason for these problems is the extremely high oxygen content of the pyrolysis products. The kinetics of catalytic reactions that remove oxygen-containing compounds from the pyrolysis liquids is being experimentally determined in a high-pressure, fixed-bed microreactor of the trickle-bed type. The reactions of interest involve the reduction of the oxygen-containing hydrocarbon with high pressure hydrogen using a solid catalyst. The catalysts which have been used are primarily commercially available hydrodesulfurization (HDS) catalysts containing molybdenum oxide with either cobalt or nickel oxide, supported on high surface area alumina matrix.

ANL BENCH-SCALE STUDIES

The emphasis of this paper will be on the bench-scale studies being performed at ANL and on the associated activities in characterizing the liquids produced in the pyrolysis reactions.

The reactor is a fixed bed contained inside a quartz tube (70-mm ID) and placed between two glass frits. The outside of the tube is enclosed in a furnace. The reactor tube can be purged from top to bottom with the desired gas(es). The reactor is operated in the nonisothermal mode with heatup rates as high as ~30°C/min. Temperatures are recorded on a multipoint recorder to allow indication of existing temperatures and temperature gradients. There is a rotameter on the inlet gas line and a dry gas meter on the outlet. Cold traps are in the gas outlet downstream of the condenser unit. These traps are filled with ice or dry ice. Downstream of the traps, the gases that do not condense are collected in plastic sampling bags for analysis by gas chromatography (GC).

Experimental results on the thermal decomposition of typical MSW components (Whatman #1, newsprint, kraft paper, cardboard,aspen, and pine) over a temperature range of 275-475°C have been gathered. The details of these experimental runs may be found elsewhere (6).

CHARACTERIZATION OF LIQUIDS

Various liquid samples produced in the bench-scale apparatus have been analyzed with GC and GC/MS. The chromatograms were qualitatively compared to each other by both measurement of peak retention times and by observation of the patterns present. As a result, six different groups were identified:

Group A - Most components elute early in the chromatogram as many sharp peaks within a small retention window.

Group B - Bulk of components elute across a 6- to 20-min retention window and are a mix of both sharp and broad peaks.

Group C - Many peaks are observed; the bulk of components elute across a 4- to 30-min retention window.

Group D - Similar to C, but most components elute across a 4- to-20-min retention window.

Group E - A few early peaks are observed, especially in the 5- to 7-min retention window.

Group F - Similar to D, but many peaks are observed in the 9- to 11-min retention window.

The gas chromatogram obtained from a typical tar sample is shown in Figure 1. This tar sample was produced from a newsprint feedstock pyrolyzed in an inert atmosphere of helium. A computer search was performed using the 31,000 component NIH/EPA library in addition to a library of compounds from Battelle Pacific Northwest Laboratory. The results of the computer search and from interpretation of various standard spectra yielded tentative identification of numerous compounds. It is apparent that many compounds of homologous series are present. Recognition of just one of the compounds in a series leads to identification of all since they will most likely differ only by 14 amu (a CH_2 group) or by 31 amu (a CH_3O group). In some cases the same identification is made for more than one compound. Actually, different isomers of the compound are probably being found. The percent found was estimated by dividing the response of the most abundant ion for a compound by the total of the responses of the most abundant ions for all compounds.

In general, the classes of compounds included furfurals (9.4%), phenols (2.5%), methoxyphenols (16.9%), cyclic compounds such as methyl cyclopentanones (10.8%), methoxy benzenes (3.8%), and the substituted propane tentatively identified for the peak at scan number 1207 (36.8%). Although the compound eluting at scan 1207 is by far in the highest concentration, insufficient information is available from its spectrum to allow a reasonable identification. The base peak observed is 75amu, and a 115 amu ion is also present at 40% abundance. These results, which are preliminary, differ somewhat from other published compositions of pyrolysis oils (3, 7-8). Future work will be required to verify the compositions observed above.

YIELDS AND ANALYTICAL RESULTS

With the Whatman No. 1 filter paper, the yields at 475°C of water vapor and gases were in the ranges 5-13 wt % and 26-34 wt %, respectively, of the original cellulose. The hydrogen balance suggests the higher water content (13%), whereas both the carbon and oxygen contents suggest a lower water yield (5 wt %).

Efforts have been made to analyze the gases collected in the sample bags. A Hewlett-Packard Gas Chromatograph is being used to identify major gas components. The preliminary GC analyses show that, for the Whatman No. 1 paper at 475°C, the gases produced are 56.6 vol% CO_2 and 43.4 vol% CO. No other gases were detected in significant quantities. The yield of CO_2 is, therefore, in the range of 18-23 wt % of the original cellulose and the yield of CO is 8-11 wt %. Roughly 25% of the energy in the feedstock is released in the gaseous products. These pyrolysis gases can be considered a low-Btu fuel.

The Whatman No. 1 paper (as received) contained 4.1 vol % moisture and 0.074 wt % ash. All results reported here are on a moisture- ash-free basis unless other wise specified. Table II summarizes some analytical results of cellulose and condensed-phase cellulose pyrolysis products. A comparison of the results in Table II for cellulose and cellulose tars indicates that the elemental composition of these two materials is very similar. (The heating

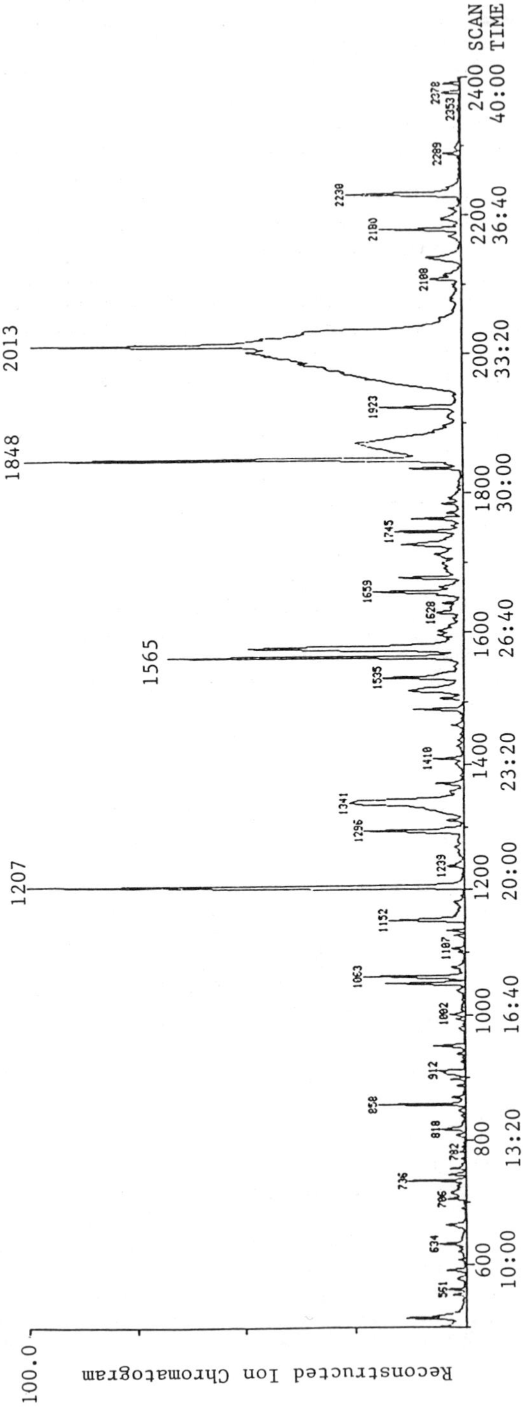

Fig. 1. Gas Chromatogram from Sample 06298402.

value of cellulose tars reported here may be low due to the loss of lower-molecular-weight products during the drying step.) Tars seem to have a slightly higher heating value than that of cellulose. These results strongly suggest that the nature of cellulose tars is similar to that of its parent cellulose.

Table II. Analytical Results of Feedstock, Tars, and Chars from Whatman No. 1 Paper (heating rate, 5°C/min; final temp., 475°C)

	Feedstock	Tar Product	Char Product
Heating Value,[a] cal/g	4170	4330	7566
Ash Content, wt %	0.074	~0.075	0.63
Moisture Content, wt %	4.1	--	--
Elemental Analysis,[a] wt %			
Carbon	44.7	42.3(43.6)[b]	81.0(27.3)
Hydrogen	5.9	6.1(47.5)	3.6(10.0)
Oxygen[c]	49.4	51.6(48.0)	15.4(4.7)
Yield, wt %	--	~46	~15

[a]Dry basis.

[b]Numbers in parentheses give percent of original material (based on the reported yields).

[c]Determined by adding together the carbon and hydrogen contents and subtracting them from 100%.

As can be seen from Table II, the cellulose chars are very different from the parent cellulose. When compared with the original cellulose, the cellulose chars have a carbon content that is roughly double, and H_2 and O_2 contents that are about one-half and one-third respectively. The richness in carbon content of the chars is indicated by their high heating value (7566 cal/g). Unfortunately, the low H_2 content of the chars make them an unlikely candidate for use as transportation fuels. Because of the high carbon content and low H_2 and O_2 content, the cellulosic chars are comparable to a low-volatile bituminous coal or a low-grade anthracite coal. However, since the chars contain no sulfur or nitrogen compounds that could form potential air pollutants upon combustion, they do have potential as a solid fuel. Tables III and IV give the analytical results for newsprint and Kraft paper feedstock.

The atomic ratios (H/C and O/C) and the heating value of cellulosic chars indicate that they are very similar to coal. The H/C ratio of cellulose tars (1.73) is comparable to that of No. 2 fuel oil (1.84). Unfortunately, the high oxygen content indicated by the O/C ratio (0.91 compared to 0.01 for fuel oil) significantly reduces the heating value of the tars.

Table III. Analytical Results of Feedstock, Tars, and Chars from Newsprint (heating rate, 5°C/min; final temp., 475°C)

	Feedstock	Tar Product	Char Product
Heating Value,[a] cal/g	4722	5573	7866
Ash Content, wt %	.95	--	4.1
Moisture Content, wt %	8.3	--	--
Elemental Analysis,[a] wt %			
Carbon	48.0	47.5(45.5)[b]	78.0(24.4)
Hydrogen	5.4	5.6(47.7)	3.7(10.3)
Oxygen[c]	46.6	46.9(46.3)	18.3(5.9)
Yield, wt %	--	~46	~15

[a]Dry basis.

[b]Numbers in parentheses give percent of original material (based on the reported yields).

[c]Determined by adding together the carbon and hydrogen contents and subtracting them from 100%.

Table IV. Analytical Results of Feedstock, Tars, and Chars from Kraft Paper (heating rate, 5°C/min; final temp., 475°C)

	Feedstock	Tar Product	Char Product
Heating Value,[a] cal/g	4445	5272	7333
Ash Content, wt %	1.3	--	4.3
Moisture Content, wt %	6.1	--	--
Elemental Analysis,[a] wt %			
Carbon	47.5	46.9(24.7)[b]	75.5(38.2)
Hydrogen	5.5	5.3(24.1)	3.9(17.0)
Oxygen[c]	47.0	47.8(25.4)	20.6(10.5)
Yield, wt %	--	~25	~24

[a]Dry basis.

[b]Numbers in parentheses give percent of original material (based on the reported yields).

[c]Determined by adding together the carbon and hydrogen contents and subtracting them from 100%.

DISCUSSION

It should be noted that the tar analysis results of this study are very similar to those obtained from vacuum pyrolysis of small cellulose samples conducted by Agrawal et al (9). Also the tar yields are comparable to those obtained by Shafizadeh using small samples of Whatman No. 1 paper under vacuum conditions (10). These findings support the assumption that limited tar decomposition takes place in the reaction bed.

Efforts are also in progress to ascertain some of the possible heat and mass transfer limitations of the pyrolysis process. Figure 1 depicts the residue and tar yields of 5-g and 15-g cellulose samples at a heating rate of 5°C/min. Increased sample weight shifts the weight-loss curve to a higher temperature by about 10°C. Figure 2 also shows that increased sample weight decreases the tar yields. Efforts to explain this effect of sample weight on product yield are in progress.

Figures 3 and 4 summarize the data for the influence of heating rate on product yields from Whatman No. 1 paper. It is seen from Figures 3 and 4 that increasing the heating rate or sample weight has a similar effect in shifting the weight-loss curve to a higher peak temperature. However, the shift in the weight-loss curve along the peak-temperature axis in the case of increased sample weight is due to mass transfer limitations, whereas, in the case of increased heating rate, this shift is due to combined effects of kinetics and heat transfer resulting in delayed decomposition. At a peak temperature of about 370°C, the product yields are essentially independent of the heating rates (Figure 4).

The results in Table V illustrate the influence of sample weight and heating rate on product yields. The data show that increasing sample size reduces the tar yields and increases the char yields. The drastic decrease in tar yields is primarily due to increased vapor residence time in the reaction bed. If the vapor residence time is reduced in the reaction bed by using a fluidized bed or an entrained flow reactor, then secondary decomposition can be significantly reduced, and the effects of sample weight will not be as drastic. Thus, data collected in the loosely packed fixed-bed reactor of the present study may represent an extreme for an operating industrial reactor.

Increasing the heating rate appears to decrease the char yields but has little influence on tar yields. This implies that gas yields increase at the expense of char yields.

Table VI summarizes elemental analyses of chars formed under various pyrolysis conditions. The elemental analyses of cellulosic chars suggest that the composition of chars is not strongly influenced by either the heating rate or sample weight.

Results to date strongly imply that, depending on the residence time, the tar yields for final pyrolysis temperatures above 300°C will be independent of heating rates. This observation is strengthened by the finding from TGA data analysis that the apparent activation energy for cellulose decomposition is similar to that for tar formation.

Olefins and other hydrocarbon gases were not detected in the pyrolysis gases. This is not surprising since these fuels are not

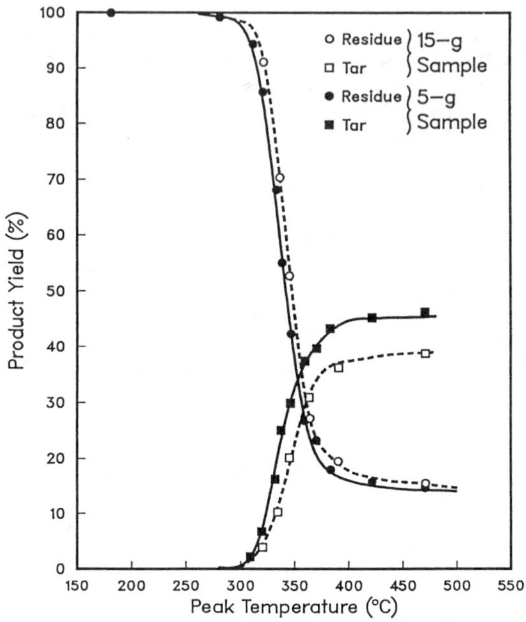

Fig. 2. Effect of Sample Weight on Product Yield for Whatman No. 1 Paper (heating rate, 5°C/min).

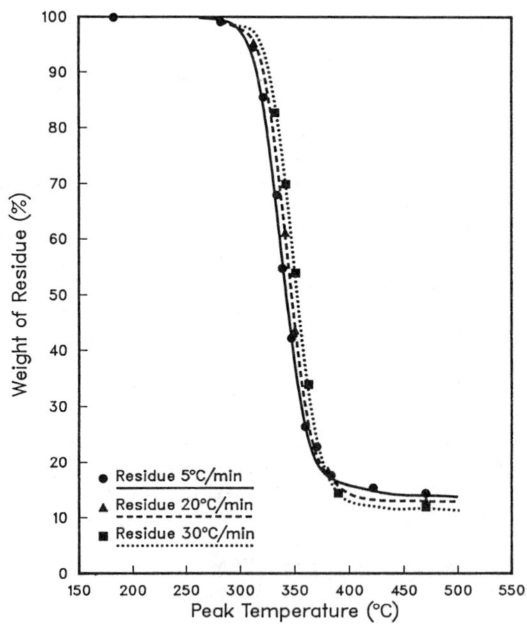

Fig. 3. Effect of Heating Rate on Weight Loss for Whatman No. 1 Paper

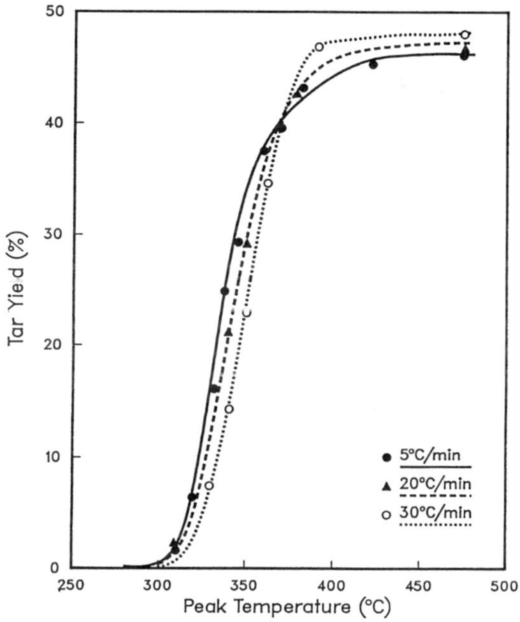

Fig. 4. Effect of Heating Rate on Tar Yields for Whatman No. 1 Paper

Table V. Effect of Sample Weight and Heating Rate on Ultimate Product Yields from Whatman No. 1 Paper

Sample No.	Pyrolysis Sample Weight, g	Heating Rate, °C/min	Yields,[a] % Tar[b]	Char
FPL08	5	5	46.14	14.55
FPL15	10	5	42.13	15.26
FPL13	15	5	38.54	15.66
FPM03	5	20	47.18	12.65
FPH05	5	30	47.76	11.93

[a]Product yields are given for a peak temperature of 475°C.
[b]Based on weight percent Whatman No. 1 paper (dry basis).

Table VI. Effect of Sample Weight and Heating Rate on the Composition of Chars from Whatman No. 1 Paper

Sample No.	Pyrolysis Sample Weight, g	Heating Rate, °C/min	Ultimate Char Yield[a] at 475°C, wt %	Chars, wt % C	H	O[b]
FPL08	5	5	14.55	82.80	3.75	13.45
FPL15	10	5	15.26	81.00	3.55	15.45
FPL13	15	5	15.66	81.90	3.45	14.65
FPM03	5	20	12.65	81.30	3.45	15.15
FPH05	5	30	11.93	80.30	3.55	16.15

[a]Yield from original Whatman No. 1 (dry basis).
[b]Derived by adding together C and H contents and subtracting from 100%.

products of primary cellulose pyrolysis (11-12). The significant yields of olefins and hydrocarbon gases from flash pyrolysis studies are most likely a result of secondary tar decomposition. Cellulose tars start to decompose at about 550°C, and most of these other studies were carried out over the temperature range of about 600-800°C. These observations suggest that results from flash pyrolysis studies are dominated by secondary tar decomposition reactions.

ACKNOWLEDGMENTS

This work was supported by the U. S. Department of Energy, Division of Biofuels and Municipal Waste Technology, under Contract W-

31-109-Eng-38. The authors wish to acknowledge the support and guidance of J. Lazar, EMW Program Manager at ANL, and K. M. Myles, Section Head in the Chemical Technology Division at ANL.

REFERENCES

1. Kuester, J. L. Thermal Systems for Conversion of Municipal Solid Waste, Vol. 5, Pyrolytic Conversion: A Technology Status Report; Argonne National Laboratory, ANL/CNSV-TM-120, (1983).
2. Helt, J. E. et al. Pyrolysis of Municipal Solid Waste; Argonne National Laboratory, ANL/CNSV-45, (1984).
3. Evans, R. J. et al. Atlas of the Primary Pyrolysis Products of Municipal Solid Waste and Selected Constituents; Solar Energy Research Institute, SERI/SP-234-2597, (1984).
4. Evans, R. J.; Milne, T. A. Mechanisms of Pyrolysis of Munici pal Solid Waste--Annual Report; Solar Energy Research Instit ute, SERI/PR-234-2852 (1985).
5. White. D. H. et al. Fundamentals of Direct Liquefaction of Municipal Solid Waste in a Semi-Continuous, Microprocessor-Controlled Autoclave; Argonne National Laboratory, ANL/CNSV-TM-180 (1986).
6. Helt, J. E. et al. Pyrolysis of Municipal Solid Waste: Annual Report July 1984-June 1985; Argonne National Laboratory, ANL/CNSV-54 (1986).
7. Shafizadeh, F. The Chemistry of Pyrolysis and Combustion; ACS Advances in Chemistry Series No. 207; American Chemical Society: Washington, DC, 1984; pp. 489-529.
8. Elliott, D. C. "Process Development for Direct Liquefaction of Biomass," Fuels from Biomass and Wastes; Ann Arbor Science Publishers: Ann Arbor, MI, 1981; pp. 435-450.
9. Agrawal, R. K., et al. J. Anal. Appl. Pyrol. 1984, 6, 325.
10. Shafizadeh, F., et al. AIChE Symp. Ser. 1979, 75, 24.
11. Antal, M. J., Jr., Adv. Solar Ener. 1983 2, 61.
12. Shafizadeh, F. J. Anal. Appl. Pyrol. 1982, 3, 283.

RECEIVED May 16, 1988

Chapter 9

Producing, Evaluating, and Upgrading Oils from Steam Liquefaction of Poplar Chips

D. G. B. Boocock, S. G. Allen, A. Chowdhury, and R. Fruchtl

Department of Chemical Engineering and Applied Chemistry, University of Toronto, Toronto M5S 1A4, Canada

Chipped poplar wood can be liquefied using only high temperature/pressure steam (300-355°C). Approximately 60 per cent of the carbon feed appears in the oil phase (acetone-soluble). The oils (≅ 45% mass yield) are just solid at room temperature but soften around 50°C (1.5% moisture). They typically contain 69% carbon, 5.8% hydrogen and about 25% oxygen. Both weight and number-average molecular weights, as measured by HPSEC, are less than 1000. The major fraction of the oil is of lignin origin. Model compound studies show that the major oil functionalities (phenolic OH and OCH_3) can be hydrodeoxygenated using typical hydrodesulphurisation catalysts.

In the early 1970's, in response to the world oil crisis, studies on the direct thermal liquefaction of biomass were initiated. These studies could be classified into those on water-based processes and those on non-water-based processes. The focus in this chapter is on the water-based processes and, in particular, on a process which uses no catalysts or reducing gases. It should be noted that actual biomass used for commercial liquefaction would certainly contain significant moisture, and water is expected as a product of the liquefaction. Despite this, the water-based and non-water-based processes are significantly different.

Two types of water-based processes were initially studied. The first of these, based on Bergstrom's earlier work (1), employed sodium carbonate as a soluble catalyst and carbon monoxide as a reducing gas (2). The second technology, also based on earlier work, used nickel metal catalysts and hydrogen (3). In all cases the substrate was powdered wood slurried in water. In both processes the role of the added chemicals was not clear. Sodium carbonate would initially dissolve any phenolic and other acidic products. However, its major role was assumed to be an aid for the addition of carbon monoxide via the formate ion. As we have shown, wood will liquefy without the addition of sodium carbonate and carbon monoxide. Researchers at the University of Saskatchewan also noted that removal

0097–6156/88/0376–0092$06.00/0

of carbon monoxide from their screw reactor system did not noticeably change the reaction, although they never went the extra step of eliminating the sodium carbonate (4).

We chose to study the nickel/hydrogen system, but here again it was unclear what role the nickel catalyst played with regard to a solid substrate. It was reasoned that gases or liquids must be formed before the catalyst could intervene. Since very little liquid was formed on slow heating in the absence of nickel and hydrogen, it was also reasoned that the catalyst/hydrogen system must stabilise the products. The early experiments were performed in batch autoclaves, but subsequently we fed water slurries of powdered wood and nickel carbonate semicontinuously to a hydrogen-containing reactor (5). The feed was heated so fast that the nickel carbonate decomposed to nickel oxide instead of reducing to nickel metal. In addition, when product was discharged from the reactor, both char and oil--the latter in 25 per cent yield--were present. We reasoned that the oil yield was initially higher, but, because of prolonged residence times and the absence of the desired stabilising system, some of the oil had recondensed and charred.

As a test of this theory, powdered wood was heated rapidly to 350°C, together with water only, in small reactors heated by a fluidised sand heater. Oil yields, up to 50 per cent by weight (based on dry wood) were obtained (6).

Until this time, all studies on water-based processes had been confined to powdered wood or sawdust, the general concern being that heat and mass transfer limitations in wood pieces would prevent liquefaction. We, therefore, studied the aqueous liquefaction of single poplar sticks (6.5 mm square cross section) in the same small reactor and showed that complete liquefaction occurred (no char) at 300°C and above (7). Although heating was fairly rapid (2 minutes) for most experiments, heat-up times were extendable to at least 11 minutes without dramatic changes in oil yield. By quenching the reaction at various temperatures and noting the change in the physical state of the sticks, as well as noting the change in heat transfer characteristics to the internal thermocouple, we were able to define the liquefaction process at the macro level, as seen in Figure 1. Initially steam enters the chip and swells it. This results in disruption of the matrix which then collapses to release condensed water. When too little water is present, the swelling phase removes free liquid water from the reactor and results in decreased heat transfer. When the matrix collapses, the reappearance of liquid water increases the heat transfer. Chemical depolymerisation then results in liquefaction. The presence of liquid water appears to retard the liquefaction slightly. It also stabilises the oil and prevents charring. Samples quenched at temperatures of 290°C and 300°C do not truly represent the swelling phase, since they have been dried (see Figure 1).

Electron scanning microscope studies of the sticks have identified the initial phases of the liquefaction at the cell level (8). On the surface, the middle lamella initially disrupts, but this is quickly followed by fusion of the adjacent cell walls. The flowing matrix then engulfs the cell cavities. Inside the stick a different sequence of events occurs. Spherical structures appear, particularly on vessel walls. These spherical structures, which result from softening matrix, eventually fill the irregular cavities formed by the breakdown of cell walls, and gas and vapour bubbles can be seen within the flowing matrix. It is assumed that this process continues, although this could not be observed, since the wood eventually had too little mechanical strength for sample preparation, even using our special freezing techniques. More recently, we have investigated the liquefaction of softwoods. The liquefaction process is slower which is probably due to decreased wood porosity. However, the spherical structures are much more prevalent during the liquefaction (Figure 2).

The Reactor

On the basis of these encouraging results, it was decided that a laboratory unit should be constructed for the purpose of studying the liquefaction of commercial-size (and larger) poplar chips (see Figure 3). A batch system was chosen employing a gravity feed and discharge system. The reactor is designed to hold 100-120 g of poplar chips and as such has an internal volume of 700 mL. A single ingot of TP 316 stainless steel was used for machining the reactor which has an internal diameter of 1.5 inches and an external diameter of 3.0 inches. The length is approximately 19 inches. At the top an Oteco hub is threaded and seal welded to the reactor body. A one inch slim line connector is threaded to the base of the reactor. A total of 9 holes are drilled in the reactor side to take ¼ inch slim line connectors. Three of these are for the steam inlet lines. Three others are for 1/8 inch thermocouples, and one of the remaining three is for the rupture disc line. A vent line and pressure gauge occupy the final two holes.

The reactor is heated by two 6 feet long heaters joined in parallel. These are coiled around the reactor and held close to it by 4 longitudinal steel strips and eight circlips. The maximum power drawn by the heaters is 4 KW. The reactor is insulated with ceramic brick which is cut and fitted to the contours of the external surface.

The inlet valve is a 1.5 inch ball type (Mogas Industries Ltd.) rated at 24.8 MPa (3600 psi) at 370°C. It is joined to the reactor by a matching hub. The controller is air operated and failure of air pressure causes the valve to close. The valve is insulated by three layers of 1.5 inch thick glass wool. A similar valve was planned for the outlet valve. However, because of cost considerations, a 0.5 inch ball valve (Crosby) rated at 18.4 MPa (2665 psi) at 370°C is currently being used. This valve operates well, but we are reluctant to operate it close to its design rating.

The cooling lock has an internal column of 300 mL and is machined from stainless steel. An outer jacket allows for a variety of coolants. The steam generator is a 2-L autoclave. A Milton-Royal high pressure pump allows for the continuous addition of water to the hot generator, ir necessary. The product collector if a pyrex vessel approximately 8 inches in diameter and 5 inches deep. Separated gas passes to a brine displacement vessel for volume measurement. A number of electrical and mechanical overrides plus barriers are designed to prevent any injury to the operator as a result of accidental discharge from the reactor or steam vessel. In addition, the control panel is located in a separate room adjacent to the reactor.

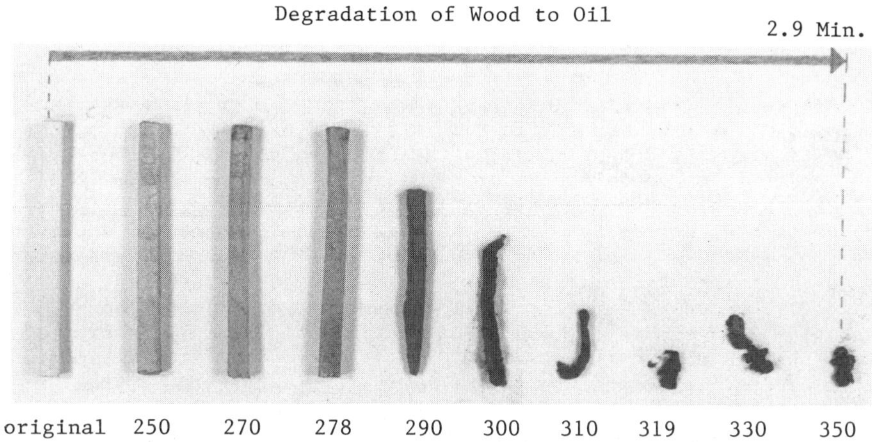

Figure 1. Liquefaction of poplar sticks in water (dried samples).

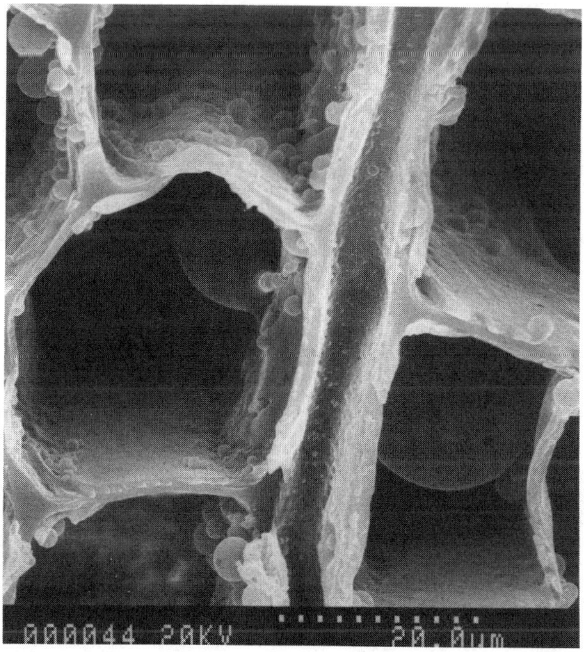

Figure 2. Scanning electron microscope picture of interior of black spruce stick (surface heated to 309°C in water).

Reactor Operation

In operation, the reactor is brought to temperature with the inlet valve open. A slow nitrogen flush through the inlet lines is also employed. The feed is contained in a cylindrical steel mesh basket which is manually dropped through the inlet valve. The purpose of the basket is three-fold. First, it allows rapid charging of the feed to the reactor. Secondly, it prevents direct contact between the chips and the reactor wall; and thirdly, it retains any unconverted wood in those experiments when large-sized feed and low reaction times and temperatures are used. After addition of the feed, the inlet ball valve is immediately closed and the steam line opened. Mechanical and electrical overrides preclude accidental ejection of steam through the inlet valve. Ordinarily, steam injection time is short (3-7 s), since the relative low temperature at the bottom of the reactor encourages flushing of the steam vessel contents. After closure of the steam valve, a further two minutes is allowed for reaction before the outlet valve is opened and the products are allowed to enter the lock. After a suitable cooling period (up to 1 minute) the products are discharged to the product collector. The basket is retrieved through the inlet valve after venting the reactor.

Results

Figure 4 shows typical temperature profiles immediately after closure of the steam lines. The temperature profile highlights the major drawback of the reactor. This is the relatively low temperature at the base of the reactor due to (a) the low rating and low thermal mass of the outlet valve; and (b) the thermal conduction to the cooling lock. Opening of the steam valve thus causes flashing of steam to this cooler area. Initially the middle and upper thermocouples are hotter than the steam, but the upper thermocouple rapidly cools as steam enters. The steam cools on expansion, and thus has a temperature less than 340°C when entering the reactor. Although heating and subsequent cooling of the bottom thermocouple is occasionally observed (particularly if the temperature is below 300°C), the most common observation is that the base temperature varies very little. After approximately two minutes the temperatures at the top and bottom of the reactor are almost equal. However, the temperature at the middle is typically 40-60°C higher after 120 s of reaction time.

Oil, water and gas exit the cooling lock and enter the collection vessel. In all, over 30 runs have been made, and of these 13 have been completely analysed in terms of carbon balance, elemental analysis and quantification of the various phases. For chipped poplar, the varied parameters have been steam temperature, injection time, water level in steam generator and total reaction time. The liquefaction of non-standard poplar in the form of dowels has also been studied.

Overall mass balances can be obtained, but these are not meaningful, given the relatively large amounts of water involved in the reaction. More important aspects are the carbon balance and the percentage of feed carbon found in the various product phases. These phases are the oil, including a separately obtained acetone wash oil from cleaning the reactor, the aqueous phase and the gas phase. In addition, a small amount of insoluble material (usually acetone-soluble) is found in the aqueous phase. Figure 5 shows a typical carbon distribution in the product phases.

The gas phase is 93% carbon dioxide, the balance being mostly carbon monoxide. About 7 litres of gas is obtained for each 100 g of chipped wood (7% moisture). Thus approximately 7 per cent of the wood carbon appears in the gas phase. The volume of the aqueous phase is difficult to control, because of the flashing of steam to the reactor. Typically the volume of aqueous phase is close to 300 mL for 100 g of chipped poplar. Interestingly, when 160 g of poplar in the form of a poplar dowel 23 inches long and 1 inch in diameter is liquefied, the volume of water collected is the same. This suggests that there is a trade off between steam, required to heat the wood, and the initial void space. The overall major finding is that, for wood chips, there is very little variation in carbon distribution amongst the various product phases for reaction times of 45-120 s and steam generator temperatures of 335-355°C. This is only true if the amount of feed is held at 100 g, and the steam vessel is relatively full of water.

Products

The Aqueous Phase. As mentioned previously, the amount of aqueous phase collected under typical operating conditions appears to be independent of void space or mass of wood in the reactor. In general, slightly less than 300 mL is usually collected for each run. The concentration of 'soluble' carbon in this water is about 45 g/L when 100 g of wood is used. This corresponds to 15 g of carbon or about 30 per cent of that in the wood. HPLC analysis has been performed for simple carboxylic acids (9) and some sugars.

A more detailed study of the aqueous phase is available for the small reactor experiments in which single poplar sticks are subjected to hot water and steam generated internally. The combined aqueous phases from single-stick experiments have been analysed for heart-wood, sapwood, and bark feeds. The analytical procedures involved extraction of the aqueous phases with relatively large amounts of ether. However, this ether extract only contained about 10 per cent by weight of the original feedstock. Since this would leave the carbon balance deficient, it is concluded that the ether did not capture all the aqueous phase. Rapid hydropyrolysis of isolated poplar lignin in the same reactor not only yielded twice the amount of oil but also produced much less ether extract from the aqueous phase. It is thus concluded that the majority of organic compounds in the aqueous phase derive from the holocelluloses. Table 1 supports this showing the major products as acetic acid, other carboxylic acids and furfurals. While hemicellulose accounds for most of the acetic acid and furfural, we have shown (10) that the cellulost yields the other furfural derivatives. The phenols obviously derive from the lignin.

The Oil. The oil yield is typically 40-45% on a dry mass basis, and the carbon content is about 59% of that in the original feed. This yield varies very little using steam temperatures of 335-355°C and

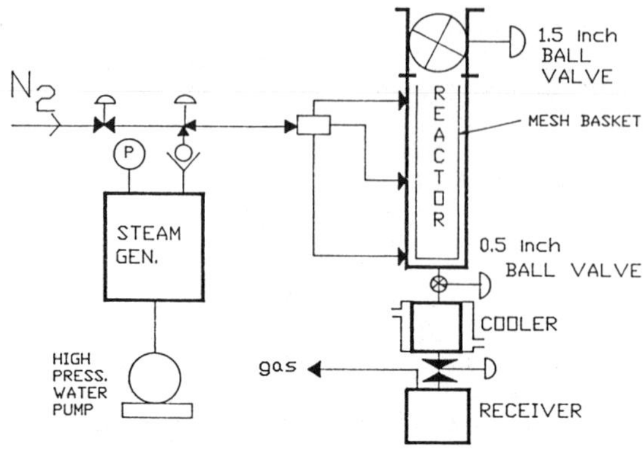

Figure 3. Schematic diagram of wood liquefaction unit.

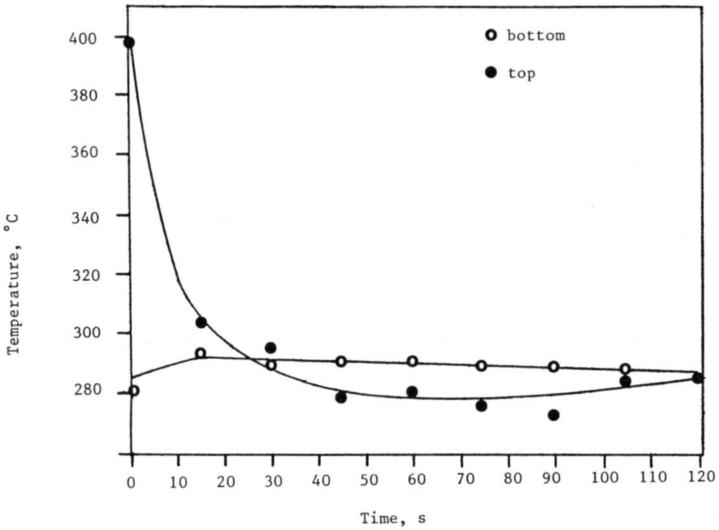

Figure 4. Temperature profiles at top and bottom of reactor.

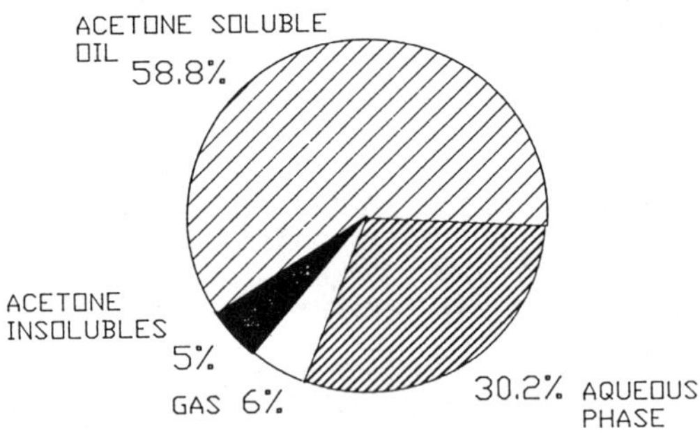

Figure 5. Carbon distribution in product phases.

Table I. Compositional Analysis of Ether-Solubles from the Aqueous Phase*

Substance	Sapwood	Heartwood	Bark	Douglas Fir
Acetic acid	30	40	20	10
Other aliphatic acids	5	10	10	5
Furfural	30	10	12	10
5-Methylfurfural	5	3	10	10
5-Hydroxymethylfurfural	3	4	5	8
Phenols (including guaiacol, syringol and alkyl derivatives)	5	5	10	5
Levulinic acid and lactones	7	10	5	20
Unidentified	15	18	28	32

* figures represent wt.% of the aqueous phase ether-solubles

reaction times of 45-150 s. The oil softens around 50°C. On average the carbon and hydrogen contents of the dry oil are 69% and 5.75%, respectively. The oxygen content is about 25% and the nitrogen is also measurable (0.3%). As mentioned previously the oil is separated easily from the bulk aqueous phase by filtration. The trapped water (up to 20%) can be separated by then melting the oil, and this lowers the water content to less than 5%. These two steps could, of course, be combined by preventing rapid cooling of the oil. However, in the laboratory we find it convenient to operate in the two step mode, although it would be relatively easy to provide heating to the product collection vessel. It should be noted that fast (non-aqueous) pyrolysis oils contain considerable quantities of water which is inseparable because of the high content of poplar compounds.

A boiling point distribution curve, together with thermal evaluation analysis data, suggests that 36% of the oil distills below 405°C which is at the upper end of the range for heavy gas oil in petroleum distillation. However, proton and carbon-13 nmr spectra clearly show that the oil is not alkane in character. A strong methoxyl signal is evident at 4.2 ppm in the proton spectrum (Figure 6) and 56 ppm in the carbon spectrum (Figure 7) and this and other evidence indicates that the oil is predominantly of lignin origin. Many of the sharp peaks found in the carbon-13 spectra of fast pyrolysis oils are absent, and there is evidence of peak broadening in both proton and carbon spectra. Spectra of co-mixtures of steam pyrolysis oil and fast pyrolysis oil show the broadening is real and is not caused by metal content in the steam pyrolysis oil (11).

Solvent-separation of the oils, using diethyl ether, chloroform and acetone, followed by analysis, support the proposed origin of the oil. Ether essentially dissolves monomeric material, while chloro- and acetone dissolve successively higher molecular weight materials. Phenols, guaiacols and syringols are found in the ether-soluble fraction along with aliphatic acids (C-5 and higher) and various benzoic acids, the latter being of lignin origin. The carboxylic acid fraction appears to be twice as prevalent as the phenolic fraction. The neutral fraction showed spectral (nmr, EIMS)

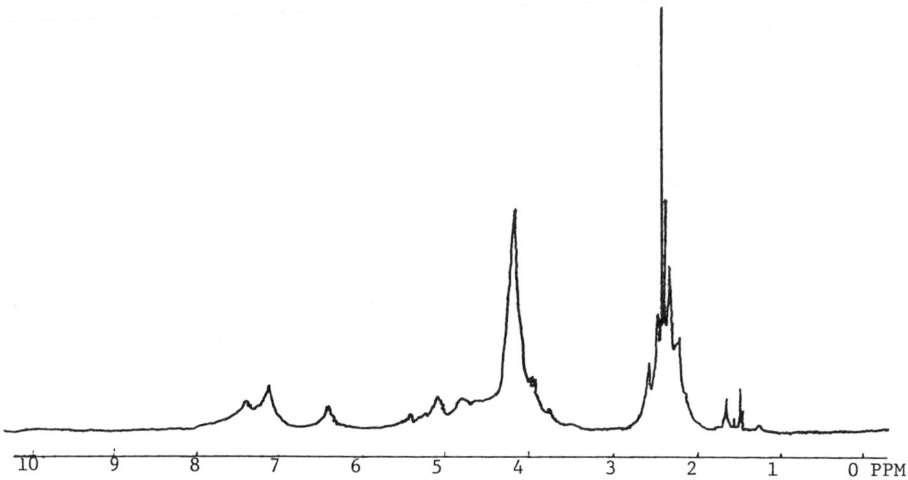

Figure 6. Proton nmr spectrum of oil.

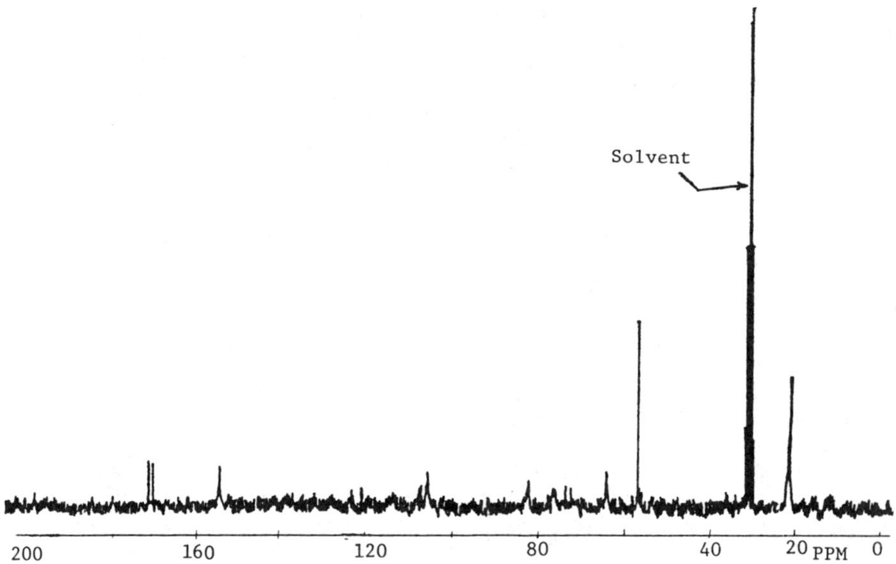

Figure 7. Carbon-13 nmr spectrum of oil.

characteristic of straight chain aliphatic protons and may contain extractives reported to be mono-, di-, and triglycerides (10).

Spectroscopic evidence shows that the chloroform soluble fraction is of both lignin and cellulose origin. In addition, the use of either lignin or cellulose as feedstock produce significant amounts of this fraction (ether-insoluble/chloroform-soluble). The ratio of syringyl and guaiacyl contributions to this fraction can be estimated from the C-13 nmr spectra. This ratio is 4:1 for heartwood, sapwood and lignin oils.

All the thermally produced oils are thermally unstable and, at sufficiently high temperatures, 'react' to form a char. This is particularly true of the fast pyrolysis oils which appear to have undergone less chemical conversion and contain more reactive functionalities. Special processes, such as entrained flow, may have to be used to upgrade all the oils.

High performance size exclusion chromatography (HPSEC) has been used to determine apparent molecular weight distributions and also to determine the effects of aging (12). The number and weight average molecular weights are less than 1000 based on polystyrene standards. Work is currently being done to quantify variations in detector response with molecular weight and sample concentration. The oils were found to age slowly at room temperature but rapidly repolymerized when heated above 100°C.

Upgrading of Oils. It is obvious from examination of the oil that it must be transformed in at least two ways. One of these transformations involves further molecular-weight reduction. This has to be achieved whilst avoiding the natural tendency of the oil to react condensatively at high temperature yielding chars. Some kind of entrainment may be necessary to obtain this desired cracking at high temperature. Our own studies have been confined to methods for removing and/or changing the phenolic OH group and methoxy groups in the product. The objective of this study was to find ways to control the transformations in specified directions and, therefore, provide maximum versatility. Although we have used many aromatic substrates, anisole and phenol have been the major focus of our work and our discussion will be limited to them.

We decided there were definite advantages to be gained if catalysts already used for hydrodesulphurisation of petroleum could be used for deoxygenation and rearrangement of functionalities. We have, therefore, studied molybdenum oxide/cobalt oxide (15%/4%) and molybdenum oxide/nickel oxide (15%/3%) (both on γ-alumina) catalysts both in sulphided and unsulphided forms. A batch reactor was used for the study and, therefore, much larger conversions were obtained than in previously used flow reactors. Disadvantages of such a system include the heat-up time required for the reactor and the non-removal of product water from the system. It was concluded from the study of phenol that the highest benzene production occurred when the sulphided cobalt catalyst was used at relatively low temperatures and hydrogen pressures (350°C, 1.4 MPa). Ratios of benzene to cyclohexane yields are 14:1 or higher. However, the reaction rate may be too low at this temperature, but raising the temperature and pressure causes the ratio to fall (2:1 at 450°C and 2.8 MPa). High conversion can be obtained at 450°C using the unsulphided cobalt catalyst. The aromatic to cycloalkane ratio is 5:1, but half the cycloalkane is methylcyclopentane. The latter product is favoured by the higher temperatures. The unsulphided nickel catalyst is slightly more active at 450°C, but the aromatic/cycloalkane ratio is not as good (3:1). If cyclohexane is the desired product, it is best to use the sulphided nickel catalyst at a relatively low temperature (350°C), when a cyclohexane/benzene ratio of 11:1 results.

Anisole readily demethylates, even in the absence of hydrogen. Even without the catalyst, some demethylation occurs. At 350°C, using the unsulphided cobalt catalyst, some of the lost methyl groups alkylate the aromatic ring to yield methyl and dimethylphenols. However, only 68% of the methyl groups can be accounted for (including 14% unconverted anisole). The same accounting holds at 450°C when deoxygenation yields toluent (21%), xylene (12%), and even some trimethylbenzenes (3%). The use of the sulphided cobalt catalyst results in a greater loss of the methyl groups (77%). Methylation of the ring is thus obviously favoured by the oxided form of the catalyst. The lost methyl groups are probably converted to methanol or dimethyl ether, but we have not analysed for these compounds. An interesting observation was that recycled (3 times) sulphided cobalt catalyst gave enhanced benzene yields (68% vs 49%). Deoxygenation was virtually complete, whereas fresh catalyst left significant unconverted phenol. Obviously the degree of sulphiding has not yet been fully exploited. Thrice used catalyst gave a ratio of aromatics to cyclohexane of 6.5:1. We suspect that similar effects would be observed if phenol were used as substrate.

Acknowledgments

We thank the Natural Sciences and Engineering Research Council of Canada, Imperial Oil and Energy Mines and Resources (Canada) for supporting various aspects of this research.

Literature Cited

1. Bergstrom, H. O.; Cederiquist, K. N. U.S. Patent 2 177 557, 1937.
2. U.S. Bureau of Mines Technical Report of Investigation #7560, Appell, H. R.; Friedman, S.; Fu, Y. C.; Wender, I.; Yavorsky, P. M., 1971.
3. Boocock, D. G. B.; Mackay, D.; McPherson, M.; Nadeau, S. J.; Thurier, R. Can. J. Chem. Eng. 1979, 47, 98.
4. Eager, R. L.; Mathews, J. F.; Pepper, J. M.; Zohdi, H. Can. J. Chem. Eng. 1981, 59, 2191.
5. Final Report of Contract File #24SU.23216-3-6143, Boocock, D. G. B.; Renewable Energy Division, Energy, Mines and Resources: Ottawa, Canada, 1984.
6. Beckman, D.; Boocock, D. G. B. Can. J. Chem. Eng. 1983, 61, 80.
7. Boocock, D. G. B.; Porretta, F. J. Wood Chem. and Technol. 1986, 6, 127.
8. Boocock, D. G. B.; Kosiak, L. Can. J. Chem. Eng. 1988, 66, 121.
9. Status Report, McKinley, J., Industrial Chemistry Division, B.C. Research: Vancouver, B.C., 1987.

10. Boocock, D. G. B.; Kallury, R. K. M. R.; Tidwell, T. Anal. Chem. 1983, 55, 1689.
11. Final Report of Contract File #38 ST.23216-6-6346, Boocock, D. G. B.; Chowdhury, A. Z., Renewable Energy Division, Energy, Mines and Resources: Ottawa, Canada, 1988.
12. Allen, S. G. B.A.Sc. Thesis, University of Toronto, Toronto, 1987.

RECEIVED May 19, 1988

Chapter 10

Oil Production by High-Pressure Thermal Treatment of Black Liquors

Process Development Studies

P. J. McKeough and A. A. Johansson

Laboratory of Fuel Processing and Lubrication Technology, Technical Research Centre of Finland, 02150 Espoo, Finland

Liquid-phase treatment of black liquors, from alkaline pulping, at 300 - 350 °C in a reducing atmosphere results in the formation of an oil-like product, which separates out from an aqueous phase containing the inorganic constituents. The process has several potential forms of application. This study was conducted in support of the development of one such application: a new recovery system for the kraft pulping process. Thermal treatment experiments were performed using different reactant gases with the main objective of elucidating the interactions between the gaseous-phase and aqueous-phase components. This information was used to advantage in the compilation of a process scheme for the recovery of the cooking chemicals from the aqueous phase.

A process producing liquid fuels from the organic matter of the black liquors from alkaline pulping is being developed at the Laboratory of Fuel Processing and Lubrication Technology, Technical Research Centre of Finland (VTT). The central operation in the process is the liquid-phase thermal treatment of black liquor at 300 - 350 °C under a reducing atmosphere (1). The treatment results in the formation of a hydrophobic oil which separates out from an aqueous phase containing the inorganic constituents and some organic matter.

The process can be applied in either of two basic ways:

- as a method to produce oil in conjunction with the kraft pulping process,
- as an entirely new system for recovering the cooking chemicals and energy from kraft spent liquors.

The first type of application exploits the favourable properties of black liquor as a feedstock for high-pressure conversion. In comparison to solid biomass, the advantages of black liquor as a feedstock include:

- no pretreatment required. Black liquor can be directly pumped into the reactor at dry solids concentrations as high as 60 %,

0097–6156/88/0376–0104$06.00/0

- black liquor contains alkaline compounds known to catalyze conversion reactions,
- cellulose, the most valuable component of wood, is not subjected to the conversion process.

In this type of application, the kraft cooking chemicals are recovered using the traditional method which is centred around the Tomlinson recovery boiler. Additional plant fuel (wood waste, peat, coal) is needed. Preliminary economic evaluations have indicated that this type of oil production process would be economic if oil prices were at their 1985 level.

In the second type of application, the Tomlinson recovery boiler is replaced by a safer and thermally more efficient recovery system, in which the oil produced by thermal treatment is used as plant fuel. This type of process is, in many respects, similar to the Hydropyrolysis Recovery Process developed by the St. Regis Paper Company, USA (2), the essential difference being the use of a reducing atmosphere in the VTT process.

The primary objectives of the present study were to elucidate the main interactions between gaseous-phase and aqueous-phase components, and, by applying this knowledge, to establish process schemes for the recovery of the cooking chemicals from the aqueous phase. Because the composition of the aqueous phase is dependent upon the composition of the reactant gas, several different gases were employed in these experiments.

Experimental

The kraft black liquor, employed in the experiments, originated from a laboratory cook of Scots pine (Pinus sylvestris). An analysis of the liquor is given in Table I.

Table I. Analysis of kraft black liquor[1].

	% of dry solids
Organic matter	78.9
Total Na	18.4
NaOH	3.0
Na_2S	5.5
Na_2SO_4	0.04
Na_2CO_3	1.1

[1] dry solids content of 30 %

A later batch of liquor, prepared in a similar way, was also analyzed for volatile acids (formic and acetic acids) and lactic acid, which are present as sodium salts in the liquor. On the basis of the analyses it can be concluded that the contents of these acids in the feed liquor used in the present experiments were approximately 6 % om, 4 % om, and 4 % om respectively, where % om denotes the percentage of black liquor organics.

The experiments were conducted in a 1-litre autoclave. In a typical experiment 500 ml of black liquor was placed in the autoclave and reactant gas was charged at sufficient pressure (5 - 9 MPa) to result in a total pressure of about 20 MPa at the reaction temperature. The autoclave was then heated at a rate of about 5 ^0C/min. A fixed time-at-temperature of 45 min was chosen for these experiments.

At the end of the reaction period, the autoclave was rapidly cooled, after which gases were released, measured, and analyzed. The organic phase and the aqueous phase were separately recovered from the autoclave and weighed. The pH of the aqueous phase was measured. In many of the experiments the aqueous phase was also analyzed for formic, acetic, and lactic acids (present as sodium salts) and for CO_2 (chemically bound in sodium carbonate or bicarbonate). The former were analyzed by gas chromatography as their benzyl esters (3), the latter by measuring the amount of CO_2 liberated upon acidification with excess mineral acid.

The primary experiments were those employing either carbon monoxide or hydrogen as reducing gas. Two temperatures, 300 ^{0}C and 350 ^{0}C were investigated. Experiments employing non-reducing carbon dioxide were also conducted in order to assess whether alkali neutralization by the reactant gas plays an important role in the thermal treatment process. One experiment was performed with a reactant gas of composition similar to that of a typical low-calorific fuel gas (producer gas) generated by gasification of biomass with air.

Results and Discussion

The main experimental results are presented in Table II.

Formation of oil phase. Although this study was not primarily concerned with the oil-forming reactions, several aspects of oil formation deserve mention. Firstly, in most of these experiments, the organic- phase product separated into two fractions, denoted as "oil" and "bitumen" in Table II. (In more recent experiments, such a separation has been avoided by using a lower temperature during pressure let-down after the reaction.) In two of the present experiments (Expts. 3 and 6), no "oil" fraction was obtained and the apparent "bitumen" yield was exceptionally high. These products were wet agglomerations of poorly-separating solids. Thus, the presence of a reducing gas was essential for obtaining an oil-like product, and, furthermore, when using hydrogen, the higher reaction temperature (350 ^{0}C) was necessary. The organic-phase product originates, to a large extent, from the lignin fraction of the black liquor. As will be discussed later, the other black liquor organics, mainly aliphatic acids, are not converted into organic-phase products (oil, bitumen) to any significant extent.

Interactions between gaseous and aqueous components. In all experiments the thermal treatment led to a decrease in the black liquor alkalinity, the final pHs being in the range 8 - 10 (Table II). The chief neutralizing agents were the gases CO and CO_2. The decrease in alkalinity was least extensive in experiments employing hydrogen as reactant gas, but it was nonetheless quite significant, particularly at 350 ^{0}C (Expts. 4 and 5). In these cases, the neutralizing agent was CO_2, a product of the thermal decomposition of the organic matter (decarboxylation).

The electrolyte systems of kraft black liquor are well known (4). In principle this data can be applied to the product mixture of the thermal treatment process, allowing the following conclusions to be drawn:

Table II. Results of autoclave experiments. Time-at--temperature: 45 min.

Experiment	1	2	3	4	5	6	7	8
Reactant gas	CO	CO	H_2	H_2	H_2	CO_2	CO_2/H_2 [1]	Producer gas [2]
Temperature, °C	300	350	300	350	350	350	345	350
Pressure, MPa	20	26	18	22	22	24	26	26
Yields, % of black liquor organics (% om)								
oil [3]	24.9	26.9	0	23.2	22.6	0	22.1	20.2
bitumen [4]	35.1	25.6	80.7	36.5	33.4	81.4	34.1	36.5
CO_2 [5]	30.1	32.1	0.4	2.4	2.4	0.6	5.2	3.9
H_2	0.5	0.7	-0.5	-0.3	-0.2	0	-0.3	-0.1
CO	-31.8	-28.6	0	0	0	0	0	-3.8
CH_x-gases	0.4	0.6	0.1	0.7	0.8	0.7	1.0	0.7
H_2S	1.4	1.1	0.3	0.5	0.4	1.0	ND	0.4
CO_2/aqueous [6]	2.0	19.0	10.0	ND	18.0	ND	19.0	ND
formic acid [7]	35.7	10.6	6.2	ND	4.6	1.6	ND	ND
acetic acid [7]	4.5	3.9	4.1	ND	3.8	3.9	ND	ND
lactic acid [7]	4.5	3.5	5.4	ND	4.0	ND	ND	ND
Aqueous-phase pH	8.5	8.5	10.2	9.5	9.4	8.9	8.4	9.1

[1] 60 % CO_2, 40 % H_2
[2] 14 % CO_2, 51 % N_2, 17 % H_2, 18 % CO
[3] organic product as oil layer
[4] heavier organic product
[5] in gaseous phase
[6] in aqueous phase as HCO_3^- or $CO_3^=$
[7] present as sodium salts
ND not determined

1. In all cases, sulphur is present as HS^- in the aqueous phase, the final pHs (8 - 10) being lower than the pK_a for HS^- (~ 13.5) and higher than the pK_a for H_2S (~ 7).
2. HCO_3^- predominates over $CO_3^=$, particularly at the lower end of the observed pH range. pK_a for HCO_3^- is ~ 10.
3. The phenolic hydroxyl groups of the lignin molecules are largely in an unionized condition, i.e. not bound to the sodium ion (pK_as: 9.5 - 11).
4. The carboxyl groups, which do not decompose during the thermal treatment, remain ionized (pK_as: 3 - 5). The three main fractions of the organic matter of the black liquor are lignin, aliphatic acids, and extractives. The aliphatic acid fraction contains the bulk of the carboxyl groups, but some are also encountered in the other fractions (5).

For the case when the neutralizing agent is CO_2, the main neutralization reactions can be written as follows:

$$NaOH + CO_2 \rightleftharpoons NaHCO_3 \quad 1)$$

$$NaHS + CO_2 + H_2O \rightleftharpoons NaHCO_3 + H_2S \quad 2)$$

$$NaO_{phenolic} + CO_2 + H_2O \rightleftharpoons NaHCO_3 + HO_{phenolic} \quad 3)$$

$$Na_2CO_3 + CO_2 + H_2O \rightleftharpoons 2NaHCO_3 \quad 4)$$

For the case when CO is the neutralizing agent, the equations become:

$$NaOH + CO \rightleftharpoons HCO_2Na \quad 5)$$

$$NaHS + CO + H_2O \rightleftharpoons HCO_2Na + H_2S \quad 6)$$

$$NaO_{phenolic} + CO + H_2O \rightleftharpoons HCO_2Na + HO_{phenolic} \quad 7)$$

$$Na_2CO_3 + 2CO + H_2O \rightleftharpoons 2HCO_2Na + CO_2 \quad 8)$$

$$NaHCO_3 + CO \rightleftharpoons HCO_2Na + CO_2 \quad 9)$$

The reactions 1 to 9 are competitive. Under different sets of reaction conditions different gas-alkali reactions will predominate. The results of Table II indicate which were the dominant reactions under the different sets of conditions employed in these experiments. At 350 °C, reactions 1 to 4 occurred to a significant extent in all atmospheres. In the CO atmosphere, a significant amount of formate was also produced (Reactions 5 to 9). At 300 °C, the gas-alkali interactions were very dependent on the composition of the reducing gas atmosphere. In the CO atmosphere, a very large quantity of formate was formed. The increase compared to the CO/350 °C experiment was presumably due to a shift in the equilibria of Reactions 8 and 9. In the H_2 atmosphere at 300 °C, carbonate formation, although still significant, was less than at 350 °C (less decarboxylation of organic matter).

Reactions of organic acid salts. A large fraction of the organic material of black liquor is comprised of aliphatic acids (5). In addition to the acids analyzed, there are significant quantities of less volatile aliphatic hydroxy acids. The acids, present as sodium salts in black liquor, are formed from wood polysaccharides during pulping. In conditions of excess alkali, these acid salts are not significantly converted into hydrophobic product (oil, bitumen) during this type of treatment (6). As confirmed in our more recent studies, three of the main acid components present after reaction are formic, acetic, and lactic acids, the less volatile hydroxy acids having undergone significant decomposition. When the yields of formic, acetic, and lactic acids (Table II) are compared to the amounts present in the feed liquor (6 %, 4 %, and 4 % of black liquor organics respectively), the following conclusions can be drawn. The treatment affected, to only a minor degree, the amounts of sodium acetate and sodium lactate present in the aqueous liquor. Any new formation of these compounds was apparently offset by loss through thermal decomposition. Sodium formate decomposed in the CO_2

atmosphere, presumably due to a shift to the left of the equilibria of Reactions 8 and 9. In the H_2 atmosphere, formate was fairly stable. In this case, the decomposition of formate to carbonate was opposed by the following reaction (7):

$$NaHCO_3 + H_2 \rightleftharpoons HCO_2Na + H_2O \qquad 10)$$

Consumption of reducing gas. From the point of view of oil formation, the most important reactions of CO or H_2 are the reduction reactions:

$$CO + O_{organic\ matter} \longrightarrow CO_2 \qquad 11)$$

$$H_2 + O_{organic\ matter} \longrightarrow H_2O \qquad 12)$$

Sodium formate can also function as a reductant in a similar way (8):

$$HCO_2Na + O_{organic\ matter} \longrightarrow NaOH + CO_2 \longrightarrow NaHCO_3 \qquad 13)$$

However, the results of the H_2/300 ^{0}C experiment suggest that the formate present in the black liquor feedstock (at 6 % om concentration) is not an effective reductant under these conditions.

In addition to its being consumed as a reductant, CO is consumed in partial neutralization of black liquor alkalinity (Equations 5 to 9) and in production of H_2 by the water-gas-shift reaction:

$$CO + H_2O \rightleftharpoons CO_2 + H_2 \qquad 14)$$

From the data of Table II, it has been possible to establish the consumption patterns of CO at 300 ^{0}C and 350 ^{0}C (Table III).

Table III. Consumption of CO.

		% of black liquor organics (% om)	
		CO/300 ^{0}C	CO/350 ^{0}C
I	Neutralization	18	3
II	Reduction	7	16
III	Shift	7	10
	Total	32	29

The total level of CO consumption, about 30 % om, is equivalent to a H_2 consumption of 2.1 % om, which is a much higher value than that observed in H_2 experiments: 0.3 - 0.5 % om (Table II).

Implications for Process Development

The aqueous phase leaving the thermal treatment reactor contains the following compounds:

- NaHS ($\rightleftarrows Na_2S$)
- $NaHCO_3$ ($\rightleftarrows Na_2CO_3$)
- Sodium salts of aliphatic acids, with formate, acetate, and lactate as three main components. The formate content is considerable when the reactant gas is CO.
- Other organic compounds in smaller amounts.

The organic acid salts and other organic compounds cannot be recycled, in their entirety, to the cooking stage. If not purged from the system, they will quickly build-up in concentration in the cooking liquor and retard the delignification reactions. Furthermore, particularly if the formate concentration is high, there may be insufficient sodium for binding to the hydroxide ion. In fact, various possibilities of converting sodium formate directly into sodium hydroxide were investigated in this study, but none of these proved to be technically feasible.

One example of a workable scheme for the recovery of chemicals is depicted in Figure 1. A part of the aqueous phase leaving the thermal treatment stage is recycled through a wet oxidation reactor where organic matter is oxidized to carbonate, CO_2, and H_2O. The heat produced is employed in heating the feed stream to the thermal treater. The rest of the aqueous phase is recausticized in the conventional way. In addition to the normal components of kraft green liquor, this liquor contains $NaHCO_3$ and some organic matter. The presence of organic matter should be an advantage because, according to the literature, a small but significant increase in pulp yield can be expected (2). The presence of $NaHCO_3$ will lead to a higher lime requirement in the recausticizing stage.

Wet oxidation is quite an economic process step in this scheme because the feed stream to it is already at the required temperature and pressure. Sulphide will be oxidized to sulphate, but it should be reduced again to sulphide in the subsequent heat treatment stage. This latter reaction will be investigated experimentally in the near future. For reasons of low gas consumption and low gas cost, producer gas is the reducing gas proposed for the scheme of Figure 1.

Preliminary economic evaluations of this scheme have indicated that it would be more economic than the conventional recovery process, particularly if a higher pulp yield were obtained. This process would also have a higher thermal efficiency than the conventional process.

Concluding Remarks

The black liquor treatment process being developed at VTT has several promising forms of application. The work described in this paper has furthered the development of one such application: a new recovery system for the kraft pulping process. Current research is being directed at gaining a better understanding of the oil-forming reactions occurring during thermal treatment.

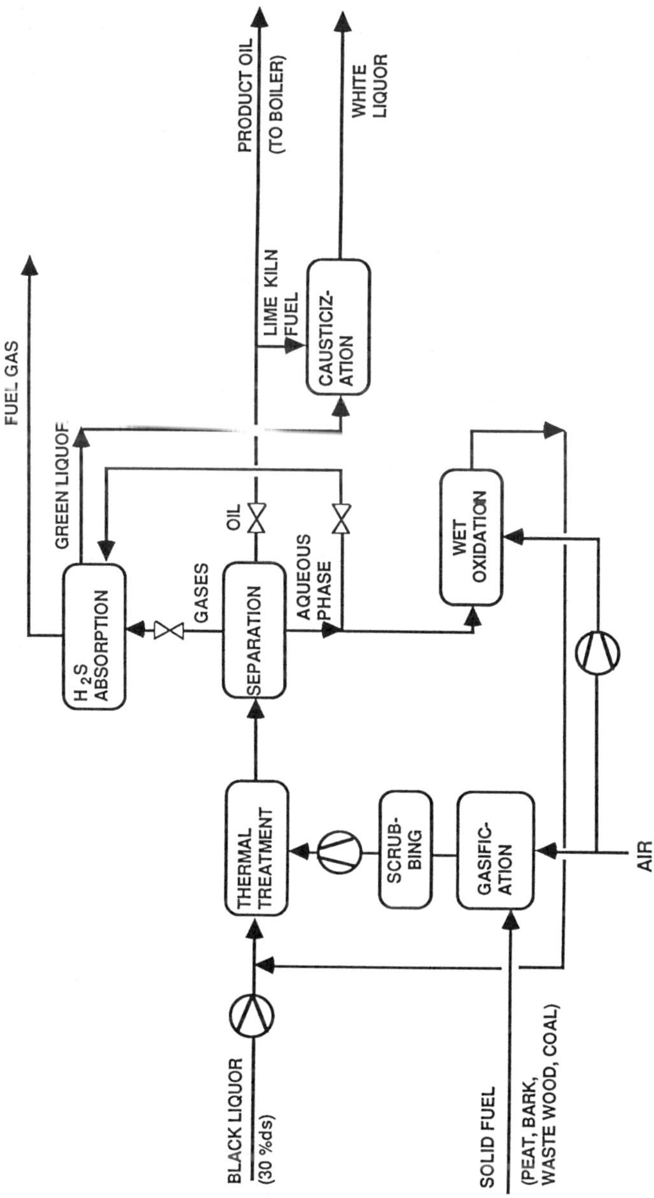

Figure 1. Prelimary scheme for a new kraft recovery process.

Acknowledgments

This work was sponsored by the Energy Department of the Finnish Ministry of Trade and Industry. The authors particularly wish to acknowledge the contribution of Paula Ylinen, M.Sc., who undertook the study of the conversion of sodium formate into sodium hydroxide. Thanks are also due to the following staff members of the Laboratory of Fuel Processing and Lubrication Technology: Dr. Raimo Alén, Virve Tulenheimo, M.Sc., and Anja Oasmaa, M.Sc.

Literature Cited

1. Johansson, A. Biomass 1984, 4. 155-60.
2. Adams, W. S.; & Maples, G. E. AIChE Symp., 1980, 76, 200, 114-19. 3. Alén, R.; Jännäri, P.; Sjöström, E. Finn. Chem. Lett., 1985, 190-92.
4. Rydholm, S. A. Pulping processes. John Wiley & Sons, New York, 1965.
5. Sjöström, E. Appl. Polym. Symp. 1983, 37, 577-92.
6. Krochta, J. M.; Hudson, J. S.; Drake, C.W. Biotech. & Bioeng. Symp., 1984, 14, 37-54.
7. Anon. Gmelins Handbuch der anorganischen Chemie. System-Nummer 21, Verlag Chemie, 1967.
8. Appell, H. R. et al. Conversion of cellulosic wastes to oil. U.S. Bureau of Mines, RI 8013, 1975.

RECEIVED June 13, 1988

Chapter 11

Formation of Aromatic Compounds from Carbohydrates

Reaction of Xylose, Glucose, and Glucuronic Acid in Acidic Solution at 300 °C

David A. Nelson[1], Richard T. Hallen[1], and Olof Theander[2]

[1]Chemical Sciences Department, Pacific Northwest Laboratory, Richland, WA 99352

[2]Department of Chemistry and Molecular Biology, Swedish University of Agricultural Sciences, S–75007 Uppsala, Sweden

The hydrothermolysis (liquefaction) of glucose, xylose, and glucuronic acid was examined in acidic solution at 300°C. Glucose provided high yields of both 5-hydroxymethyl-2-furaldehyde and 2-furaldehyde. The major products, catechol, 2,3-dihydroxyacetophenone, and 3,8-dihydroxy-2-methylchromone, as well as 2-furaldehyde, were quantitatively determined in the acidic reaction mixtures of xylose and glucuronic acid. Considerable variation of these components occurred within pH 1.7-4.0. Such products, potentially produced during biomass hydrolysis, may have an inhibiting effect on subsequent fermentation processes.

For several years our respective groups have investigated the formation of aromatic compounds from carbohydrates in aqueous solution at various pH-values under reflux or hydrothermolytic conditions. For instance, previous papers(1-6) in this series concerned the degradation of hexoses, pentoses, erythrose, dihydroxyacetone, and hexuronic acids to phenolic and enolic components. Of particular interest were the isolation and identification of catechols, an acetophenone, and chromones from pentoses and hexuronic acids at pH 4.5 (1,2). The formation of these compounds, as well as reductic acid(7),was found to be more pronounced than that of 2-furaldehyde(**2**) under acidic conditions. The aromatic precursors of **4** and **5** were also isolated from these reaction mixtures. This is in contrast to the high yields of **2** obtained from pentoses(8) and hexuronic acids(9) at very low pH. Similar products were obtained in lower yield from glucose and fructose under acidic conditions(10). However, the predominant product of these hexoses was 5-hydroxymethyl-2-furaldehyde (**1**) as would be expected from prior work(11). Surprisingly, similar products are noted at neutral and even alkaline pH with glucose and

0097–6156/88/0376–0113$06.00/0

xylose(12). Previous hydrothermolytic studies of cellulose indicated that certain aromatic products could be obtained when the pH was maintained in the range of 4-11(13). This suggested that aldol condensation, a prime route for the production of aromatics from saccharides, could function under moderately acidic conditions.

The current research was initiated to study the competition between the formation of phenolic compounds (aldol involvement) and that of furans (dehydration and cyclization). Hydrothermolytic (liquefaction) conditions, 5-7.5 minutes at 300°C, were chosen to examine the effect on potential biomass materials while exposed to mild acid. Xylose and glucuronic acids were previously found to provide higher yields of phenols than glucose. It is also of increasing interest for those involved with the hydrolysis of biomass, including steaming and autohydrolysis under slightly acidic conditions at 170-250°C, to obtain substrates for various fermentation processes or as a pretreatment for other uses. It is very likely that the aromatic products, particularly those formed from pentosans and polyuronides, may have an inhibiting effect on fermentation processes. More information, therefore, is needed concerning the formation of aromatic components and their precursors from the exposure of cellulosic material to aqueous, acidic processing of biomass. Experimentation with this material at high temperature may provide adequate direction toward this important chemistry.

Experimental

A series of 3.0 mL capacity tubing autoclaves (316 stainless steel) were used. Each tube was 0.6 x 9 cm and sealed with Swagelok fittings. The tubes were charged with 0.27 g sodium glucuronate, 0.19 g D-xylose, or 0.22 g D-glucose, respectively. Buffered acid solutions (2.0 mL) were added to the tubes. For instance, sodium acetate-acetic acid buffer was used for the pH 3 to 4 reactions, while a potassium chloride-hydrochloric acid buffer was used for the pH 1.7-1.9 reactions. The void space of each tube was swept with nitrogen prior to insertion into a 300°C sand bath. Interior tube temperature reached 300°C within 2.5 minutes, while quenching to below 100°C required only 0.1 minute. After cooling the solutions contained minimal or no precipitate. However, the solutions were dark brown after quenching. The tube contents were extracted with ethyl acetate. This extract was dried, and the solvent was removed. After the solids were weighed the extract was reconstituted with 1.0 mL of ethyl acetate for GC analyses. Gas chromatographic analyses were obtained with a Hewlett-Packard 5880A instrument using a DB-5 capillary column (30 m x 0.25 mm). The temperature program involved an initial temperature of 40°C for 5 minutes followed by an increase to 300°C at 10°C/minutes.

Results and Discussion

The yields of the solvent free extracts are presented in Table I. Column A shows the standard wt.% yields. Column B was formulated to show a loss of three moles of water for glucose and xylose and a loss of one mole of carbon dioxide for glucuronic acid. This represents the theoretical conversion of carbohydrates to furan or phenolic components. The standard yields (column A) give mixed

Table I. Yields of Ethyl Acetate Extracts After Acidic Treatment of Glucose, Xylose, and Glucuronic Acid at 300°C

	pH	Time(min.)	A*	B**
Glucose	1.7	5	37	52
Xylose	1.7	5	27	42
	3.6	5	40	62
	3.6	7.5	38	59
Glucuronic Acid	1.9	5	20	41
	3.0	5	22	45
	3.6	5	20	41
	3.6	7.5	31	63
	4.0	5	15	31

*A equals wt.% based on the amount of carbohydrate.
**B equals wt.% based on glucose or xylose minus 3 H_2O, and glucuronic acid minus $3H_2O$ and CO_2.

results when pH is compared; i.e., xylose shows higher yields at higher pH, while glucuronic acid does not. This may reflect two different mechanisms, however. These solvent extracted yields are rather similar to those obtained under basic conditions(13).

There was some change in pH after the acidic hydrothermolysis of glucose, xylose, and glucuronic acid. The pH of the aqueous phase of glucose and xylose increased from 1.7 to about 2.6 after 5 min at 300°C. Those reactions of xylose buffered at pH 3.6 held that acidity level rather well. The pH of the glucuronic acid reactions tended to increase more than those of xylose regardless of buffer; i.e., pH 1.9 to 3.2, 3.0 to 3.4, 3.6 to 3.8, and 4.0 to 5.2. This may be partially attributed to the decarboxylation of the glucuronic acid.

Table II presents the quantitative results of those components volatile enough for GC analysis. At low pH the furan compounds predominate when both glucose and xylose are exposed to 300°C. This is not unexpected since all pentoses form 2-furaldehyde(**2**) in high yield when exposed to aqueous acid solution(14). However, the presence of **2** in the glucose reaction mixture is of interest. The major product obtained from hexoses at elevated temperatures and aqueous acid is 5-hydroxymethyl-2-furaldehyde(**1**) with minor amounts of 2-(hydroxyacetyl)furan(15). The 2-furaldehyde has been detected after acidic treatment of fructose(16), glucose(15,17), and is a major component after the thermolysis of cellulose in distilled water(13). One plausible explanation for the formation of **2** may involve loss of formaldehyde(18) from glucose with consequent pentose formation. It should be noted that the pyrolysis of 1 does produce a small amount of **2**(19). However, the reaction conditions are sufficiently different to suggest a different mechanism for hydrothermolysis.

The xylose results are also notable with the increase of **3** and **5** at pH 3.6 and longer time. In contrast, **2** decreased with increased pH and time. During previous work(1) with xylose in refluxing acid at pH 4.5, catechol (**3**) was not detected. However, 3 has been detected after xylose was exposed to refluxing caustic solution(12). The presence of **3** in basic solutions of xylose was

Table II. Major Identified Components of Glucose, Xylose, and Glucuronic Acid After Hydrothermolysis at 300°C with Various Times and pH

Component**	Glucose* pH 1.7 5 min	Xylose pH 1.7 5 min	Xylose pH 3.6 5 min	Xylose pH 3.6 7.5 min	Glucuronic Acid pH 1.9 5 min	Glucuronic Acid pH 3.0 5 min	Glucuronic Acid pH 3.6 5 min	Glucuronic Acid pH 3.6 5 min	Glucuronic Acid pH 4.0 5 min
1	19.9	-	-	-	-	-	-	-	-
2	8.4	46.5	6.9	2.9	2.7	0.7	-	-	-
3	-	-	3.5	6.5	3.8	4.2	16.7	4.0	8.5
4	-	-	-	-	0.5	0.5	4.1	0.9	0.6
5	-	0.4	6.3	8.5	2.3	3.3	-	0.7	-

* Values are reported as mole%; oil yields are reported in Table I; those values not reported are <0.1%.
**The components refer to those compounds identified in Figure 1.

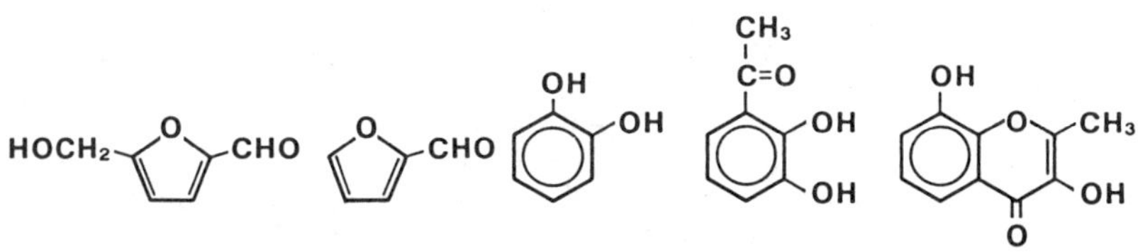

1 2 3 4 5

Figure 1. Major components isolated from carbohydrates exposed to 300°C and acidic conditions.

attributed to retro-aldol and re-aldol reactions since both xylose and glucose yielded the same type of products. Unfortunately, this does not explain the presence of **3** in the acidic hydrothermolysis product, but it has been shown that the aldol reaction can occur at pH 4.0(13). Detection of 3,8-dihydroxy-2-methylchromone(5) has been noted previously in xylose solutions at pH 4.5(1,20). Since 5 is a ten-carbon product, it is presumed that at least two moles of xylose were necessary for its composition. Thus, the mole% values of 5 (Table II) should be doubled to reflect this. Further support for this was recently demonstrated by E. Olsson, N. Olsson, and O. Theander in unpublished work during the preparation of 5 at pH 5 and 100°C from 1-^{13}C-pentose (prepared by a Kiliani synthesis involving erythrose and $K^{13}CN$). The major distribution of the ^{13}C-label was at the 2-methyl and C-8a positions.

The results from glucuronic acid do not appear quite as informative as those from xylose. The 2-furaldehyde content decreased with increasing pH; none was observed beyond pH 3.0. Correspondingly, the amount of phenolic components (**3** and **4**) increased with pH, but reached a maximum at pH 3.6. The decrease of

3 and **4** after 7.5 min at pH 3.6 may be due to instability of those components toward the thermal conditions, however degradation of **3** at pH 4.5 was negligible at 100°C(1). The results at pH 4.0 do suggest a contribution toward less **3** due to the decreasing acidity. The decrease of **1** with increasing pH is also of interest since it had previously been isolated from glucuronic acid exposed to pH 4.5 and 100°C(1,20). The Table II values obtained for **3**, **4**, and **5** from glucuronic acid may also represent only 50% of the mole percentage since each component may require more than one mole of glucuronic acid for its preparation.

Several unidentified components in the ethyl acetate fraction were also observed in the reaction mixtures of hydrothermolyzed glucuronic acid and xylose. Unfortunately, isolation attempts were not successful for these products although two were found in moderate amounts. The mass spectra (m/e 164 and 162 from glucuronic acid and xylose, respectively) were not definitive. However, both appeared to be related isomers of the methylchromone(**5**). When both are compared to **5**, one compound (m/e 162) is missing a mass of 18 (possibly water), while the other (m/e 164) is missing a mass of 16 (possibly oxygen). Several other products remained in the aqueous phase after solvent extraction. We presumed that they represented a low molecular weight fraction comprised of formaldehyde, acetaldehyde, crotonaldehyde, acetol, and pyruvic acid, for example (18,21).

It is evident from the results of this research that phenolic products, especially **3**, may be obtained by the acidic hydrothermolysis of xylose and glucuronic acid containing materials. The phenolics and 2-furaldehyde could contribute toward the inhibition of fermentation organisms if acidic pretreatment procedures are not carefully controlled.

Acknowledgments

This research was supported by the U.S. Department of Energy, Office of Basic Energy Sciences, under contract DE-AC06-76RLO 1830. O.T. thanks the Northwest College and University Association for Science (NORCUS) for the opportunity to conduct research at the Pacific Northwest Laboratory.

Literature Cited

1. Popoff, T.; Theander, O. Carbohydr. Res., 1972, 22, 135.
2. Popoff, T.; Theander, O. Acta Chem. Scand., 1976, B30, 705.
3. Popoff, T.; Theander, O.; Westerlund, E. Acta. Chem. Scand., 1978, B32, 1.
4. Olsson, K.; Pernemalm, P.-A.; Theander, O. Acta. Chem. Scand., 1978, B32, 249.
5. Theander, O.; Westerlund, E. Acta. Chem. Scand., 1980, B34, 701.
6. Miller, R.; Olsson, K.; Pernemalm, P.-A. Acta. Chem. Scand., 1984, B38, 689.
7. Feather, M. S. J. Org. Chem., 1969, 34, 1998.
8. Hurd, C. D; Isenhour, L. L. J. Am. Chem. Soc., 1932, 54, 317.
9. Stutz, E., Deuel, H. Helv. Chim. Acta, 1956, 39, 2126.

10. Popoff, T.; Theander, O. Acta. Chem. Scand., 1976, B30, 397.
11. Moye, C. J.; Krzeminski, Z. S. Aust. J. Chem., 1963, 16, 258.
12. Forsskahl, I.; Popoff, T.; Theander, O. Carbohydr. Res., 1976, 48, 13.
13. Nelson, D. A.; Molton, P. M.; Russell, J. A.; Hallen, R. T. I&EC Product Res. Develop., 1984, 23, 471.
14. Scott, R. W.; Moore, W. E.; Effland, M. J.; Millett, M. A. Anal. Biochem., 1967, 21, 68.
15. Miller, R. E.; Cantor, S. M. J. Am. Chem. Soc., 1952, 74, 5236.
16. Bonn, G.; Bobleter, O. J. Radioanal. Chem., 1983, 79, 171.
17. Smith, P. C.; Giethlein, H. E.; Converse, A. O. Solar Energy, 1982, 28, 41.
18. Rice, F. A. H.; Fishbein, L. J. Am. Chem. Soc., 1956, 78, 1005.
19. Kato, K. Agr. Biol. Chem., 1967, 31, 657.
20. Lindgren, G.; Pernemalm, P.-A. J. Liq. Chrom., 1980, 3, 1737.
21. Hayami, J. Bull. Chem. Soc. Jpn., 1961, 34, 924.

RECEIVED April 18, 1988

Chapter 12

Kinetics of Alkaline Thermochemical Degradation of Polysaccharides to Organic Acids

John M. Krochta, Joyce S. Hudson, and Sandra J. Tillin

Western Regional Research Center, Agricultural Research Service, U.S. Department of Agriculture, Albany, CA 94710

Thermochemical degradation of starch and cellulose in alkaline solution can be described by second-order kinetics, with the hydroxide ion concentration determined by the stoichiometry of polysaccharide conversion to organic acids. The thermochemical degradation activation energy for both starch and cellulose is 39.5 Kcal/mol. The production of organic acids proceeds more slowly for starch compared to cellulose. Ultimately, the yields are quite similar for both. Formic, acetic, glycolic, lactic, 2-hydroxybutyric, 2-hydroxyisobutyric and 2-hydroxyvaleric acids are produced in significant amounts. The maximum yield of lactic acid is 20% and the maximum total yield of the acids identified is 45% for the conditions investigated.

Very little research has been done to explore commercial application of the thermochemical degradation of polysaccharide materials to organic acids in alkaline solution. Much of the work on alkaline degradation of polysaccharides has been conducted at 100°C or lower to investigate mechanisms of degradation, reaction termination and the effects of molecular structure (1-4). However, little conversion of polysaccharides to acids occurs at these low temperatures. Most investigations at higher temperatures have involved study of the alkaline pulping process at approximately 170°C (5,6), where polysaccharide degradation is limited and not desired. Chesley et al. (7) proposed a commercial method for producing formic, acetic, glycolic and lactic acids from cellulosic materials. More recent work has provided additional information on the thermochemical degradation of cellulose to organic acids at reaction conditions of practical interest (8-10). No data has been reported for the important polysaccharide, starch, at reaction conditions of practical interest.

The object of this study was to identify additional organic acid products and to obtain and compare kinetic data for the

Published 1988 American Chemical Society

thermochemical degradation of starch and cellulose in alkaline solution. Reaction conditions leading to substantial or complete degradation in short times were selected so that the results could have practical application to the production of organic acids. Both starch and cellulose were studied to determine if they react with alkali at similar rates and produce the same products. Data for the formation of identified acids were also collected to allow determination of reaction conditions leading to optimum yields. The reaction system used in a previous study (8) was modified to allow rapid attainment of reaction temperatures, thus making kinetic data more easily analyzed.

Experimental

Materials. Starch used in experiments was commercial, food-grade starch (CPC International Inc., Englewood Cliffs, NJ) having a moisture content of 13%. Cellulose was purchased commercially as a highly purified, finely powdered product (Cellulay-Cellulose, United States Biochemical Corp., Cleveland, OH) and had a moisture content of 6%.

Formic acid (88%) (Fisher Scientific Co., Fair Lawn, NJ), glacial acetic acid (J. T. Baker Chemical Co., Phillipsburg, NJ), glycolic acid, L(+)-lactic acid (grade L-1), DL-α-hydroxybutyric acid (sodium salt), α-hydroxyisobutyric acid and DL-α-hydroxyvaleric acid (sodium salt) (Sigma Chemical Co., St. Louis, MO) were used as standards in high performance liquid chromatography and gas chromatography. Glutaric acid (Mallinckrodt, Inc., St. Louis, MO) was used as an internal standard in the GC analyses. Boron trifluoride (14% in propanol) (Eastman Kodak Co., Rochester, NY) was used as a derivatizing agent to produce propyl esters of the organic acids for GC analyses.

Reactions. All reactions were performed under nitrogen in a 1-L magnetically stirred autoclave equipped with a cooling coil (Model AFP 1005, Autoclave Engineers, Erie, PA). Each experiment was run by first adding 167g of aqueous solution containing 10g of NaOH to the autoclave and heating the sealed autoclave . The autoclave contents were heated to a temperature approximately 10°C above the desired reaction temperature. A well-mixed slurry of 10g of starch or cellulose (moisture free) in 85g of water was then added to an adjacent Kuentzel vessel (Vessel KD-19.3-SS11, Autoclave Engineers, Erie, PA) which had tubing with a ball valve leading from its bottom to the stirred autoclave. The tubing extended through the head of the stirred autoclave to below the surface of the alkaline solution. The Kuentzel vessel was quickly sealed, pressured to 100 psi above the pressure in the stirred autoclave with N_2, and starch or cellulose slurry was injected into the hot alkaline solution in the stirred autoclave by opening the ball valve. Resulting solution volume was 250ml and it contained 10g of starch or cellulose (0.25M based on glucose monomer) and 10g of NaOH (1.00N). Solution temperature dropped below the target reaction temperature for a short time, but returned to the desired temperature in 1-3 min. The reaction was timed from the moment of injection.

Experiments were conducted at temperatures between 180 and 300°C (145-1246 psia), and the reactants were held at the selected temperature for varying times. The temperature dropped quickly upon flow of water through the cooling coil and removal of the autoclave heater. The time required to cool the reactants to below 100°C was 5-15 min.

After cooling to room temperature, the autoclave was opened and the reaction products were suctioned out. In the case of cellulose, any unreacted material appeared as a solid residue in the reaction solution. Any residue was filtered from the reaction solution and then extracted with water. The filtrate was combined with extract, and an aliquot of the combined solution was titrated for total organic acids. The solid residue was dried at 40°C for 24 hr. in a forced-air oven and then at 98°C for 24 hr. in a vacuum oven. In the case of starch, any unreacted starch was soluble in the alkaline product solution. An aliquot of the solution was titrated for total acids. Unreacted starch was precipitated quantitatively with acetone (11). A 5.0g aliquot of the product solution was neutralized with 2.0N HCl to pH 7 and 25ml of acetone was added. After sitting overnight, the starch precipitate was separated by centrifugation, washed with 20ml of a 1:3 acetone-water solution, separated again by centrifugation, and finally dried in the same manner as for cellulose solid residue.

High-Performance Liquid Chromatography. Filtered product solutions from reactions with starch and cellulose were analyzed using a Waters Associates high-performance liquid chromatograph model ALC 201 equipped with a refractive index detector model R-401 (Waters Associates, Milford, MA) to identify water-soluble products. Product identification was accomplished by matching retention times of product compound peaks with retention times of standards. A 300 mm x 7.8 mm Aminex HPX-87H organic acid analysis column (Bio-Rad Laboratories, Richmond, CA) was used. Elution was carried out at 60°C using 0.002N H_2SO_4 at a flow rate of 0.5 ml/min. Data acquisition from the chromatographic system was by the Hewlett-Packard integrating recorder model 3388A (Hewlett Packard Co., Palo Alto, CA).

Gas Chromatography. To verify the compounds identified by HPLC, both standards and product solutions were converted to propyl esters with BF_3-propanol according to the method of Salwin and Bond (12). Analysis of the esterified compounds was performed on a Hewlett Packard 5880A gas chromatograph equipped with a flame ionization detector and a bonded Superox FA (Alltech Assoc., Inc. Deerfield, IL) fused-silica capillary column (25m x 0.25mm I.D., 0.2 µm film thickness). The temperature program was 100°C to 240°C at 5°C/min and 10 min at 240°C. The injector temperature was 225°C and the detector temperature was 300°C. Carrier flow rate (He) was 1.0 ml/min. Injection volume was 1µl with a split ratio of 80:1.

Results and Discussion

Degradation of Starch and Cellulose. An analysis of the data was performed to determine whether starch and cellulose degradation in alkaline solution could be described by second-order kinetics according to the equations:

$$\text{polysaccharide} + \text{N}\cdot\text{alkali} \longrightarrow \text{products} \quad (1)$$

$$\frac{dC_p}{dt} = -k_a C_p C_a \quad (2)$$

$$\ln \frac{C_a}{C_p} = \ln M + C_{p_o}(M-N)k_a t \quad (3)$$

$$M = C_{a_o}/C_{p_o} \quad (4)$$

$$N = (C_{a_o} - C_a)/(C_{p_o} - C_p) \quad (5)$$

where k_a is the reaction rate constant for starch or cellulose in alkaline solution, C_p is the concentration of starch or cellulose, C_{p_o} is the concentration of starch or cellulose at t=0, C_a is the concentration of alkali, C_{a_o} is the concentration of alkali at t=0, t is the reaction time, and N is the stoichiometric reactant ratio between hydroxide ion and polysaccharide.

The catalytic effect of the hydroxide ion would normally be represented as part of the reaction rate constant (k_a) for each temperature, because catalyst concentration normally remains constant. However, in the case of alkaline degradation of starch or cellulose, organic acids are produced which are converted to their salts by the alkali present, thus reducing the hydroxide ion concentration. Therefore, it seemed that this degradation reaction could be represented by second-order kinetics, with the hydroxide ion concentration determined by the stoichiometry of conversion of starch or cellulose to organic acids.

Figures 1 and 2 show experimental data plotted according to Equation 3 to determine applicability of second-order kinetics. The linear plots indicate that the results conform to second-order kinetics quite well. Table I shows the reaction parameters determined for each temperature plotted in Figures 1 and 2.

Table I. Reaction Parameters for Alkaline Degradation

	Starch		Cellulose	
T °C	N	k_a liter/mole min.	N	k_a liter/mole min.
180	1.44	0.0026	2.09	0.0034
200	1.17	0.0243	1.85	0.0194
220	1.16	0.1334	1.71	0.1189
240	1.16	0.3733	1.59	0.6000

Interestingly, although $M = C_{a_o}/C_{p_o} = 4.0$ for both starch and cellulose, extrapolating the data plotted in Figures 1 and 2 back

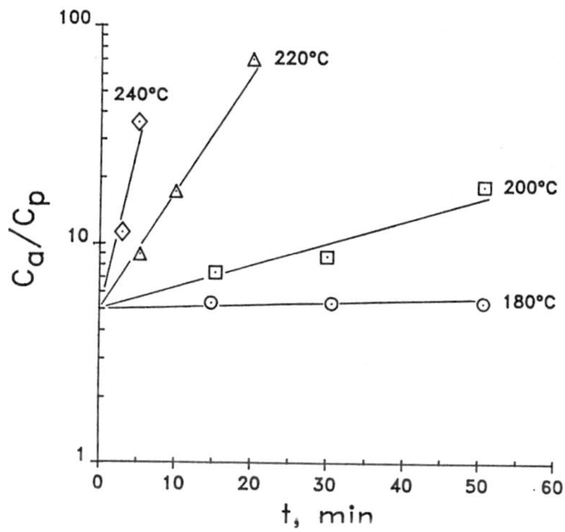

Figure 1. Effect of time on thermochemical degradation of starch in alkaline solution according to Equation 3.

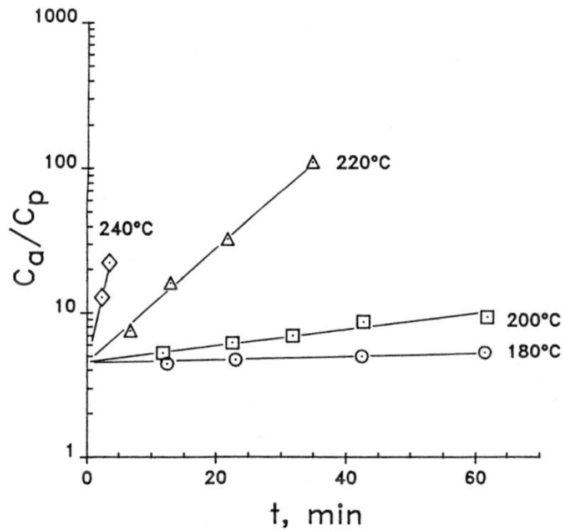

Figure 2. Effect of time on thermochemical degradation of cellulose in alkaline solution according to Equation 3.

to t = 0 gives M = 5.0 for starch and M = 4.6 for cellulose. Thus, M increased from 4.0 to 5.0 for starch and 4.6 for cellulose in very little or no time instead of the expected finite time. Considering that M eventually reaches infinity for the systems studied, this represents a small fraction of more easily degraded polysaccharide. The results of this paper apply to the more resistant fractions of starch and cellulose, which constitute the bulk of those polysaccharides.

Figure 3 shows the reaction rate constants determined from the slopes of the lines in Figures 1 and 2 plotted against the reciprocal of absolute temperature. The slope of the line through the points allows determination of the reaction activation energy according to the Arrhenius equation: $k = k_0$ e exp (-E/RT). The value of E determined in this manner is 39,500 calories/mole, which appears to apply equally well to the degradation of starch or cellulose.

Production of Organic Acids. The values of N calculated for starch and cellulose are shown in Table I. They reveal that a smaller number of organic acid molecules were formed from each glucose monomer in starch compared to cellulose. This means that the organic acids produced initially from starch had a greater molecular weight than those produced from cellulose. This is illustrated clearly in Figure 4, which shows equivalents of organic acids produced from the degradation of starch or cellulose as a function of time. Although starch and cellulose degrade at the same rate, the rate of total organic acid formation was less for starch, and fewer equivalents of organic acids were produced in a given time from starch than from cellulose. The relative difference decreased as the temperature increased, and at higher temperatures the difference in total equivalents of organic acids produced disappeared at short times. A maximum concentration of approximately 0.45 equivalents/L of total organic acids was produced for both starch and cellulose for the reaction systems studied. Apparently, in starch degradation, the initially larger molecules of organic acids eventually break down into a larger number of smaller organic acids. Therefore, the total organic acid formation for starch ultimately equals that for cellulose. Since the starting concentration of starch and cellulose was 0.25 mol/L based on the glucose monomer, the average equivalent weight for the organic acids produced is (0.25)(180)/(0.45) = 100g/eq. This assumes that all the polysaccharide material degrades to acids.

The HPLC chromatograms of the alkaline degradation products of starch and cellulose were quite similar. Based on retention times of standards, formic, acetic, glycolic, lactic, 2-hydroxybutyric and 2-hydroxyvaleric acids were identified. The GC chromatograms of the propyl esters of the starch and cellulose chromatograms were also quite similar. Since GC gave much better separation of degradation products than HPLC, the GC chromatograms were used to calculate the yields of glycolic, lactic, 2-hydroxybutyric and 2-hydroxyvaleric acids. Formic and acetic esters were hidden under the GC solvent peak; thus, their yields were determined from the HPLC chromatograms. In addition, 2-hydroxyisobutyric, which was hidden under the HPLC lactic peak, was identified and quantified

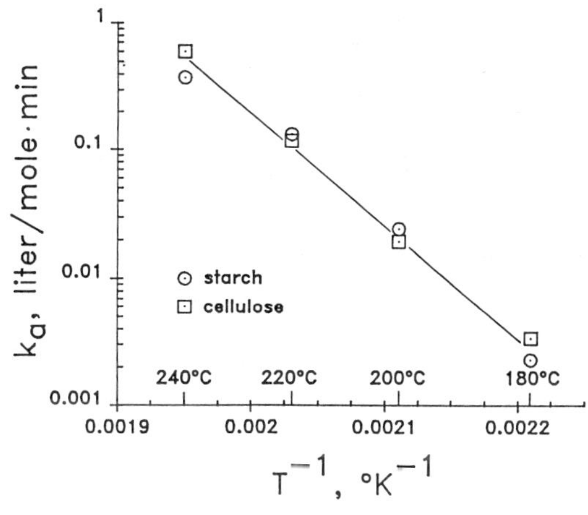

Figure 3. Relation of second-order reaction rate constants for starch and cellulose to temperature according to Arrhenius equation.

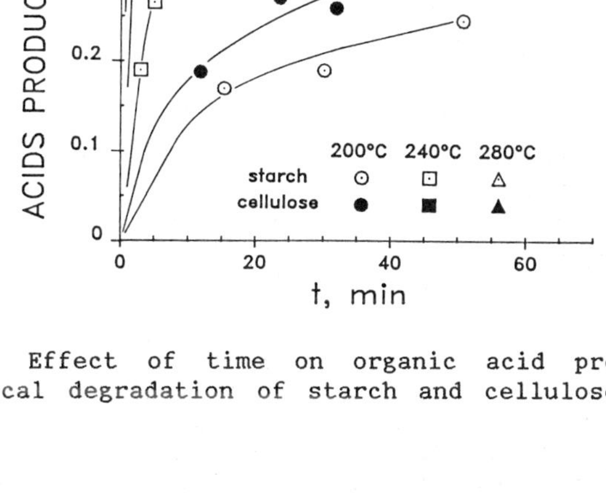

Figure 4. Effect of time on organic acid production from thermochemical degradation of starch and cellulose in alkaline solution.

with GC. Tables II and III give the maximum yields of these acids at several temperatures. The corresponding times at temperatures to achieve maximums are included in the headings.

Table II. Maximum Organic Acid Yields from Degradation of Starch in Alkaline Solution

	Yields[1] %			
	240°C 30 min.[2]	260°C 30 min.	280°C 20 min.	300°C 5 min.
Formic	10.5	10.9	10.5	10.2
Acetic	1.6	1.9	2.3	2.1
Glycolic	3.4	4.5	5.3	5.6
Lactic	16.9	17.5	19.1	19.0
2-Hydroxybutyric	2.3	2.9	3.6	3.6
2-Hydroxyisobutyric	2.4	2.6	2.7	2.5
2-Hydroxyvaleric	1.1	1.4	1.6	1.3
Total	38.2	41.7	45.1	44.3

[1]Based on starting dry weight of starch
[2]Maximum not yet attained

Table III. Maximum Organic Acid Yields from Degradation of Cellulose in Alkaline Solution

	Yields[1] %			
	240°C 32 min.	260°C 18 min.	280°C 6 min.	300°C 3 min.
Formic	10.8	10.9	10.7	10.2
Acetic	1.6	1.9	1.7	1.7
Glycolic	4.6	4.8	5.5	5.2
Lactic	19.8	19.9	20.4	18.5
2-Hydroxybutyric	3.4	3.3	3.5	3.4
2-Hydroxyisobutyric	2.1	2.1	2.1	1.9
2-Hydroxyvaleric	1.3	1.6	1.4	1.3
Total	43.6	44.5	45.3	42.2

[1]Based on starting dry weight of cellulose

Consistent with the results for rates of total acids produced shown in Figure 4, it took a longer time to reach the maximum yields starting with starch compared to cellulose. The maximum yields of the identified product acids at each temperature were quite similar for starch and cellulose starting materials. The time required to reach the maximum yields decreased as temperature increased. The trade-off is the higher operating pressure at higher temperature. The total yields of the identified acids appear to reach an optimum at approximately 280°C for both starch and cellulose.

The similarity in maximum yields for the identified product acids from starch and cellulose suggests similar or identical reaction pathways. The slower rate for acid formation from starch could be explained by dissolved starch polymer or partially degraded starch polymer interfering with additional degradation of initial conversion products. The greater ease with which starch polymer dissolves in alkaline solution compared to cellulose is consistent with this possibility.

Although significant amounts of other organic acids are produced, lactic acid is the major product for the reaction conditions studied. As well as having numerous direct uses, lactic acid has the potential of being an important intermediate in the production of other valuable products (13, 14). Preliminary economic analysis indicates that production of lactic acid from alkaline degradation of cellulose has economic merit (9). It is also conceivable that all the hydroxy acids produced could be used together without separation to produce polyester materials.

Conclusions

Starch and cellulose can both be thermochemically degraded in alkaline solution to water-soluble compounds of relatively low molecular weight. (A parallel study examines more completely the nature of these compounds (15).) Both starch and cellulose degradation processes can be described by second-order kinetics, with the hydroxide ion concentration determined by the stoichiometry of polysaccharide conversion to organic acids. The thermochemical degradation activation energy in alkaline solution for both starch and cellulose is 39,500 calories/mole.

The production of organic acids from starch proceeds more slowly than from cellulose. Ultimately, however, the yields of acids are quite similar from both. Formic, acetic, glycolic, lactic, 2-hydroxybutyric, 2-hydroxyisobutyric and 2-hydroxyvaleric acids are produced in significant amounts. The maximum yield of lactic acid is approximately 20%, and the maximum sum total yield of all identified acids is approximately 45% for the conditions investigated.

Increasing the yields of the desirable acids produced from the thermochemical, alkaline degradation of polysaccharides will require additional understanding of the reaction mechanisms and kinetics involved. Additional research should also proceed to determine the effect of other bases and supplementation of alkali catalysis by other catalytic materials.

Literature Cited

1. Whistler, R. L.; BeMiller, J. N. Adv. Carbohydr. Chem. 1958, 13, 289-329.
2. BeMiller, J. N. In Starch: Chemistry and Technology; Whistler, R. L.; Paschall, E. F.; BeMiller, J. N.; Roberts, H. J., Eds.; Academic: New York, 1965; Vol. 1, Chap. XXI.

3. Richards, G. N. In Methods in Carbohydrate Chemistry; Whistler, R. L.; Green, J. W.; BeMiller, J. N.; Wolfrom, M. L., Eds.; Academic: New York, 1963, Vol. III, Chap. 27.
4. Molton, P. M.; Demmitt, T. F. Plym. Plast. Technol. Eng. 1978, 11(2), 127-157.
5. Lowendahl,L.; Peterson, G.; Samuelson, O. Tappi, 1976, 59(9), 118-121.
6. Sjostrom, E. Tappi 1977, 60(9), 151-154.
7. Chesley, K. G.; Montgomery, C. W.; Sandborn,L. T. U. S. Patent 2 750 414, 1956.
8. Krochta, J. M.; Hudson, J. S.; Drake, C. W. Biotechnology and Bioengineering Symp. No. 14, Wiley: New York, 1984, p. 37.
9. Allen, B. R.; Dawson, W. J.; Jenkins, D. M. Report No. DOE/ID/12519-T1, Battelle Columbus Laboratories, Columbus, Ohio, 1985.
10. Niemela, K.; Sjostrom, E. Biomass 1986, 11, 215-221.
11. Taylor, T. C.; Salzmann, G. M. J. Am. Chem. Soc. 1933, 55, 264-275.
12. Salwin,, H.; Bond, J. F. J. Assoc. Off. Anal. Chem. 1969, 52, 41-47.
13. Lipinsky, E. S. Science 1981, 212, 1465-1471.
14. Lipinsky, E. S.; Sinclair, R. G. Chem. Eng. Prog. 1986, 82(8), 26-32.
15. Krochta, J. M.; Tillin, S. J.; Hudson, J. S. J. Appl. Polym. Sci. 1987, 33, 1413-1425.

RECEIVED March 31, 1988

Chapter 13

Direct Liquefaction of Wood by Solvolysis

J. M. Bouvier, M. Gelus, and S. Maugendre

Chemical Engineering Department, University of Technology, B. P. 649, F–60206, Compiegne, France

Direct liquefaction of wood can be based on a two steps scheme : solvolysis then upgrading.
In this paper, experimental results on solvolysis of wood in an acidified water-phenol mixture are presented. The process conditions -temperature, time, liquid to solid ratio, solvent composition-, are investigated and it is shown that mild conditions allow to directly solvolyse wet chips. Solvolysis products are : gases (few percent), water, light products and heavy oil. Analysis of material balances shows that wood dissolution involves a condensation reaction between phenol and cellulose.
Upgrading of the solvolysis products is necessary to recycle them as a solvent. Hydrotreatment seems to be very convenient. No water is produced but phenol, which was linked to solvolysis products, is made free. Such observations explained why upgrading increases the ability of the oil to dissolve wood. Optimization of this step is underway in order to design a two steps liquefaction process.

Since the early seventies, several authors have tried to convert lignocellulosic materials into liquid fuels or chemicals. This can be done either indirectly by use of gas synthesis or directly by thermochemical conversion. This second way, so-called liquefaction, has been first carried out in one-step processes, which are grouped in the first generation liquefaction processes (1). They combined drastic conditions : high temperatures (300 - 350°C), high pressures (5 - 15 MPa), catalysts and reducing gases (H_2, CO, H_2 + CO). More recently, a second generation of liquefaction processes appeared, because of technological problems

0097–6156/88/0376–0129$06.00/0

related to too severe conditions. They are based on a two steps scheme : solvolysis and upgrading. In the firt step, wood polymers are partly converted under mild conditions (200 - 280°C and low pressure) to complete their dissolution into specific solvents like phenols or alcohols. The resulting solution is then upgraded often by conventional refining processes.

Solvolysis processes were introduced first by the LAWRENCE BERKELEY LABORATORY, (2, 3). YU showed (3) that pine wood can be completely dissolved in acid-added phenol, with a H_2SO_4/phenol ratio from 0.1 up to 0.25 wt %, at 240 - 260°C and atmospheric pressure ; the residence time is close to 20 minutes. He also showed that similar result can be obtained with a water-phenol mixture (water/phenol = 1) providing the weight solvent/wood ratio is not below 4. Later, CHORNET and al (4) proposed the UDES-S process which combines a highly shearing pretreatment step with dissolution in creosote oil or ethylene glycol. MAUGENDRE (5) proposed a solvolysis process carried out in phenol and water with sulphuric acid, for combining hydrolysis with alcoholysis reactions. Finally, some authors attempted to use organic solvents, like acetone, methanol, tetrahydrofuran for instance at supercritical conditions to avoid non desirable side reactions, as proposed by Mc DONALD (6) or KALIAGUINE (7).

From material balances, it is generally concluded that solvent plays a keyrole in the solvolysis step ; it avoids intermediate components to recondense, preventing production of highly condensed molecules giving char. Besides, water is produced whilst the production of gases is negligible. Analysis of the material balance allows only an overall description of the solvolysis process and the mechanism of the solvolysis step is not well established, in relation to specific characteristics of wood polymers.

In this paper, a solvolysis process of wood is presented using phenol as solvent medium. Its characteristics are examined in term of material balances, the role of phenol in the solvolysis reaction as well as the origin of the water produced. Then, the solvolysis product is upgraded through a hydrotreatment, the purpose of which is to increase the proportion of light fractions as well as to produce a satisfactory solvent to be recycled in the solvolysis step.

EXPERIMENTAL PROCEDURES AND MATERIAL BALANCES

The first step is solvolysis and, as proposed by YU, we used phenol or phenol based solvents to perform the dissolution of wood, which should be complete, in order to avoid a difficult solid-liquid separation. So we carried out a lot of experiments in order to find the optimal conditions, i.e., a quantity of phenolic solvent as small as possible, low temperature and low pressure. Analysis of material balances of different runs shows that it is necessary to keep the weight solvent/wood above 4. To complete the dissolution of wood, the minimum amounts of water and phenol in the solvent mixture are respectively 20 wt % and 25 wt % (Figures 1.a and 1.b). In this case, the liquid phase is completed

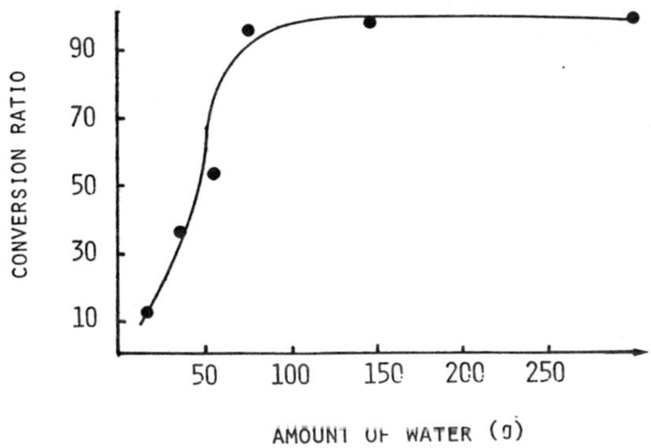

Figure 1-a. Conversion ratio versus amount of water (Oak wood = 100 g ; phenol = 100 g ; water + oil = 300 g ; H_2O = 1 g ; T = 250°C ; t = 40 - 60 minutes).

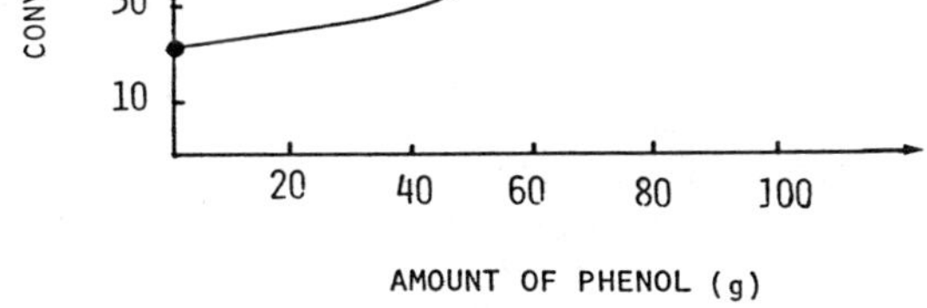

Figure 1-b. Conversion ratio versus amount of phenol (Oak wood = 100 g ; water = 80 g ; oil = 220 g ; H_2O = 1 g ; t = 250°C ; t = 40 - 60 minutes).

by an aromatic oil (aromatics : 63 wt %, saturated : 35 wt %, polars : 2 wt %).

The nature of the solvent influences the solvolysis rate. At 240°C, if solvent is phenol itself, wood is completely dissolved during the heating-up period (10 minutes), with no coke formation, even though the reaction medium is maintained at 240°C during several hours. Conversely, in a water (20 wt %) - phenol (25 wt %) - aromatic oil (55 wt %) the rate of dissolution is slower. The conversion ratio reaches 75 % during the heating-up period and complete conversion is achieved within 20 - 25 minutes as reaction time.

The material balance is set up through the analytical procedure summarized in Figure 2. Five fractions were separated and evaluated : water, light oil, heavy oil, free phenol and linked phenol, which cannot be separated by physical methods. The crude mixture is first distilled, giving two fractions. The light one contains water, phenol and light products. The heavy one while anhydrous, contains free phenol, the concentration of which is evaluated by gas chromatography.

Then, by steam carrying the heavy oil, an aqueous fraction is obtained containing free phenol, light products and a heavy oil, which is solid at room temperature. The aqueous fraction is settled, then distilled and the light fraction is separated from free phenol. A typical material balance is shown in Figure 3, whilst the effect of steam carrying is illustrated in Figure 4, through the thermogravimetric analysis of the heavy fraction. Clearly, the steam carrying eliminates light products and free phenol.

SOLVOLYSIS CHARACTERISTICS

It is obvious that the rate of dissolution increases with the temperature, but the effect of temperature depends on the composition of the solvent. The mass balance of the solvolysis reactions is greatly influenced by the temperature. Below 280°C, the production of gases is quite low, approximately 2 wt % of dry wood. As the temperature increases to 340°C, the gas production increases up to 4-5 % (Figure 5). The production of water also depends upon the temperature as shown in Figure 6. Below 220°C, produced water is approximately 20 wt % of initial dry wood. At 280°C, the amount of water reaches 30 %. The amount of water produced strongly depends also upon the nature of the wood, even though its variation with temperature shows the same behaviour. In fact, pine wood gives more water than oak wood and cellulose still produces much more water. To a first approximation, it is believed that water entirely comes from carbohydrates. Moreover, linked phenol is apparently linked to carbohydrates too, but not to lignin. Therefore, solvolysis is not only a simple dissolution process, but it involves reactions based on covalent exchanges between phenol and carbohydrates leading to approximately 3 phenol molecules per glucose. In the same time, about 3 molecules of water are produced, meaning that the dehydration process is enhanced by phenol, which could stabilize the reactive sites.

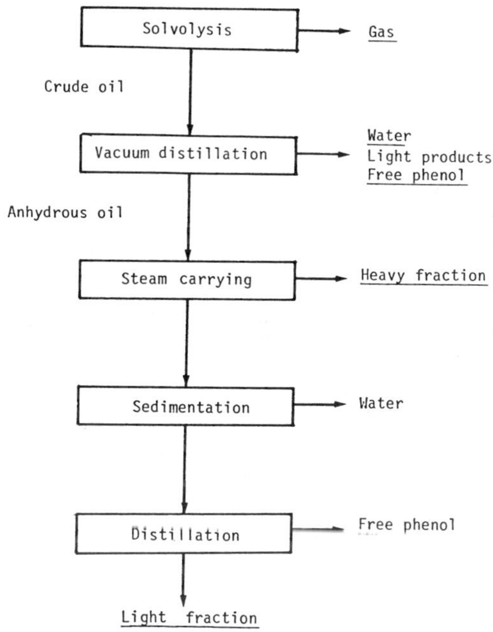

Figure 2. Scheme of the analytical procedure.

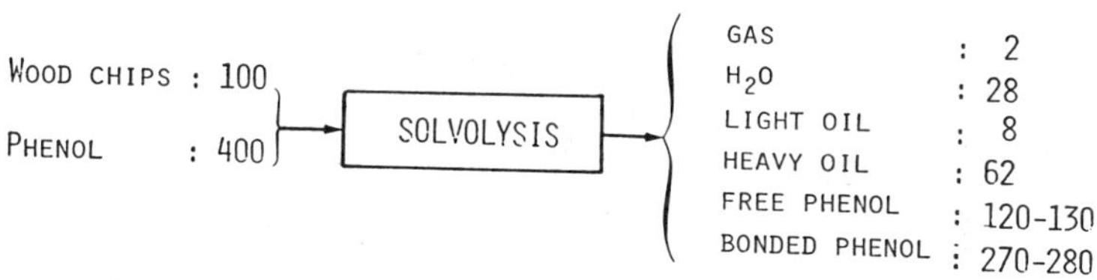

Figure 3. Material balance of the solvolysis conversion.

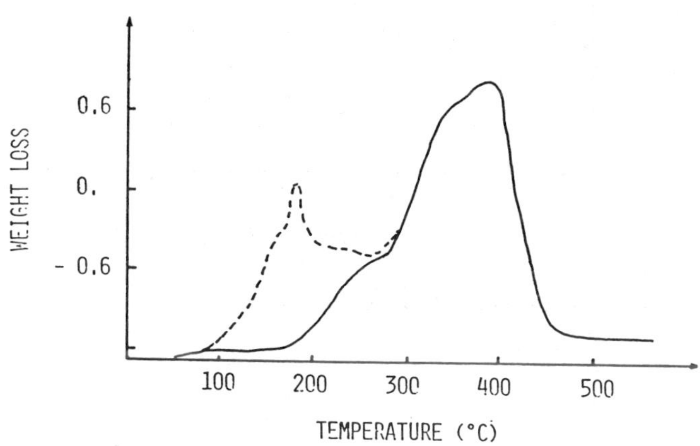

Figure 4. Thermogravimetric analysis of solvolysis oil ($10^{\circ}C$/minute ; gas carrier : argon)
.... before steam carrying
—— after steam carrying

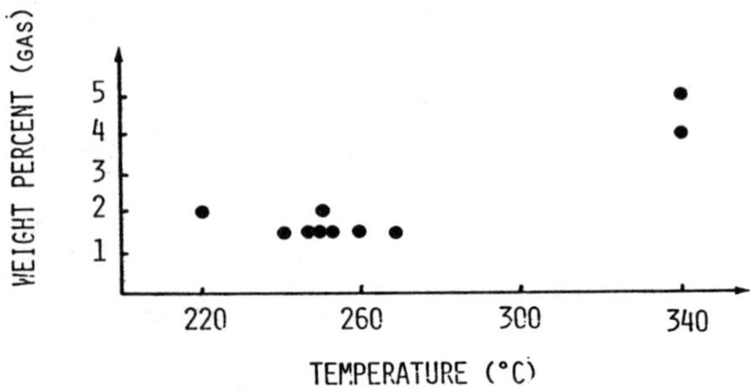

Figure 5. Weight percent (100 g dry wood) gas in solvolysis.

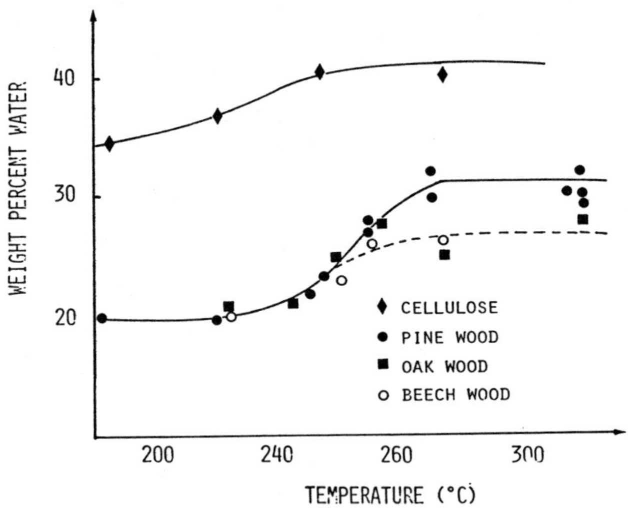

Figure 6. Weight percent (100 g dry wood) water in solvolysis.

Besides, it seems that lignin is simply dissolved in phenol rather than covalently linked to. This result was confirmed by experimenting with deuterated phenol, even though the analysis is quite complex due to exchanges of deuterium with hydrogen of sulphuric acid and water.

The solvolysis product is already a complex mixture ant it is important to identify the major components which are present into it. This was done by High Resolution Mass Spectrometry (HRMS).

The light fraction (8 % on initial dry wood) contains a large amount of benzofuran (5 %), methoxyphenol (1,8 %), methylbenzofuran (0,8 %), methylphenol (0,4 %). It contains also traces of diphenol, hydroxybenzofuran, methoxybenzofuran, dimethylbenzofuran , methyl-dibenzofuran and methylphenylketone.

Analyzing the heavy fraction is quite cumbersome, but allows us to identify some basis molecules as aromatic etheroxydes on condensed polyphenols, with molecular weights in the range of 200 to 300.

Therefore, we can draw some partial conclusions : wood can be dissolved in pure phenol or in a mixture of water and phenol with addition of sulphuric acid. Without such adition, dissolution of wood is not completed. That is of great importance from a practical point of view : it is possible to use wet chips as a raw material. Water is formed during the reaction, and this formation is attributed to the condensation of carbohydrates with phenol. It seems that one molecule of glucose, the cellulose monomer, involves bonding of three molecules of phenol and formation of three molecules of water. Phenol prevents recondensation of solvolysis products, allowing stability of the oil.

RECYCLING AND UPGRADING OF SOLVOLYSIS OIL

It is unrealistic to design a solvolysis process consuming a large proportion of new phenol. In a continuous process, it will be necessary to recycle the final product as a solvent for solvolyzing wood.

In a first attempt, successive solvolysis operations were carried out starting from a first solvolysis operation which solution is recycled as the solvent of the second solvolysis operation and so on, each operation being characterized by the temperature (260 °C), the residence time (30 minutes), the solvent/wood ratio (4/1) and the presence of sulphuric acid (1 % dry wood basis). It is shown that the conversion ratio stays close to 100 %, but the viscosity increases sharply, step by step. In fact, the viscosity of the solvolysis oil is 130 cp, 6 10^3 cp and 1,6 10^5 cp for the first, the second and the third operation respectively. The fourth operation leads to a solid mixture at room temperature.

Now, it is clear that the solvolysis oil must be upgraded in order to improve its solvent ability as well as to convert heavy fraction to a lighter one. A classical hydrotreatment was chosen following these conditions : temperature (330°C), residence time (30 minutes), H_2 initial pressure (2 MPa), commercially available Co/Mo catalyst. Hydrotreatment does not produce water while the viscosity of the produced oil decreases from 130 cp to 35 cp. The upgraded oil was recycled as solvent in a solvolysis operation followed by an upgrading treatment similar to the previous hydrotreatment.

Again, the resulting upgraded oil was recycled until seven solvolysis-upgrading cycles are realized, without adding new phenol. The overall material balances show that :

- the amount of water produced is 30 % (dry wood weight basis) ;
- the quantity of produced gas remains at the same level during the solvolysis but increases in the hydrotreatment as the number of cycles increases. In addition, the CO_2 tends to increase too, while the other-CO and CH_4 - did not vary in a large extent, except for the last step ;
- the oil viscosity increases too, tending to become very high after the last cycle (Figure 7) ;
- the free phenol content linearly decreases till the 5th cycle and then remains constant (20 % wt initial phenol basis). This means that the upgrading generates free phenolics (Figure 7) ;
- the overall material balance can be established after the seven cycles. It is based on initial dry wood and shown in Figure 8.
- hydrogen consumption is small, less than 1.5 wt % related to dry wood.

These experimental facts show clearly that the upgrading step is not the optimal one. Experimental work is underway to improve the hydrotreatment, by applying it, not on the crude solvolysis products, but after extraction of water and elimination of remaining sulphuric acid. Simultaneously, we use more severe conditions by increasing both H_2 initial pressure and temperature and we test other catalysts, like sulphurized oxydes of Mo and Ni.

ECONOMICAL EVALUATIONS

With the informations drawn from our experimental work, a preliminary conceptual design has been developed. Then, we used the same assumptions as proposed by SRI, which published a rough economical analysis of the PERC Process (8), in order to perform also economical analysis of a two steps liquefaction process.

The capacity of the plant is 400 t a day of wet wood feed (humidity : 50 %). The major products are coke (10 t/d), gases (20 t/d), water (260 t/d) and oil (112 t/d). The heating value of oil is 14,150 Btu/lb, with a 16,5 % oxygen content. Within these assumptions the calculated price of the oil is 13,1 $\$/10^6$ Btu or 65 $/bbl.

In order to compare the two steps process with the PERC and the LBL ones, the three plants were costed out with the same capacity. Results are given in Table I,
and it appears that a two steps process is less expensive. This difference comes in a major part from the investment costs, which are very lower for this process than for the two others. However, the price of oil from biomass is, in a very large extent, higher than fossil one.

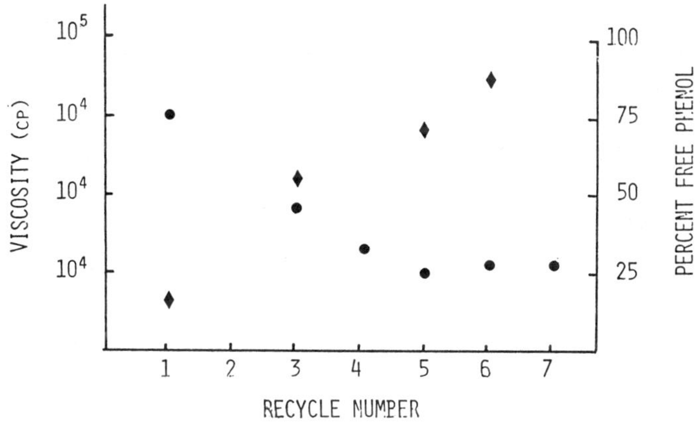

Figure 7. Oil viscosity (♦) and free phenol content (●) when upgraded oil is recycled.

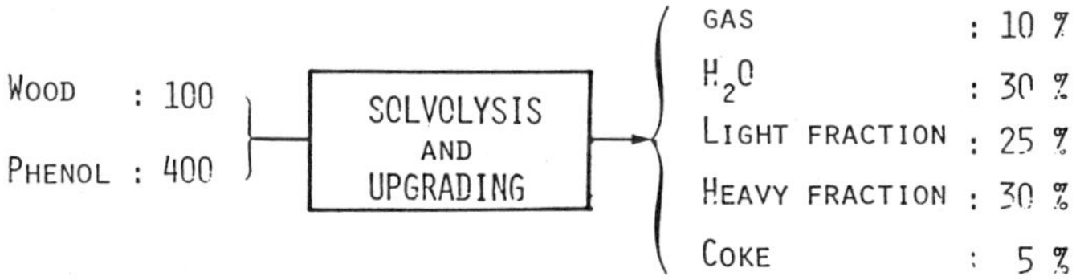

Figure 8. Overall material balance over 7 recycles.

Table I. Comparison of oil prices

	PERC process	LBL process	TWO STEPS process
Capacity (wet wood/day)	1815	1815	1815
Produced oil (bbl/day)	1728	1372	3280
Oxygen content	16,6	10,8	16,5
Oil price ($/bbl)	74,9	77,6	47,4

Acknowledgments

The Company SHELL RECHERCHES S.A. is acknowledged for its contribution to the analysis of the chemical characteristics of solvolysis oils.

Literature Cited

1. Bouvier, J.M.; Gélus, M.; Maugendre S. Applied Energy, accepted, to be published.
2. Figueroa, C.; Schaleger, L.L.; Davis, H.G. Proc. Energy from Biomass and Wastes VI Symp., 1982, p. 1097.
3. Yu, S.M. Ph. D. Thesis, Lawrence Berkeley Laboratory, 1982.
4. Chornet, E,; Vanasse C,; Overend, R.P. Entropie, 1986, 130/131, 89-99.
5. Maugendre, S. Ph. D. Thesis, University of Technology, Compiegne, France, 1986.
6. Mc Donald, E.C.; Benett, B.; Howard, J. Proc. 3rd Specialists Mtg. on Biomass Liquefaction, Sherbrooke, 1983.
7. Poirier, M.G., Ahmed, A., Grandmaison, J.L., Kaliaguine, S.C.G. Ind. Eng. Chem. Res., 1987, 26, 1738-1743.
8. Wilhem, D.J.; Kam, A.Y. Proc. 12th Biomass Thermochemical Conversion Contractors Mtg., 1981, Washington, Paper T.

RECEIVED May 19, 1988

Chapter 14

Solid Residues from Supercritical Extraction of Wood

Characterization of Their Constituents

J. L. Grandmaison, A. Ahmed, and S. Kaliaguine

Department of Chemical Engineering, Université Laval, Quebec G1K 7P4, Canada

The systematic analysis of the solid residues of the supercritical methanol extraction of Populus tremuloides was performed for samples prepared at temperatures varying from 250 to 350°C and pressures from 3.4 to 17.2 MPa, using such analytical techniques as wet chemistry, chromatography, thermogravimetric analysis, diffuse reflectance FTIR spectrometry and photoelectron spectroscopy. The results allow monitoring of the continuous changes in chemical composition of samples ranging from partly extracted wood samples to highly recondensed polyaromatic structures.

In the recent studies of thermal and thermochemical processes of wood liquefaction, considerable progress has been reported in the analysis of gaseous and liquid products (1). Some attention has been given to the composition of the solid products by wet chemistry analysis (2,3).

For the last twenty years much work has been done in the study of the thermal stability of lignocellulosic materials by thermal analytical methods. Since these materials are complex mixtures of organic polymers, thermogravimetric (TG) analysis causes a variety of chemical and physical changes depending on the nature of the sample and its treatment prior to analysis. These problems have been reviewed recently (4).

Lignocellulosic material can also be analyzed by IR spectrometry. This analytical method was used for characterization of modified lignin and cellulose in various ways (5-13). Quantification by infrared spectrometry has been reported, for example, in analysis of the three basic constituents in sweetgum and white oak chips pretreated at temperatures ranging from 140 to 280°C(5) using the diffuse reflectance FTIR spectrometry (DRIFT). The technique is simple and applicable to powdered solids and dark samples(14) and

0097–6156/88/0376–0139$06.00/0

can be used for the characterization of the chemical bonds and their modifications by thermal processes.

In this paper we report our efforts to characterize the solid residues produced in a series of experiments with the semicontinuous extraction of Populus tremuloides (aspen wood) in supercritical methanol[15], at temperatures ranging from 250 to 350°C (Supercritical Extraction residues or SCE residues), by using wet chemistry and chromatographic[16], thermogravimetric and spectral methods such as DRIFT[17] and ESCA[18].

Experimental Procedure

The solid residues analyzed here were produced by supercritical extraction with methanol of Populus tremuloides in a tubular reactor[15]. The analytical procedures used for these residues were described previously as elemental analysis, Klason lignin test, thioglycolic acid lignin test, recondensed material and carbohydrates[16], thermogravimetric (TG/DTG) and FTIR[17] and ESCA[18]. Table 1 reports results obtained using these procedures as well as conditions of extraction for each SCE residue.

Results and Discussion

Wet Chemistry and Chromatography. For the analysis of wood, the Klason lignin test, performed in concentrated sulfuric acid, is the accepted method for the determination of lignin content. We performed similar tests using also trifluoroacetic acid (TFA), the results of which are almost identical to those of the Klason tests. TFA has the advantage of allowing further analysis of the saccharides in the acid soluble fraction, as it can easily be evaporated from the solution.

The acid insoluble fraction, usually designated as "Klason lignin", is referred to in this work as Klason residue. Figure 1 shows that for the most severe extraction conditions the SCE residue is almost entirely constituted of Klason residue. The fact that this Klason residue cannot be considered as lignin has been established through elemental analysis and IR spectroscopy in KBr pellets[16].

In order to determine if the solid residues still contain lignin, the old method of forming a soluble lignin derivative with thioglycolic acid was used. This reagent reacts by displacement of α-hydroxyl and α -alkoxyl groups in lignin, and the derivative so produced can be solubilized by alkali and recovered. Results are reported in Table 1 and Figure 1 showing that thioglycolic acid lignin (TGAL) decreases from 15.6% in wood to 3.3-5.9% in the samples prepared at 350°C. It was shown by IR spectrometry that the TGAL keeps the characteristic features of lignin even for SCE temperatures of 350°C[16].

This confirms our belief that the thioglycolic acid test is a suitable method for the determination of uncondensed lignin in SCE residues. In spite of the fact that: 1) the Klason test induces some condensation reactions, and 2) the thioglycolic acid test may only extract those lignin fragments containing benzyl alcohol groups or aryl ether groups[19], we would like to suggest that: 1) the

Table I. Supercritical extraction (SCE) residues from Populus tremuloides: extraction conditions and analysis

Sample	SCE Temperature °C	SCE Pressure MPa	SCE Flowrate L/h	Residue yield wt%(a)	Klason residue wt%(b)	TGAL wt%(b)	Recond. material wt%(b)	TFA soluble wt%(b)	Glu wt%(b)	Xyl wt%(b)	%C (b)	%H (b)	%O (b,c)
Wood	–	–	–	100.0	17.6	15.6	–	78.5	47.6	27.2	46.1	6.50	47.4
MP-22	250	3.4	1.5	74.8	26.1	14.3	11.8	70.8	45.7	18.0	50.0	5.84	44.2
MP-6	250	10.3	0.5	55.4	30.5	22.2	8.3	64.1	38.1	12.8	53.5	5.29	41.2
MP-17	250	10.3	2.5	69.7	10.6	6.5	4.1	89.9	38.0	19.2	45.0	6.49	48.6
MP-15	250	17.2	1.5	68.4	7.6	5.4	2.0	90.6	26.9	12.3	44.1	6.97	48.9
MP-16	300	3.4	0.5	40.5	42.2	12.1	30.1	46.4	25.9	7.9	64.9	5.82	29.2
MP-20	300	3.4	2.5	31.0	78.1	12.3	65.8	–	0.0	0.0	75.9	5.12	19.0
MP-9	300	10.3	1.5	16.4	92.1	4.3	87.8	5.0	0.0	0.0	78.2	4.96	16.8
MP-12	300	10.3	1.5	15.9	92.5	3.8	88.7	1.6	0.6	0.0	80.4	4.62	15.0
MP-21	300	10.3	1.5	15.7	89.2	4.5	84.7	2.4	0.0	0.0	79.8	5.35	14.8
MP-13	300	17.2	0.5	18.4	65.8	6.0	59.8	27.2	10.3	3.3	64.8	5.47	29.7
MP-13	300	17.2	2.5	15.0	86.7	5.3	81.3	4.0	0.7	0.0	73.3	5.17	21.6
MP-27	330	17.2	1.2	8.4	71.4	4.8	66.7	19.8	7.1	0.0	74.1	6.08	19.8
MP-14	350	3.4	1.5	26.9	90.0	5.2	85.1	11.1	4.8	1.5	74.8	4.34	20.8
MP-11	350	10.3	0.5	18.2	92.3	3.8	88.5	0.8	0.0	0.0	83.5	4.89	11.6
MP-8	350	10.3	2.5	10.2	95.1	5.9	89.2	1.2	0.0	0.0	92.8	3.84	3.33
MP-24	350	17.2	1.5	6.1	80.3	3.3	77.0	3.1	0.0	0.0	76.3	5.00	18.8

(a) Dry wood basis.
(b) Dry SCE residue basis.
(c) Calculated by difference.

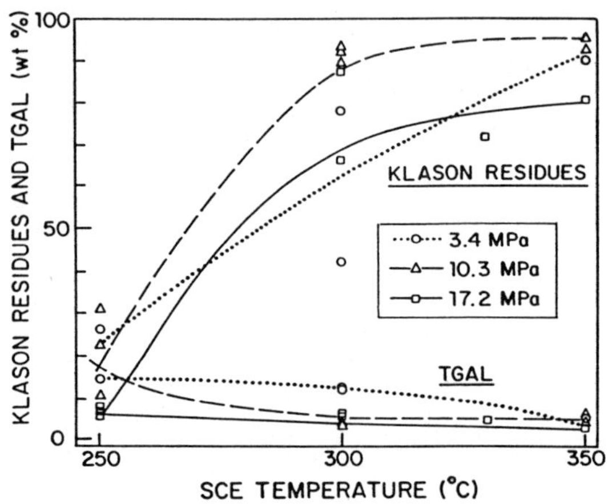

Figure 1. Klason residue and thioglycolic acid lignin (TGAL) in SCE solid residues (expressed on dry SCE residue basis).

thioglycolic acid lignin represents a good estimate of unconverted lignin, and 2)the Klason residue represents the summation of unconverted lignin and of condensation products formed by pyrolysis reactions during the SCE process.

As a consequence, we suggest that the difference between the Klason residue and the thioglycolic acid lignin is representative of recondensed material (RM) in SCE residues. The calculated values for recondensed material in these residues are reported in Table 1. Figure 2 gives the values for the percent recondensed material expressed on a dry wood basis.

Figure 3 shows the percentage of recondensed material, expressed on dry wood basis, plotted as a function of lignin conversion. This graph suggests different condensation reactions at 250°C and at 300-350°C. At 250°C in particular, the condensation seems to be a secondary reaction of lignin conversion. As also shown on the figure, for several experiments the percents of recondensed material are higher than the value which would be calculated assuming that all converted lignin is transformed to recondensed material (line A). It is believed that this indicates that the condensation reaction involves not only products of degradation of lignin but also some of carbohydrates.

The glucose and xylose contents were determined in the soluble TFA acid hydrolysis fraction by liquid chromatography using a cation exchanged resin (Ca^{++} form) column. The results are reported in Table 1. Most of the samples prepared at 300-350°C show only minor amounts of hydrolyzed material, except for samples MP-16, MP-13 and MP-14 which were prepared at low pressure or low flow rate. The percents of glucose and xylose for these samples as well as those for the samples prepared at 250°C, expressed on dry wood basis, are plotted on Figure 4. The rather well defined curve indicates that cellulose and hemicellulose are simultaneously degraded at or below 250°C.

Thermogravimetric Analysis. Thermogravimetric analysis (TG and DTG) under nitrogen atmosphere was performed for aspen wood and the 16 partially converted wood residues. The TG and DTG curves are reproduced in Figure 5 for untreated wood and for 4 selected representative SCE residues.

The examination of TG and DTG curves, shows that:

a) aspen wood loses weight starting near 230°C (pyrolysis of hemicellulose (21)), and between 350 and 420°C, with a maximum rate of weight loss at 385°C (cellulose and lignin pyrolysis (21)); the weight lost at 700°C is 89.4%.

b) the SCE residues can be classified according to their temperature of extraction.

For the residues of type I prepared at 250°C (e.g. sample MP-6), the weight loss takes place between 350 and 420°C, with a maximum rate at 375-390°C. The weight lost at 700°C is between 82.5 and 94.6%.

For the type II residues produced at 300°C (e.g. sample MP-12), a continuous weight loss is observed from 300 to 600-700°C, with a maximum rate at temperatures ranging from 380°C to 510°C. The total weight loss at 700°C is less important than for samples of the previous type, ranging from 27.2 to 57%.

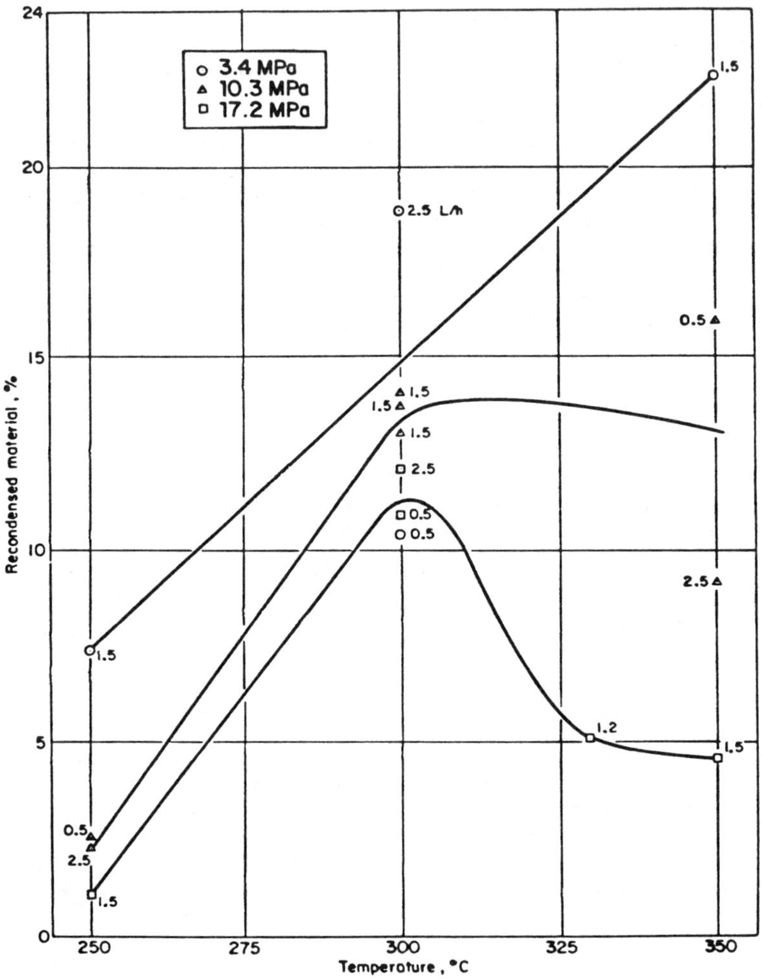

Figure 2. Effect of SCE conditions on recondensed material (expressed on dry wood basis).

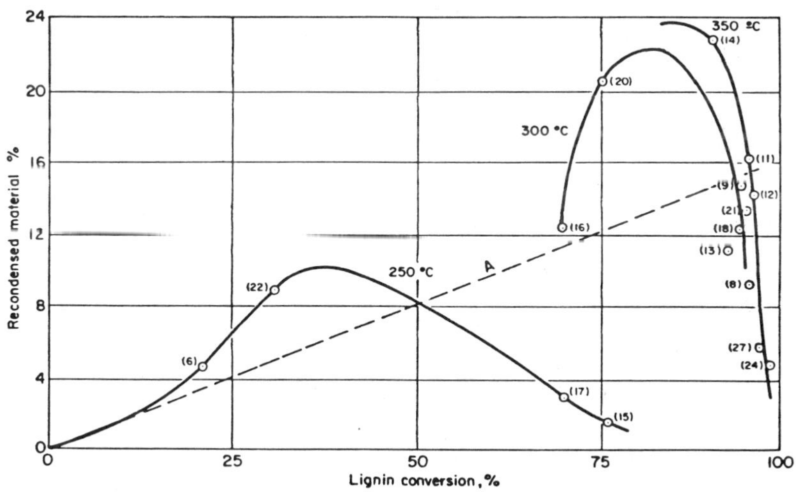

Figure 3. Recondensed material (expressed on dry wood basis) as a function of lignin conversion (experiment number shown in parentheses).

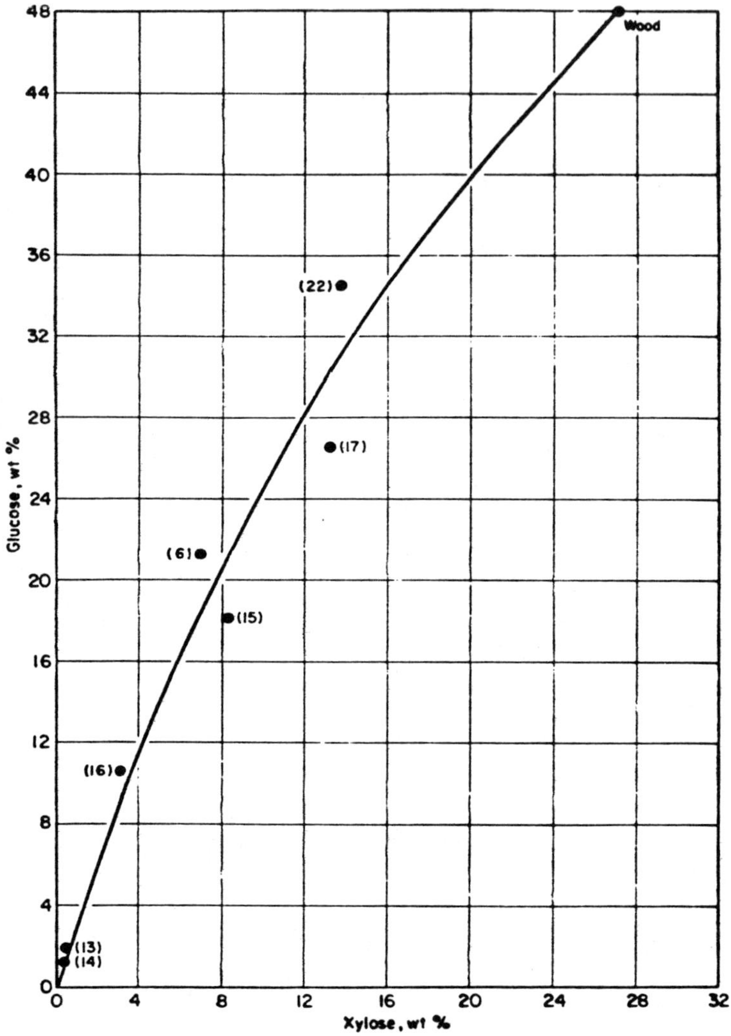

Figure 4. Residual hydrolyzed glucose as a function of residual hydrolyzed xylose (both expressed on dry wood basis).

A

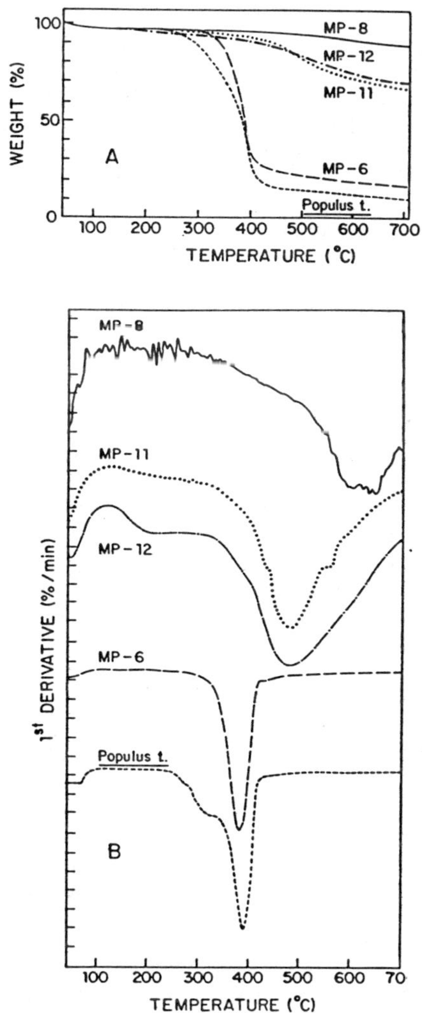

B

Figure 5. Thermogravimetric analysis of Populus tremuloides and of four SCE residues; TG(A) and DTG(B) in flowing nitrogen.

For samples of type III prepared at 350°C (e.g. samples MP-11 and MP-8), the weight loss is slower than for those of type II but happens roughly in the same temperature range (300 to 600-700°C) and with maximum rates occuring at higher temperatures, from 380 to 620°C. The weight loss at 700°C is also small, ranging from 8.2 to 34.2%.

A closer analysis of these curves shows that there is a continuous change in the shape of the thermogram of the residues as the SCE pressure is increased for experiments at the same SCE temperature. As shown in Figure 6, the temperature T_{max} corresponding to a maximum on the DTG curve shows a continuous evolution with the parameters of extraction. Two maxima are observed at the lower SCE pressure of 3.4 MPa (with only the overall maximum being shown in Figure 6), showing that when the extraction is performed less efficiently, some of the unconverted lignin and cellulose is still present at relatively high content in the residue prepared at 350°C.

The smooth evolution in the temperature of the high temperature DTG peak reflects a change in the nature of the volatile fraction of the recondensed material.

A very good correlation (0.97) was found between weight lost between 200 and 420°C and the weight of trifluoroacetic acid soluble plus unconverted lignin previously determined by wet chemistry. These data suggest that the material still not volatilized at 420°C would be identical with what we defined as the recondensed material. It was indeed verified that the correlation between recondensed material and weight % of the solid not volatilized at 420°C is also excellent. In fact, minor corrections can be made taking into consideration the amounts of char formed during TG analysis from polysaccharides and lignin still present in the SCE residue. These corrections affect only very slightly the slope of the correlation curves. From thermogravimetric data the recondensed material in a given SCE residue can thus be further characterized by the weight fraction of RM volatilized between 420 and 700°C.

Diffuse Reflectance Infrared Spectrometry. DRIFT spectra in KCl were obtained for the 17 afore mentioned samples, and the spectra of the five representative samples used to present the TG/DTG data are reported in Figure 7.

The spectrum of aspen wood (Figure 7.I) shows the presence of the three fundamental wood constituents. A band for cellulose is at 895 cm^{-1} (β-anomer in pyranose ring) (22). The band at 1740 cm^{-1} is due to uronic acid and acetyl groups in hemicellulose (22). The bands from 1235 to 1605 cm^{-1}, specially the one at 1505 cm^{-1}, are representative of lignin (19), while those from 950-1100 cm^{-1} are due, in part, to carbohydrates (C-O bonds in alcohols).

As shown by the other spectra in Figure 7, these bands are significantly modified by the SCE treatment.

Spectral region 2870-3050 cm^{-1}. A band at 3050 cm^{-1} (aromatic and/or alkene C-H stretching) becomes evident at 300°C (MP-12, Figure 7-III) and dominates this region at 350°C (MP-8, Figure 7-V). The band in the 2900 cm^{-1} region (aliphatic C-H stretching) which is broad in the initial wood sample, is progressively resolved in three separate bands (2870, 2930 and 2955 cm^{-1}) as the SCE temperature is

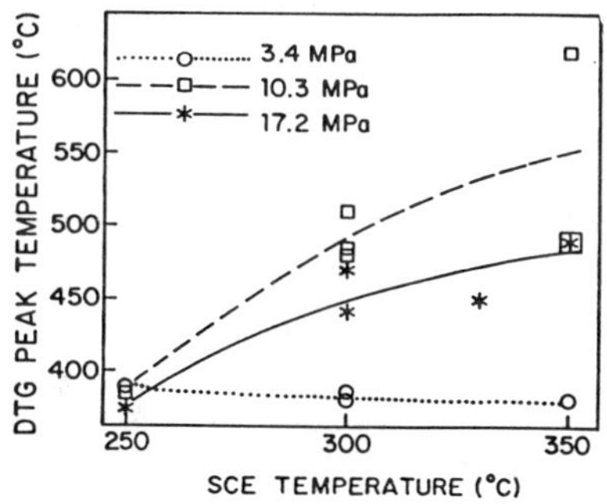

Figure 6. DTG peak temperature as a function of SCE temperature and pressure.

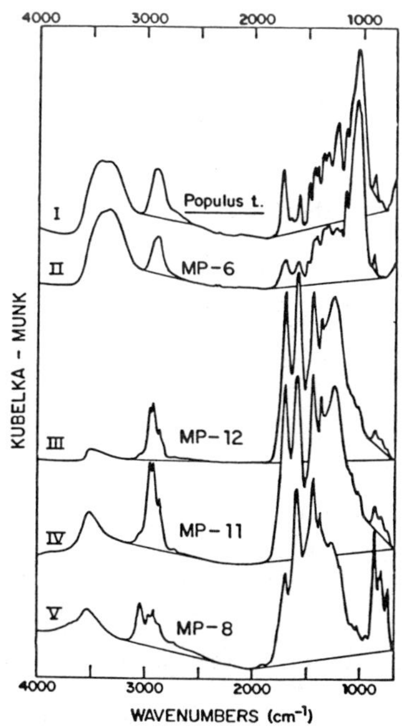

Figure 7. DRIFT spectra (in KCl) of Populus tremuloides and of four SCE residues.

increased (MP-12, MP-11, and MP-8, Figures 7III, IV and V). The overall pattern in Figure 7V corresponding to the most carbonized sample is similar to the ones reported for higher rank bituminous coal [23] and for vitrinite [24], with the 3050 cm^{-1} even more intense in our MP-8 sample. As it was shown earlier that this sample contains 89.2% of recondensed material, it may be concluded that this material has a coal-like polyaromatic nature. This is supported by the changes in the next spectral region.

Spectral Region 700-950 cm^{-1}. The band at 895 cm^{-1} is discernible in wood and MP-6 (SCE temperature 250°C) but disappears from spectra of samples treated at higher temperatures where saccharides analysis has also shown the absence of cellulose. As carbonization proceeds, the out-of-plane bending below 900 cm^{-1} (at 870-881, 800-833 and 752-766 cm^{-1}) increases in intensity; the position for some of the bands (800 and 750 cm^{-1}) can suggest condensed ring systems.

The band at 950 cm^{-1}, which is visible in MP-8, is probably due to elimination reaction giving t-alkenes[19].

Spectral Region 1450-1595 cm^{-1}. The characteristic aromatic ring vibration at 1505 cm^{-1}, clearly visible in the spectrum of wood, is gradually hidden with an increase in the SCE temperature. This corresponds to the progressive deqrease in lignin content of the residue. Inversely, two bands near 1450 and 1595 cm^{-1} become very intense and dominate in the spectra of residues produced at 300 and 350°C (Figures 7-III, IV and V). These changes parallel the modifications in the 2870-3050 cm^{-1} region. The band at 1450-1460 cm^{-1} can be attributed to methyl and methylene bonding and also to aromatic ring modes [24,19]. The band at 1595 cm^{-1} is also assigned to aromatic ring stretching. Its high intensity in spectra 7-III, IV and V, could possibly be given the three following explanations [24]: 1) aromatic ring stretching in combination with a chelated conjugated carbonyl structure, 2) aromatic ring stretching mode, with possible intensity enhancement due to phenolic groups, 3) aromatic ring stretching of aromatic entities linked by methylene and possibly ether linkages.

Spectral Regions 1035-1370 and 1705-1740 cm^{-1}. The bands from 1035 to 1171 cm^{-1} (C-O bonds in polysaccharides, lignin alcohols and lignin ethers), present in wood and samples obtained at 250°C, drop in the spectra of SCE residues produced at and above 300°C. The aromatic ethers bands (up to 1330 cm^{-1}), and the phenolic stretching near 1250 cm^{-1}, decrease also.

The hemicellulose band, at 1740 cm^{-1}, present on untreated wood almost disappears in residues prepared at 250°C. The unconjugated carbonyl and/or carboxyl and/or ester of conjugated acids at 1710-1720 cm^{-1} from original lignin is still visible at 250°C when hemicellulose is partly removed, but at higher SCE temperatures it is hidden by the highly intense 1705 cm^{-1} band. This last band can be attributed to a conjugated carbonyl or carboxyl structure, but it would be surprising that carboxyl could resist at severe SCE conditions. Further study is necessary for definite assignment of this band.

Quantification. Schultz et al[5] reported correlations of FTIR absorbance ratios with such variables as percent glucose, xylose and Klason lignin for wood chips pretreated by the RASH process at temperature ranging from 140 to 280°C. These correlations do not fit correctly our data, so we developed our own equations by non-linear least squares regression. For quantitative evaluation of absorbances, baseline was defined as shown on spectra on Figure 7. These equations are described in Table II. In equations 1-5, the intensities (I) are expressed in Kubelka-Munk units. Selection of the independent variables in equations 1-5 was done by first selecting 16 significant bands in the 17 DRIFT spectra. All linear regression equations describing the five dependent variables, as function of all possible combinations of intensity ratios, were obtained using two to six of these variables. A preliminary selection was made on the basis of R^2 values and F test significance levels. A final selection was performed, among the statistically equivalent regression equations, on the basis of the analytical significance of the bands selected, yielding equations 1-5.

ESCA

ESCA is a surface sensitive technique, based on the measurement of kinetic energies of photoelectrons ejected from a given atomic energy level under the action of a monoenergetic X-ray beam. It provides quantitative information on the elemental composition as well as on the chemical environment of each atom (bonding and oxidation state).

The kinetic energy of photoelectrons (E_K), as measured with respect to the vacuum level, is expressed as:

$$E_K = E_X - (E_B + \phi + E_C) \quad (6)$$

where E_X is the energy of the incident photon, E_B is the binding energy of the electron on its original level, ϕ is the work function of the spectrometer, and E_C is the energy lost in counteracting the potential associated with the steady charging of the surface. ϕ and E_C are essentially corrections. ϕ depends on the spectrometer and is not liable to be modified between experiments. E_C is high on low conductivity samples and can be made lower by the use of a flood gun.

ESCA spectra corresponding to carbon 1s peaks of Populus tremuloides and 3 samples isolated at three different SCE temperatures are illustrated in Figure 8. There is a general agreement in the literature on the assigment of components C_1, C_2 and C_3 in wood derived materials: C_1 corresponds to carbon linked to H or C, C_2 has one link to oxygen, whereas C_3 has two. In the solid phase, C_1 is referenced at 285.0 eV and C_2 and C_3 are usually close to 287.0 and 289.5 eV [25].

In all SCE residues, a fourth C_{1s} component is found on the low binding energy side of the spectrum, shifted from the C_1 component by 1.4 ± 0.5 eV. This is thereafter designated as the C_0 component. As the temperature of extraction is increased from 250 to 300°C, the C_0 component increases continuously whereas the general trend of C_1, C_2 and C_3 components in a continuous decrease.

Table II. Regression equations to calculate five parameters in SCE residues

$$\%\ \text{Klason residue} = 17.2 + \left(97.7 \times \frac{I_{3050}}{I_{1505}}\right) - \left(77.7 \times \frac{I_{2870}}{I_{1505}}\right) + \left(51.1 \times \frac{I_{1370}}{I_{1505}}\right) \quad (1)$$

$R^2 = .952$

$$\%\ \text{Unconverted lignin} = -10.6 + \left(31.1 \times \frac{I_{2955}}{I_{1595}}\right) - \left(62.2 \times \frac{I_{2870}}{I_{1595}}\right) + \left(22.8 \times \frac{I_{1505}}{I_{1595}}\right) + \left(12.5 \times \frac{I_{1160}}{I_{1595}}\right) \quad (2)$$

$R^2 = .879$

$$\%\ \text{Recondensed material} = 166.1 + \left(114.1 \times \frac{I_{3050}}{I_{1370}}\right) - \left(69.5 \times \frac{I_{2930}}{I_{1370}}\right) - \left(111.1 \times \frac{I_{1505}}{I_{1370}}\right) - \left(181.2 \times \frac{I_{895}}{I_{1370}}\right) \quad (3)$$

$R^2 = .967$

$$\%\ \text{Glucose} = -42.8 + \left(20.0 \times \frac{I_{1505}}{I_{1595}}\right) + \left(41.8 \times \frac{I_{1260}}{I_{1595}}\right) - \left(54.9 \times \frac{I_{895}}{I_{1595}}\right) \quad (4)$$

$R^2 = .940$

$$\%\ \text{Xylose} = -23.2 + \left(35.3 \times \frac{I_{1260}}{I_{1595}}\right) - \left(9.1 \times \frac{I_{1210}}{I_{1595}}\right) - \left(33.2 \times \frac{I_{895}}{I_{1595}}\right) \quad (5)$$

$R^2 = .961$

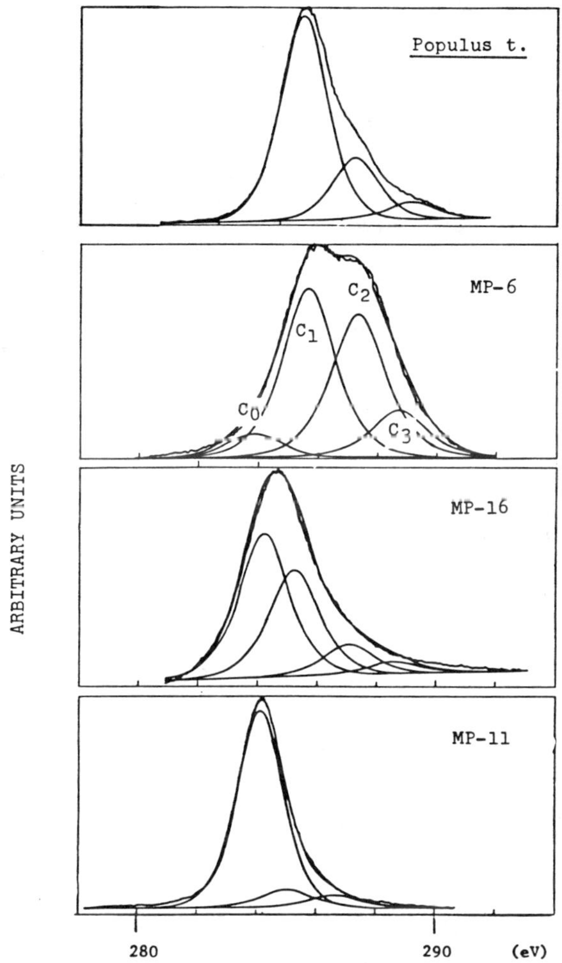

Figure 8. ESCA spectra of C_{1s} peak for Populus tremuloides and for 3 SCE residues.

It is interesting to note that the uncorrected experimental C_{1s} binding energy for dibenz (a,h) anthracene and for a sample of high rank commercial coal is very close to the binding energy for this C_o peak. As polyaromatics are electrical conductors, the charging is expected to be low and E_c close to 0. On this basis, C_o component is assigned to carbon in polyaromatics. This complementes the infrared conclusion about the polyaromatic nature of recondensed material.

The ratio C_{RM}/C_{SR} (where C_{SR} is the carbon content of the whole solid residue as determined by elemental analysis, whereas C_{RM} is the calculated mass of carbon in the recondensed material contained in a given sample) was calculated for 6 samples. Figure 9 shows that this ratio is correlated to the C_o fraction of the C_{1s} peak for ground chars, but also that both values are almost the same for each sample; this represents the case where residues are well ground (black dots on the figure). For two samples, MP-22 and MP-16, the ESCA analysis were ran on partially ground samples. When the grinding is less severe (MP-22-a and MP-16-a), the fibers are only separated and not broken. The C_o area value is the same as for non-ground samples. Otherwise, by increasing the efficiency of the grinding (MP-22-b and MP-16-b), the C_o% value decreases and, ultimately is equal to the bulk % carbon in recondensed material. It can be concluded that in partially converted material, the recondensed material is located at the external surface of the fibers but uniformally distributed within the raw samples.

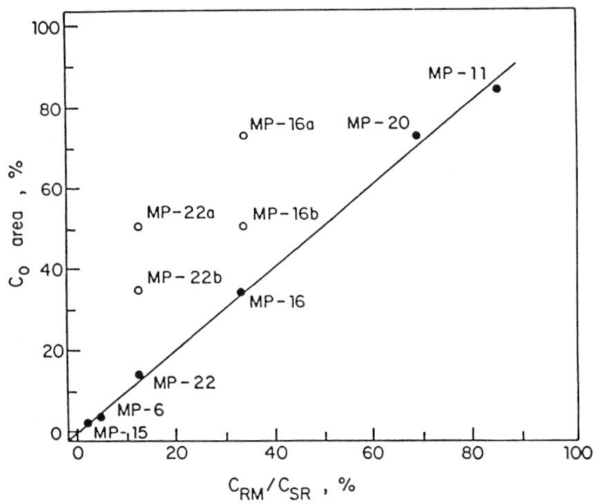

Figure 9. Relation between carbon in recondensed material of some SCE residues and ESCA C_o peak.

Therefore it may be concluded that the ESCA technique provides a simple means for the determination of the extent of recondensation reactions by a mere determination of the proportion of the C_o component in the C_{1s} band of the solid residue.

Acknowledgements

We thank A. Adnot for performing ESCA spectra; P. Chantal (CANMET, Ottawa) for running thermogravimetric analysis; G. Chauvette (DREV,

Valcartier) for having allowed the use of his FTS-40 spectrometer for running some infrared spectra and J. Thibault for calculating the regression equations.

Literature Cited

1. Fifth Canadian Bioenergy R & D Seminar, Ed. Hasnain, S.; Elsevier, New-York, 1984.
2. McDonald, E.; Howard, J.; Fourth Bio-Energy R & D Seminar, Winnipeg, Canada, March 29-31, 1982.
3. Biermann, C.; Schultz, T.P.; McGinnis, G.D. J. Wood Chem. Technol. 1984, 4, 111-128.
4. Nguyen, T.; Zavarin, E.; Barrall II, E.M. J. Macromol. Sci.-Rev. Macromol. Chem a) 1985, C20, 1-65; b) 1985, C21, 1-60.
5. Schultz, T.P.; Templeton, M.C.; McGinnis, G.D. Anal. Chem. 1985, 57, 2867-2869.
6. Chua, M.G.S.; Wayman, M. Can. J. Chem. 1979, 57, 2603-2611.
7. Marchessault, R.H.; Coulombe, S.; Morikawa, H. Can. J. Chem. 1982, 60, 2372-2382.
8. Nassar, M.M.; Mackay, G.D.M. Wood and Fiber Sc. 1984, 16, 441-453.
9. Meier, D.; Larimer, D.R.; Faix, O. Fuel 1986, 65, 916-921.
10. Lephardt, J.O.; Fenner, R.A. Appl. Spectrosc. 1981, 35, 95-101.
11. Haw, J.F.; Schultz, T.P. Holzforshung 1985, 39, 289-296.
12. Dollimore, D.; Hoath, J.M. Thermochimica Acta 1981, 45, 103-113.
13. Shafizadeh, F.; Sekiguchi, Y. Carbon 1983, 21, 511-516.
14. Fuller, M.P.; Griffiths, P.R. Anal. Chem. 1978, 50, 1906-1910.
15. Poirier, M.; Ahmed, A.; Grandmaison, J.L.; Kaliaguine, S. I & EC Research, 1987, 26, 1738-1743.
16. Ahmed, A.; Grandmaison, J.L.; Kaliaguine, S. J. Wood Chem. Technol. 1986, 6, 219-248.
17. Grandmaison, J.L.; Thibault, J.; Kaliaguine, S.; Chantal, P.D. Anal. Chem., 1987, 59, 2153-2157.
18. Ahmed, A.; Adnot, A.; Kaliaguine, S. J. Appl. Polym. Sci., 1987, 34, 359-375.
19. Lignins. Occurence. Formation. Structure and Reactions; Ed. Sarkanen, K.V.; Ludwig, C.H.; John Wiley and Sons; 1971.
20. Grandmaison, J.L.; Kaliaguine, S.; Ahmed, A. in Comptes Rendus de l'atelier de travail sur la liquéfaction de la biomasse, Sherbrooke, Canada, NRCC 23130, Sept. 29-30, 1983.
21. Shafizadeh, F. J. Anal. and Appl. Pyrolysis 1982, 3, 283-305.
22. Nakanishi, K.; Infrared Absorption Spectroscopy-Practical; Holden-Day, Inc.: San Francisco, 1962.
23. Gerson, D.J.; McClelland, J.F.; Veysey, S.; Markuszewski, R. Appl. Spectrosc. 1984, 38, 902-904.
24. Painter, P.C.; Snyder, R.W.; Starsinic, M.; Coleman, M.M.; Kvehn, D.W.; Davis A. Appl. Spectrosc. 1981, 35, 475-485.
25. Gelius, U.; Heden, P.F.; Hedman, J.; Lindberg, B.J.; Manne, R.; Nordberg, R.; Nordling, C.; Siegbahn, K. Physica Scripta, 1970, 2, 70.

RECEIVED March 31, 1988

Chapter 15

Some Aspects of Pyrolysis Oils Characterization by High-Performance Size Exclusion Chromatography

David K. Johnson and Helena L. Chum

Chemical Conversion Research Branch, Solar Energy Research Institute, 1617 Cole Boulevard, Golden, CO 80401

The utilization of biomass pyrolysis oils or isolated fractions of these feedstocks requires a fast overall characterization technique. Gas chromatographic techniques typically analyze only the volatile fraction (5%-50%) of underivatized oils. With proper choice of solvent and detector systems the HPSEC, on polystyrene-divinylbenzene copolymer gels, of the whole oils can provide valuable information on their apparent molecular weight distributions and changes that occur upon aging or chemical fractionation. Several pyrolysis oils have been analyzed as well as fractions isolated by solvent elution chromatography. In order to better understand the observed low-molecular-weight region, a number of model substances of the main classes of compounds found in pyrolysis oils have been investigated. While hydrogen bonding between the phenolic groups and tetrahydrofuran occurs, solute-solute interactions can be kept very small by operating at very low concentrations of solute; solute-gel interactions do occur with polycyclic aromatic compounds. HPSEC provides very good information on the shelf life and reactivity of pyrolysis oils, and can be used to compare oils produced under different process conditions.

Many biomass pyrolysis processes convert 55%-65% of dry biomass to a very inexpensive pyrolysis oil (1-3). Costs of the oils will range from \$0.02-\$0.08/lb of oil, depending on the biomass feedstock cost (\$10-\$40/dry ton biomass). Therefore, these inexpensive oils, rich in phenolic fractions, acids, and furan-derivatives could be feedstocks for further upgrading or could be used because of their reactivity, in applications such as thermosetting resins and other wood-bonding methods. One of the

0097-6156/88/0376-0156\$06.00/0

important considerations for this use or further processing is the stability of the oil. Fast techniques to determine such properties become necessary. We present a method of characterization of pyrolysis oils and chemically isolated fractions using high-performance size exclusion chromatography (4-5), a technique commonly employed in the determination of the molecular weight distribution of polymers. We discuss the potential of the method and its limitations. Classes of model compounds commonly found in these oils have been investigated in the low-molecular-weight range to shed light on interactions between solute and solvent, solute and gel material (polystyrene-divinylbenzene), and solute-solute which can be kept to a minimum by operating at very dilute conditions.

Experimental

High performance size exclusion chromatography was performed on Hewlett-Packard 1084 and 1090 liquid chromatographs using HP1040A diode array and HP-1037A refractive index detectors. Data were stored on a HP 85 microcomputer. The columns (300 x 7 mm) used in this study were purchased from Polymer Laboratories Inc. and were a PL Gel 100 Å (10 μ particles) and a PL Gel 50 Å (5 μ particles). The solvent employed was tetrahydrofuran (Burdick and Jackson, chromatographic grade) used as received.

Details on the preparation of pyrolysis oils at SERI in the entrained-flow, fast ablative pyrolysis reactor can be found in a report by Diebold and Scahill (2). The oils in Figure 1 were obtained from two runs, #40 (c and d) and #41 (a and b), the oils being collected from a packed scrubber (a and c) and a cyclone scrubber (b and d). The oil obtained from the packed scrubber in run 41 was subjected to sequential elution by solvents chromatography (SESC) according to the method of Davis et al (7). The HPSEC of fractions 1 through 6 appear in Figure 3. Fraction 1 was eluted with 15% toluene in hexane (yield 0.4%), Fraction 2 was eluted with chloroform (yield 1.5%), Fraction 3 was eluted with 7.5% ether in chloroform (yield 7.5%), Fraction 4 was eluted with 5% ethanol in ether (yield 19.5%), Fraction 5 was eluted with methanol (yield 38.1%), and Fraction 6 was eluted with 4% ethanol in THF (yield 3.1%). The oils displayed in Figures 4 and 5 were produced from run 66.

The details of the methods of preparation of the oils shown in Figure 2 can be obtained by writing to their suppliers. Briefly, Oil 1 was supplied by D. S. Scott of the University of Waterloo and was produced by flash pyrolysis of a poplar-aspen hybrid. Oils 2, 3 and 6 were supplied by C. Roy of the University of Sherbrooke. Oil 2 was produced by vacuum pyrolysis of Avicel at 360° C, Oil 3 by vacuum pyrolysis of an aspen-poplar hybrid at 534° C and 2.2 torr, and Oil 4 by vacuum pyrolysis of aspen at 315° C and 0.7 torr. Oil 4 was supplied by J. Howard, of B.C. Research, Vancouver and was produced by extraction of aspen using supercritical acetone. Oil 5 was supplied by S. Kaliaguine of the University of Laval and was produced by extraction of aspen with supercritical methanol at 350° C and 1500 psi.

The aromatic hydrocarbons, acids and naphthalenes used for Figure 6 were purchased from the Aldrich Chemical Co. and were toluene (1), propyl benzene (2), s-butyl benzene (3), naphthalene (4), 1,4-dimethyl naphthalene (5), 1-phenyl naphthalene (6), vanillic acid (7), and 4-hydroxy-3-methoxy cinnamic acid (8).

The lignin model compounds used for Figure 7 were prepared by J. A. Hyatt (6) and were a 5, 5'-biphenyl tetramer hexaol, $C_{42}H_{54}O_{14}$ (A), a β-O-4 tetramer heptaol, $C_{41}H_{52}O_{15}$ (b), and a β-O-4 dimertriol, $C_{17}H_{20}O_6$ (C). The phenols used for Figure 7 were purchased from the Aldrich Chemical Co. and were phenol (1), p-cresol (2), 2-propyl phenol (3), guaiacol (4), syringyl alcohol (5) and acetovanillone (6).

Results and Discussion

Comparison of Pyrolysis Oils Obtained from Various Sources. The HPSEC of four wood pyrolysis oils obtained from the entrained flow, fast ablative pyrolysis reactor at SERI are shown in Figure 1. The oils were obtained from two separate runs and collected from two different scrubbers. The apparent molecular weight distributions of the four oils are very similar, indicating little selectivity on the basis of molecular weight distribution. Figure 2, however, shows the HPSE chromatograms of a number of other pyrolysis oils obtained under a variety of conditions from many different sources. Clearly, some of the oils contain components of high apparent molecular weight even to the extent that some are excluded from the pores of the column polymer, indicated by the peaks at about 4.5 minutes in the chromatograms. The oils also have varying amounts of more sharply resolved components at lower apparent molecular weight. Thus, HPSEC may be used to characterize pyrolysis oils obtained from different sources, and comparisons may be drawn regarding their relative apparent molecular weight distributions as long as the analyses were carried out under the same chromatographic conditions.

The wood oil obtained from the packed scrubber in Run 41 at SERI was also subjected to fractionation by sequential elution by solvents chromatography (SESC) according to the method of Davis et. al. (7). The fractions obtained were analyzed by HPSEC and the chromatograms are shown in Figure 3. The HPSEC shows a general trend to higher apparent molecular weight as the polarity of the eluting solvent was increased up to methanol. A number of the fractions appear to contain relatively large amounts of distinct components (the sharp peaks) of lower apparent molecular weight. The sixth fraction was produced by going back to a less polar solvent. A seventh fraction was produced using a more polar eluant of 10% acetic acid in methanol which could not be analyzed by HPSEC because it was insoluble in tetrahydrofuran. About three-quarters of the oil was found in Fractions 3, 4, and 5, the last being the major fraction. If the chromatograms in Figure 3 were combined taking into account the yields of the various fractions then, as expected, a close comparison could be made with the chromatogram in Figure 1 of the unfractionated oil.

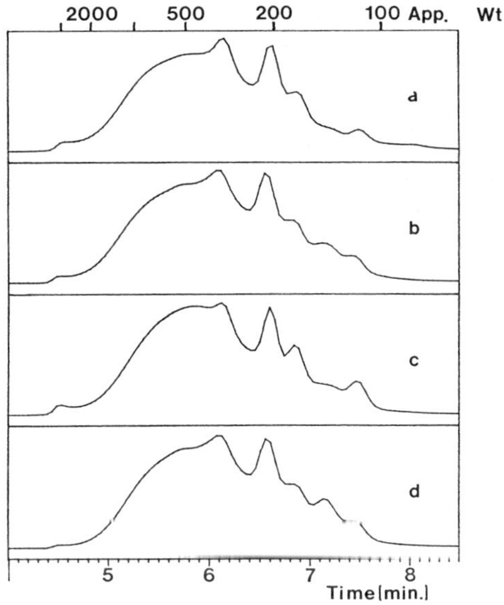

Figure 1. HPSEC of wood pyrolysis oils from the SERI entrained-flow, fast ablative pyrolysis reactor. Analysis on 100 Å, 10 μ GPC column using THF at 1 ml/min^{-1} with detection at 330 nm (bandwidth 140 nm), see Experimental for a description of the oils.

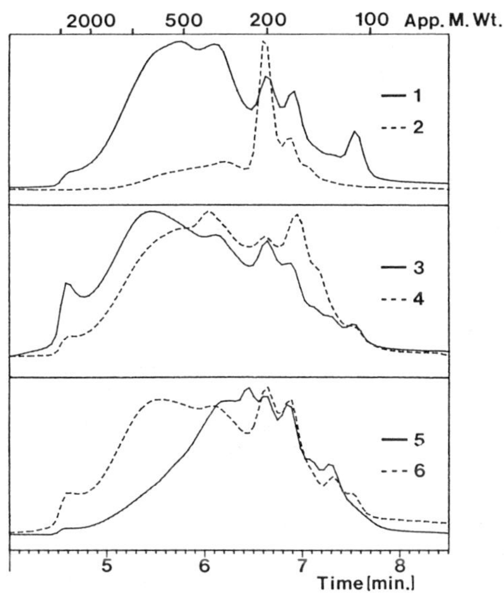

Figure 2. HPSEC of wood pyrolysis oils from a variety of sources. Analysis conditions as per Figure 1, see Experimental for a description of the oils.

Doubts have been expressed that these pyrolysis oils could have molecular weights as high as indicated by these chromatograms as they are obtained by condensation of the primary vapors from pyrolysis. Analysis by techniques requiring revaporization of the oils consistently do not detect high-molecular-weight components, possibly because they are difficult to vaporize and also because they may be thermally degraded to either higher or lower molecular weight components (8) or both. It has been suggested that the high apparent molecular weights observed by HPSEC are the result of solute-solute or solute-solvent associations producing high-molecular-weight complexes. To verify the results obtained by HPSEC, the three major fractions and the original unfractionated oil were subjected to proton NMR analysis. The spectra of Fraction 5 and the original oil contain broad peaks characteristic of irregular polymers such as lignin, while the spectrum of Fraction 3 contains sharp peaks indicative of a mixture of simpler, low-molecular-weight compounds; Fraction 4 is intermediate between 3 and 5. Thus, the HPSEC and proton NMR spectra appear to be in general agreement in that this pyrolysis oil contains mixtures of possibly higher molecular weight polymeric components and simpler low-molecular-weight compounds.

Many of the chromatograms shown here are of samples whose history of handling and age are not known in detail. It has been suggested that because of the very reactive and acidic nature of these oils that the high molecular weights observed are produced as the oils get older and are exposed to ambient conditions. Consequently, a study has been started to examine the effects of aging and the conditions under which pyrolysis oils are stored. A pyrolysis oil was produced in the SERI entrained-flow, fast ablative pyrolysis reactor. In this reactor, the primary vapors are scrubbed out with water such that about 90% are dissolved out. One sample of this aqueous solution of pyrolysis oil was stored at 4°C and the other was analyzed by HPSEC. The sample to be analyzed was made up by dissolving a small amount of the aqueous solution in tetrahydrofuran. Storage was under ambient conditions. The THF solution was analyzed several times over the period of a week to look for changes in its HPSE chromatogram as shown in Figure 4. At the end of this period, the sample kept at 4°C was also analyzed to determine the effect of aging on the oil. Actually, a physical change took place in the aqueous sample stored at 4°C in that a small amount of tar separated out on the bottom of the vial. Consequently, two samples were made up in THF from the cooled sample, one from the aqueous part and one from the tar. Figure 5 compares the HPSEC of the sample kept at ambient conditions to those of the cooled samples. The HPSEC of the tar sample shows it consists of relatively much larger amounts of material higher in apparent molecular weight. The aqueous fraction of the cooled sample appears very similar to the sample stored at 25°C, although the latter does appear to contain a slightly larger relative amount of apparently higher molecular weight material. The degree to which storage at lower temperature has prevented any increase in molecular weight of the pyrolysis oil with time is difficult to ascertain because of the

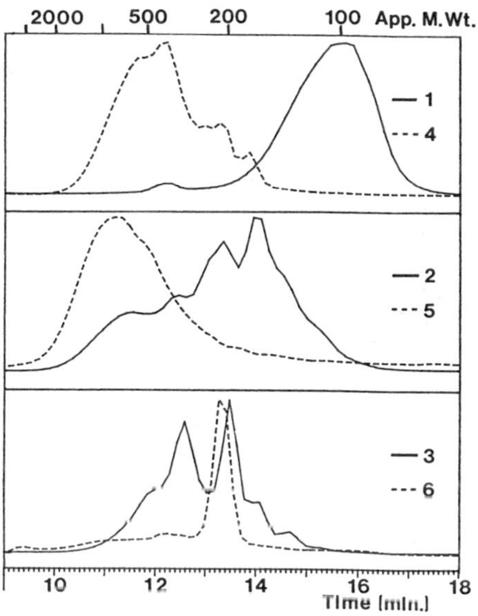

Figure 3. HPSEC of SESC fractions from a SERI pyrolysis oil. Analysis on a 100Å 10 μ GPC column using THF at 0.5 ml/min^{-1} with detection at 330 nm (bandwidth 140nm), see Experimental for a description of the fractions.

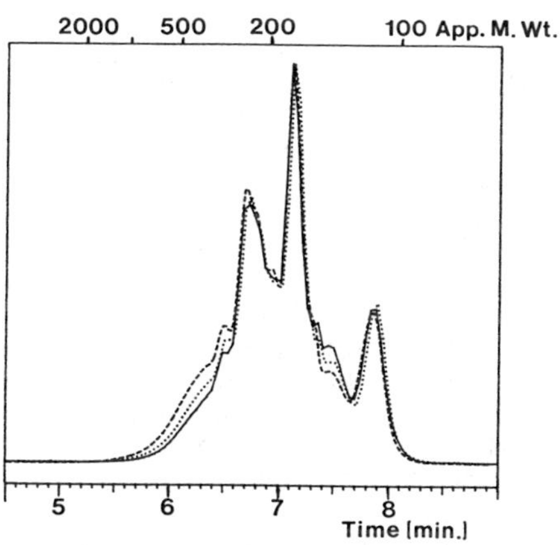

Figure 4. HPSEC of a SERI pyrolysis oil kept under ambient conditions, after 5 hours (———), 3 days (.....), and 7 days (-----). Analysis on a 50 Å 5μ GPC column using THF at 1.0 ml/min^{-1} with detection at 270 nm (bandwidth 10 nm).

fractionation of the refrigerated sample. The sample kept at ambient conditions did not have the opportunity to fractionate because of the solvent it was dissolved in.

The HPSEC of the unrefrigerated sample (Figure 4) does indicate that the pyrolysis oil "aged" over the period of a week with increasing amounts of apparently higher molecular weight components being produced with time. Most of the samples obtained from outside of SERI are much older than one week. Pyrolysis oils are generally very reactive so that unless they are effectively stabilized in some way, increasing molecular weight should be expected as they get older.

What are the Limitations of HPSEC as a Technique When Applied to Pyrolysis Oil Characterization. One of the major advantages of HPSEC as a technique is that with the proper choice of solvent to dissolve the sample, the whole of the sample may be analyzed under very mild conditions. Because HPSEC is an isocratic technique, differential refractometers may be used as detectors so that, again, all of the sample may be detected. This is not a great concern when applied to pyrolysis oils, as they tend to absorb quite strongly in the ultraviolet. With a modern UV-visible diode array detector, a number of wavelengths can be monitored to ensure all the components of the oil are monitored. However, the eluting solvent must be chosen such that all the sample is dissolved, and as pyrolysis oils are fairly polar and often contain water, the solvent will also need to be fairly polar. The combination of polar solutes and polar solvents means that solute-solvent interactions through hydrogen bonding must be a concern. Tetrahydrofuran, probably the most popular solvent for HPSEC, can form hydrogen bonds with certain species such as phenols producing a complex molecule exhibiting greater molecular size and lower retention volume than would be expected (9). When nonpolar solvents are used such as toluene or chloroform, the molecular size should be relatively unaffected, but oil solubility then becomes very limited. The use of solvents of greater solvating power, such as dimethyl formamide, also generates problems (10) due to solute-solute association, interaction between polystyrene standards and the column gel and column gel-solvent interactions.

The other major limitation of HPSEC as a technique comes from the desire to correlate solute elution time with molecular weight. As stated in its name, this is a method of separation based on molecular size. HPSEC columns contain a polymer gel of polystyrene-divinylbenzene produced with a controlled pore size distribution. Solutes of different size are separated by the different degrees of their penetration into the pores of the gel. The parameter that can be obtained from HPSEC is effective molecular length; e.g., material excluded from a column containing gel with 100 Å pores should have an effective molecular length of 100 Å or greater. To correlate retention times to molecular weight, it is necessary to use calibration standards similar in structure to the solute whose molecular weight is being determined. The most common calibration materials used are polystyrenes of low polydispersity. Others used include straight chain alkanes, polyethylene glycols, and the related materials

IGEPALS™ that are 4-nonylphenyl terminated. If a column were calibrated with straight chain alkanes, it is unlikely to be much good for obtaining molecular weights of aromatic solutes, as a benzene ring is only about as long as propane, and anthracene is only about as long as hexane. When dealing with much larger molecules, it is difficult to estimate what their size might be in three dimensions in solution. Although pyrolysis oils have a high level of aromatic components, especially phenolics, they are a very complex mixture of components, and so it is unlikely that any one set of calibration standards would do a very good job. Despite these limitations, HPSEC can give an idea of the molecular weight distribution of an oil and certainly can be used in comparing oils. Establishing molecular weights for low-molecular-weight components is probably the most difficult task. Figures 6 and 7 compare the actual molecular weights of a variety of different types of compounds with their apparent molecular weights calculated from their retention times on a 50 Å 5 μ HPSEC column calibrated with polystyrenes and IGEPALS. If the calibration was good for all compounds, then they should all fall on the straight lines. The aromatic hydrocarbons follow the calibration, but the aromatic acids and naphthalenes deviate greatly and in opposite directions. The aromatic acids contain both carboxylic and phenolic groups and so probably have higher apparent molecular weights than their actual molecular weights because of hydrogen bonding with the solvent tetrahydrofuran. The naphthalenes have lower apparent molecular weights than actual not only because their condensed structure makes them relatively small for their molecular weight, but also because of interactions between these solutes and the column gel. Philip and Anthony (9) observed retention volumes that were longer than expected for anthracene, benzopyrene, and coronene, considering their molecular size. They attributed this behavior to interaction of these highly aromatic solutes with the phenyl groups of the polymer chains of the gels.

The phenols and lignin model compounds follow the calibration quite closely, tending to show slightly higher apparent molecular weights than they actually have, probably because of association with the solvent. This is encouraging for the HPSEC of pyrolysis oils as these types of compounds are more likely to be present. Heavily cracked oils, however, can be rich in polynuclear aromatics.

Solute-solute association has not been observed for any of these molecules or for others when using tetrahydrofuran as solvent. Retention time changes of less than 0.01 minutes were observed in changing sample concentrations in the mg/mL range (~4 mg/mL) to the μg/mL range (~3 μg/mL) when injecting 5 μl of these solutions. HPSEC of pyrolysis oil samples made up in this concentration range should also be free of solute-solute association which would artifically increase the apparent molecular weight of the oils.

Conclusions

HPSEC has been shown to be a useful method of characterizing pyrolysis oils because it examines the whole of the oil. When

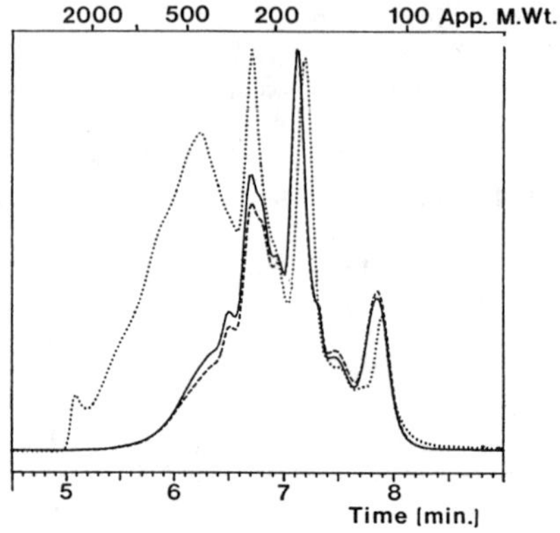

Figure 5. HPSEC of a SERI pyrolysis oil after 7 days kept under ambient conditions (———), tar fraction of refrigerated sample (.....) and aqueous fraction of refrigerated sample (-----). Analysis conditions as per Figure 4.

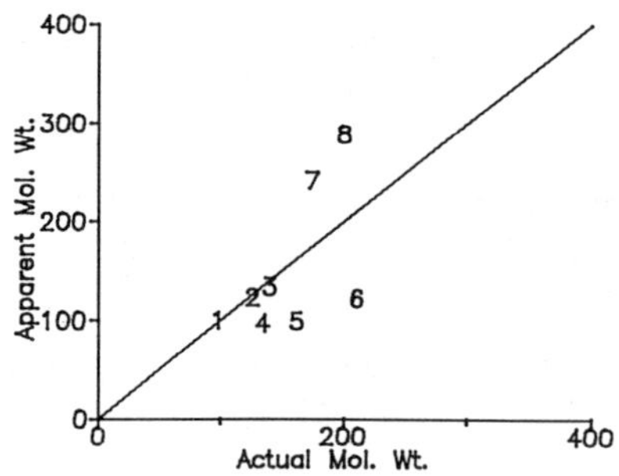

Figure 6. Applicability of calibration using polystyrenes and IGEPALS™ for aromatic hydrocarbons, acids, and naphthalenes, see Experimental for a description of the compounds.

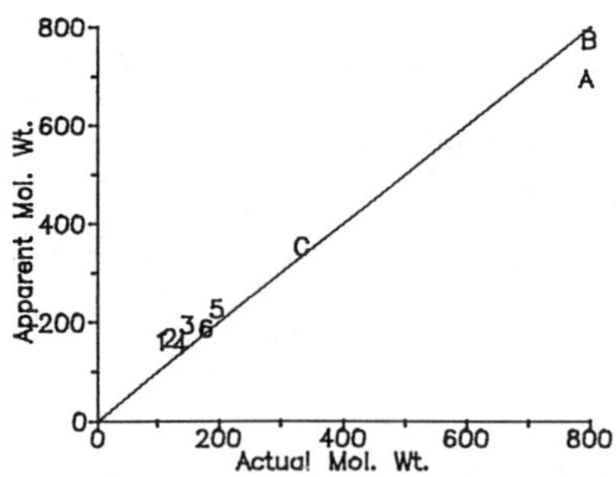

Figure 7. Applicability of calibration using polystyrenes and IGEPALS™ for phenols and lignin model compounds, see Experimental for a description of the compounds.

using polystyrene-divinyl benzene polymer gel columns, tetrahydrofuran as solvent and polystyrenes and IGEPALS™ as calibration standards a good indication of molecular weight distribution can be obtained for oils from a variety of sources. The high apparent molecular weights observed appear to be real, and some corroboration is seen in proton NMR spectra. Although some solute-solvent association can be expected, use of phenolic model compounds has shown that HPSEC can give a good indication of molecular weight. However, if the oils contained large amounts of either much more polar compounds or condensed aromatic compounds, then interpretation of HPSEC on the basis of molecular weight would be much more difficult. Pyrolysis oils are reactive materials and an awareness of the length of time and conditions under which they are kept must be maintained and is important for further processing.

Acknowledgments

This work was supported by the Office of Industrial Programs of the U.S. Department of Energy (FTP 587). Thanks are due to Mr. A. Schroeder, Program Manager, and to all producers of pyrolysis oils: J. Diebold, T. Reed, J. Scahill, C. Roy, S. Kaliaguine, J. Howard, and to R. Evans and T. Milne for profitable discussions. Special thanks to J. A. Hyatt for supplying three of the lignin model compounds employed.

Literature Cited

1. Diebold, J. P., Ed. Proc. Specialists Workshop on Fast Pyrolysis of Biomass, Copper Mountain, SERI/CP-622-1096, Solar Energy Research Institute: Golden, Colorado, October 1980.
2. Diebold, J. P.; Scahill, J. W. Entrained-Flow Fast Ablative Pyrolysis of Biomass, SERI/PR-234-2665, Solar Energy Research Institute: Golden, Colorado, 1985.
3. Overend, R. P.; Milne, T. A.; Mudge, L. K., Eds. Fundamentals of Thermochemical Biomass Conversion; Elsevier: New York, 1985.
4. Provder, T., Ed. Size Exclusion Chromatography: Methodology and Characterization of Polymers and Related Materials; ACS Symposium Series No. 245; American Chemical Society: Washington, D.C., 1984.
5. Yau, W. W.; Kirland, J. J.; Bly, D. D. Modern Size Exclusion Chromatography: Practice of Gel Permeation and Gel Filtration Chromatograhy; Wiley: New York, 1979.
6. Hyatt, J. A. Holzforschung 1987, 41, 363-70.
7. Davis, H. G.; Eames, M. A.; Figueroa, C.; Gansley, K. K.; Schaleger, L. L.; Watt, D. W. In Fundamentals of Thermochemical Biomass Conversion; Overend, R. P.; Milne, T. A.; Mudge, L. K., Eds.; Elsevier: London, 1985, pp. 1027-1038.
8. Evans, R. J.; Milne, T. A. Energy and Fuels, 1987, 1, 123-137.

9. Philip, C. V.; Anthony, R. G. In Size Exclusion Chromatography; Provder T., Ed.; ACS Symposium Series No. 245; American Chemical Society: Washington, DC, 1984, 257-270.
10. Chum, H. L.; Johnson, D. K.; Tucker, M. P.; Himmel, M. E. Holzforchung 1987, 41, 97-107.

RECEIVED March 31, 1988

Chapter 16

Composition of Oils Obtained by Fast Pyrolysis of Different Woods

J. Piskorz, D. S. Scott, and D. Radlein

Department of Chemical Engineering, University of Waterloo, Waterloo, Ontario N2L 3G1, Canada

Liquids obtained by fast pyrolysis of four different woods were analysed. On addition of excess water they separated into water-soluble and water-insoluble fractions. The former which is principally of carbohydrate origin was shown by HPLC analysis to consist of sugars, anhydrosugars and low molecular weight carbonyl compounds. The latter was shown by ^{13}C NMR to be a "pyrolytic lignin". In this way 81% to 92% of the organic content of the liquids has been characterised.

During the last six years a fluidized bed fast pyrolysis process for biomass has been developed at the University of Waterloo (The Waterloo Fast Pyrolysis Process). This process gives yields of up to 70% of organic liquids from hardwoods or softwoods, which are the highest yet reported for a non-catalytic pyrolytic conversion process. A fluidized sand bed is used as a reactor and optimum liquid yields are normally obtained in the range of 450° to 550°C at about 0.5 seconds gas residence time with particles of about 1.5 mm diameter or smaller. Two units are in use, one with a throughput of 20 to 100 gms/hr, and another with a throughput of 1 to 4 kg/hr. Details of the process have been published by the authors in several publications [1], [2], [3].

Several studies have been published describing results from the flash pyrolysis of biomass. Most of these studies were carried out at higher temperatures and were intended to promote biomass gas production. However, the work of Roy and Chornet [4] reported high liquid yields from biomass pyrolysis under vacuum conditions. More recently, Roy et al. [5] have described a vacuum pyrolysis system for the production of liquids from biomass, based on a multiple hearth type of reactor. Knight et al. [6] have developed an upward flow entrained pyrolyzer for the production of liquids from the thermal pyrolysis of biomass.

0097–6156/88/0376–0167$06.00/0

However, few details of the chemical nature of the pyrolytic oils produced from wood or other biomass have been reported. Some recent studies of the composition of pyrolysis oils obtained from poplar wood were carried out by workers of the Pacific Northwest Laboratories of Battelle Institute [7] and the Universite de Sherbrooke [8]. Methods of quantitative determination of functional groups in the pyrolytic oils from wood were tested in our laboratory by Nicolaides [9]. However, more detailed characterisation exists in the literature for the products of the thermal degradation of cellulose [10,11,12,13,14].

Results

In previously reported tests (3), yields have been classified as gases, organic liquids, char and product water. Mass balances were generally close to 95% or better. Elemental analyses were also reported for many runs, but identification of individual compounds was done only for non-condensable gases and for some volatile organics such as methanol, acetaldehyde, furan, etc. However, more detailed analyses of liquid products have recently been carried out, and some of these preliminary results which are of particular significance are reported here.

All experimental results given were obtained at conditions of close to optimal feed rate, particle size and residence time for maximum liquid yield at the stated temperatures, as determined by over 200 bench scale runs and 90 pilot scale runs. These optimal conditions are those shown in Table I, that is, 480° to 520°C, 0.5 to 0.7 seconds apparent residence time and a maximum particle size of 1 mm.

Table I shows the experimental yields of products from selected runs for four different woods whose properties are given in Table II.

High organic liquid yields are characteristic of all four materials when undergoing fast pyrolysis in our process. The total liquid yield, including water of reaction, varies from 70% to 80% of the dry biomass fed, all of which can be directly used as a substitute fuel oil if desired. The liquids are single-phase, homogeneous fluids, which pour readily, and which contain from 15% to 25% water depending on the feed material and its moisture content.

These liquids are quite stable at room temperature. The water content, in a period of twelve months, was found to increase slightly presumably due to the slow processes of condensation-polymerization going on even at room temperature. At higher temperatures, 120°C and above, the oils become increasingly unstable and decompose with evolution of gas, and finally charification of a polymeric residue. All the pyrolysis vapors produce an oil mist following rapid quenching, but their polar character allows the easy utilisation of electrostatic precipitation in the recovery system of a pyrolysis process.

Preliminary small scale combustion tests carried out with the pyrolysis oil as produced (20% water content) showed that it burns readily in a furnace with a conventional pressure atomizing burner, providing the combustion box is preheated. If an air atomizing nozzle is used with a pilot flame, no pre-heating of the combustion chamber is necessary. Larger scale tests for extended periods have not yet been done, but preliminary work shows that the pyrolysis oil has potential as a substitute fuel oil.

Table I

Pyrolysis Yields from Different Woods

	Brockville Poplar			White Spruce			Red Maple			IEA Poplar		
Run #	51	58	50	42	43	45	14	63	81	27	59	A-2*
Temperature, °C	497	504	555	485	500	520	532	508	515	500	504	497
Moisture content, wt%	+	5.2	+	+	7.0	+	3.8	5.9	9.5	6.2	4.6	3.3
Particle Top Size, μm	+	1000	+	+	1000	+	590	590	1000	590	1000	590
Apparent Residence Time, s	0.64	0.47	0.57	0.70	0.65	0.62	0.69	0.47	0.44	0.55	0.48	0.46
Feed Rate, kg/hr	2.24	2.10	4.12	2.07	1.91	1.58	2.16	1.98	1.32	2.24	1.85	0.05
Yields, wt% of m.f. wood												
Organic liquid	62.9	62.9	59.0	63.1	66.5	66.1	67.3	67.9	65.0	65.8	66.2	65.7
Water	10.3	9.7	10.2	10.7	11.6	11.1	7.4	9.8	10.0	9.3	10.7	12.2
Char	14.4	16.5	10.6	16.3	12.2	12.3	9.0	13.7	12.1	12.1	11.8	7.7
Gas:												
H_2	0.02	0.02	0.07	0.04	0.02	0.01	0.04	0.01	0.02	0.02	0.01	--
CO	4.95	4.71	8.40	4.16	3.82	4.01	6.96	4.12	4.83	5.32	4.44	5.34
CO_2	6.14	5.89	7.06	3.38	3.37	2.69	4.02	4.89	5.36	6.30	5.75	4.78
CH_4	0.45	0.44	0.97	0.34	0.38	0.43	0.75	0.36	0.57	0.48	0.37	0.41
C_2H_4	0.22	0.19	0.40	0.16	0.17	0.16	0.24	0.16	0.21	0.20	0.13	0.19
C_2H_6	0.06	0.05	0.11	0.02	0.03	0.05		0.04	0.07	0.04	0.05	--
C_3	0.06	0.07	0.10	0.03	0.04	0.06	0.04	0.07	0.10	0.09	0.08	0.09
C_4	0.13	0.16	0.28					0.14	0.41		0.19	3.10
Total Gas	12.0	11.5	17.4	8.1	7.8	7.4	12.1	9.8	11.6	12.4	11.0	10.8
Overall recovery wt%, m.f.	99.7	100.5	97.2	97.8	97.7	96.7	95.7	101.2	98.7	99.7	99.8	96.4

* Bench Scale Unit

Table II

Properties of Feed Materials

	Whole Tree Poplar	White Spruce	Red Maple	IEA Poplar Wood
Source	Brockville Plantation, Whole Tree Except Leaves and Roots Ontario MNR Clone C-147	Eastern Canada Clean Wood Sawmill Sawdust and Mill Scrap		Clean Wood only Ontario MNR Clone D-38
Moisture, wt%	5.2	7.0	3.8 - 25.2	4.6
Ash, % mf	1.19	0.50	0.40	0.46
Elemental Analysis, %				
C	49.06	49.63	48.5	49.45
H	6.23	6.36	6.1	6.05
O	43.6	43.1	--	44.4
N	1.08	0.2	--	0.07

Some properties of "wet" liquids as produced are given in Table III. The elemental analyses of the pyrolytic liquids as given in Table III are very similar to those of the starting materials - wood. One could probably fairly accurately describe these liquids, therefore, by the term "liquid wood". Two major characteristics of these liquids are high oxygen content and high density (much higher than wood). Another specific property is a limited water solubility. In the case of pyrolytic sirups produced by the Waterloo process, water is dissolved in the organic phase. The addition of more water to the level of about 60% by weight causes a phase separation and this behaviour has been utilized in our work for analytical purposes, in that both fractions were analyzed separately after dilution of the original oil product.

The details of the HPLC technique developed for analysis of the water soluble fraction are given below:

Column	:	Aminex HPX-87H, high performance cation exchange resin in hydrogen form 300 x 7.8 mm from Biorad
Eluant	:	H_3PO_4, 0.007 N
Flow Rate	:	0.80 ml/min, isocatic
Temperature	:	65°C
Detector	:	Waters R 401 Differential Refractometer
Internal Standard	:	n-propanol

The quantitative data obtained by the HPLC technique are presented in Table IV. A typical HPLC-chromatogram is shown in Figure 1.

To obtain relative response factors and retention times, the pure compounds were fed and then eluted, although some of them, such as cellobiosan and 1,6 anhydro-β-D-glucofuranose, had to be synthesized in-house [13]. Confirmation of compound identification was obtained by GC-MS [Hewlett-Packard 5970 Mass Selective Detector coupled to 5890A Gas Chromatograph]. For GC-MS analysis, sugars and anhydrosugars were first trimethylsilylated to the corresponding ethers. Small amounts of simple phenols and of furanoid compounds were also detected by GC-MS in the water soluble fraction. These components were not quantified by HPLC.

The yields of the water insoluble fraction are given also in Table IV. This fraction separated as a dark brown viscous liquid which solidified during drying into a hard, black, easily powdered material.

The carbon-13 NMR spectrum of this pyrolytic product is shown in Figure 2 together with similar spectra published by Marchessault et al. [15] for milled wood lignin (MW) and steam exploded lignin (EXW). The similarity of the spectra of the steam exploded lignin and our pyrolytic lignin is quite striking. It appears that the oils produced by the Waterloo Fast Pyrolysis process contain a significant fraction which is apparently derived from the natural wood lignin. This "pyrolytic lignin" represents nearly 80% of the original content of wood lignin. Evidence for this conclusion was first reported from the work of H. Menard [16] using thermogravimetric and infrared analysis.

The Waterloo NMR spectrum was recorded using a 9% solution in DMSO-d_6 at 62.9 MHz and 50°C with broad-band proton decoupling. Delay time between pulses was 10 seconds.

Table III

Properties of Pyrolitic Liquids

	Brockville Poplar			White Spruce			Red Maple	IEA Poplar
Run #	51	58	50	42	43	45	63	59
Yields, wt% of wood as fed	74.1	73.3	70.5	75.2	79.2	78.5	77.9	77.6
Water content, wt%	19.8	18.7	21.7	21.9	22.4	21.8	18.0	18.6
pH	2.6	2.4	2.8	2.1	2.1	2.3	2.4	2.4
Density, g/cc	1.19	1.20	1.18	1.22	1.22	1.22	1.19	1.23
Elemental analysis, wt%, m.f.								
Carbon	54.1	54.7	55.6	53.5	54.0	56.6	54.7	53.6
Hydrogen	7.1	6.9	6.9	6.6	6.8	6.9	6.4	7.0

Table IV

Analysis of Liquid Products

	Brockville Poplar	White Spruce	Red Maple	IEA Poplar
Run #	58	43	63	A-2
Temperature	504	500	508	504
Yields, wt % of feed, m.f.				
Organic Liquid	62.9	66.5	67.9	62.6
1. Oligosaccharides	-	-	-	0.70
2. Cellobiosan	1.11	2.49	1.62	1.30
3. Glucose	0.55	0.99	0.64	0.41
4. Fructose	1.34	2.27	1.51	1.32
5. Glyoxal	1.42	2.47	1.75	2.18
6. Methylglyoxal	2.52	3.96	2.84	0.65
7. Levoglucosan				3.04
8. 1,6 anhydroglucofuranose	-	-	-	2.43
9. Hydroxyacetaldehyde	6.47	7.67	7.55	10.03
10. Formic Acid	5.40	7.15	6.35	3.09
11. Formaldehyde	-	-	-	1.16
12. Acetic Acid	6.30	3.86	5.81	5.43
13. Ethylene Glycol	0.87	0.89	0.63	1.05
14. Acetol	1.70	1.24	1.15	1.40
15. Acetaldehyde	-	-	-	0.02
Water-Solubles - Total Above	27.7	33.0	29.9	34.2
Pyrolytic Lignin	24.8	20.6	20.9	16.2
Amount not accounted for (losses, water soluble phenols, furans, etc.)	10.5	12.9	17.1	11.91
by G.C.				
Methanol	0.63	1.11	0.77	0.12
Furfural	0.46	0.30	-	-
Methylfurfural	0.18	0.05	0.42	-

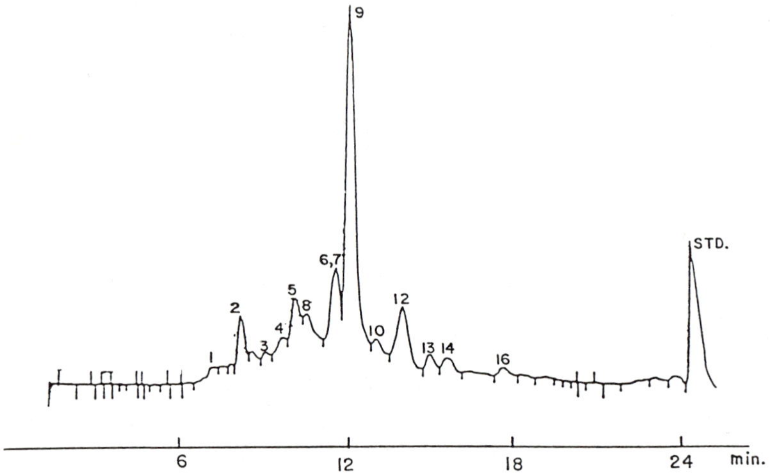

Figure 1. Typical HPLC chromatogram for water soluble fractions of pyrolytic oil from poplar wood. Peak numbers correspond to those in Table IV.

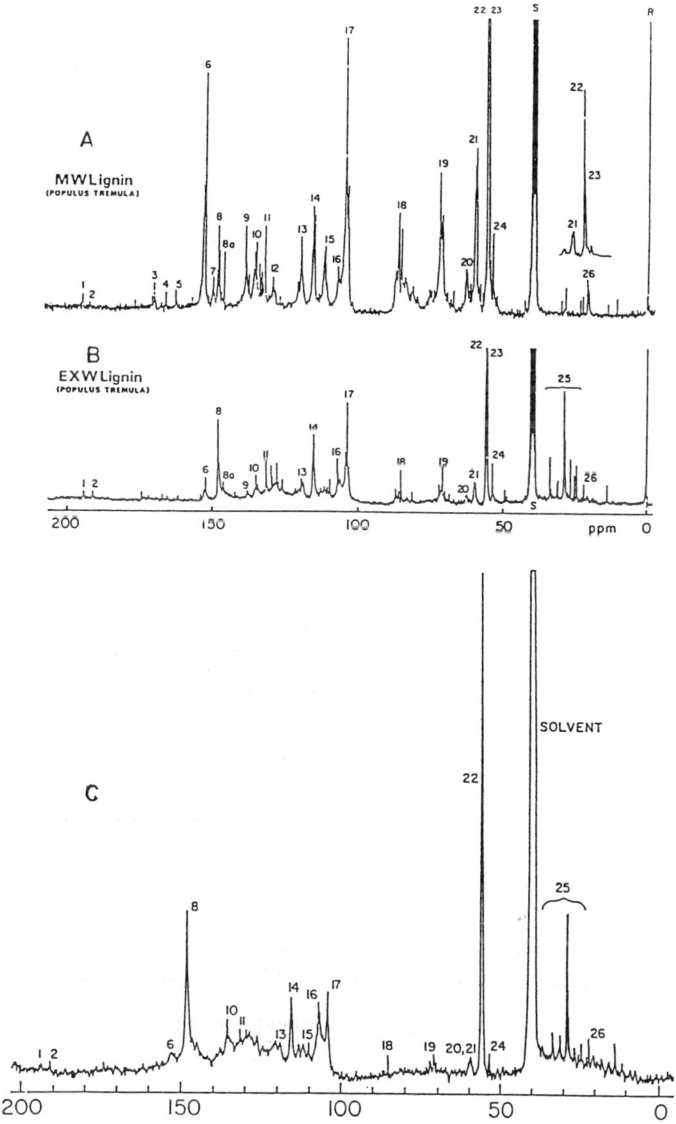

Figure 2. ^{13}C NMR spectra of A) milled wood lignin; B) steam exploded lignin (from Marchessault et al [15]); C) Waterloo pyrolytic lignin. All from poplar wood.

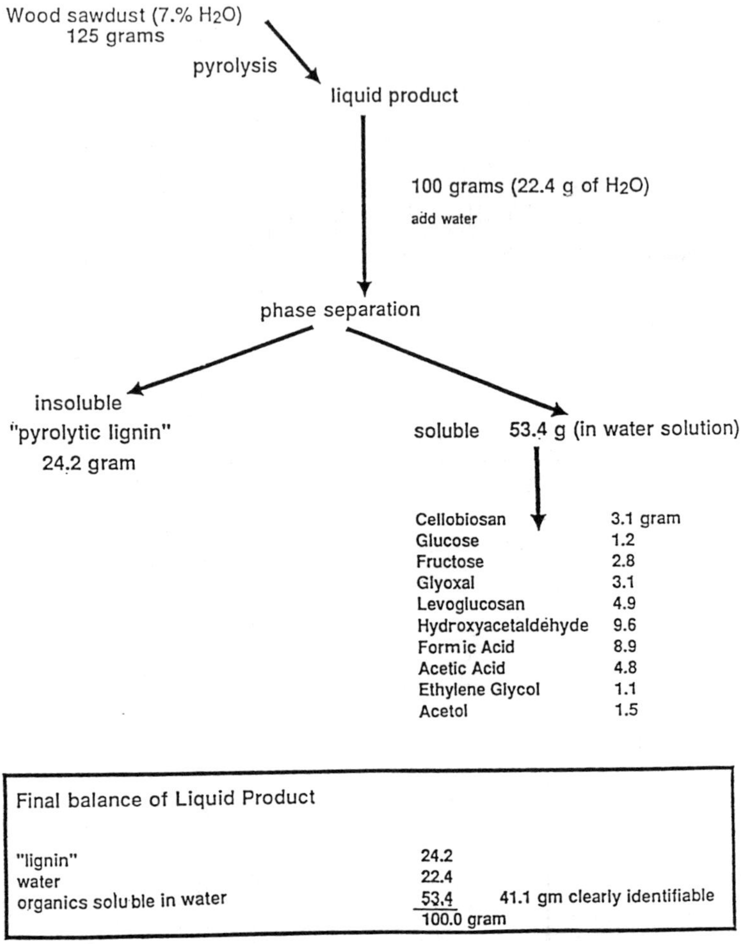

Figure 3. Material balance for separation of liquid pyrolysis product from eastern spruce wood.

Conclusions

Four different woods under conditions of fast pyrolysis yielded very similar liquid products. Analyses show that this product is a complex mixture of chemicals.

The pyrolytic "wet" tar-sirup can be readily separated into two principal fractions by water extraction. The water-insoluble fraction is derived from lignin while the water-soluble fraction is carbohydrate in origin. Analytical results indicate large amounts of low molecular weight (~100) lactones and aldehydes, and a significant fraction of these are multifunctional in nature. Four major classes of chemicals can be differentiated,

1. sugars and anhydrosugars
2. carbonyl and hydroxycarbonyl compounds
3. acids - formic, acetic
4. "pyrolytic" lignin.

Results in Table IV show that 81% to 92% of the content of the pyrolysis oils produced in this work from wood has been quantitatively identified. A schematic of the material balance for a typical separation is shown in Figure 3 to illustrate the yields obtained for the various fractions.

Detailed analysis of pyrolytic oils is needed in order to allow possible mechanistic or kinetic models to be formulated which can explain the various observed aspects of fast biomass thermal degradation. A knowledge of chemical composition of these oils may also assist in the eventual future utilization, upgrading or separation of these compounds as higher value products.

Acknowledgments

The financial supportfor this work of the National Research Council of Canada and of the Natural Sciences and Engineering Research Council of Canada are gratefully acknowledged by the authors. The assistance of P. Majerski and A. Grinshpun with experimental measurements is also acknowledged with pleasure.

Literature Cited

1. Scott, D. S.; Piskorz, J. Can. J. Chem. Eng., 1982, 60, 666.
2. Scott, D. S.; Piskorz, J. Can. J. Chem. Eng., 1984, 62, 404.
3. Scott, D. S.; Piskorz, J.; Radlein, D. Ind. Eng. Chem. Process Res. Devel., 1985, 24, 581.
4. Roy, C.; Chornet, E. Proc. 2nd World Congress Chem. Eng., Montreal, 1985, Vol. II, p 315.
5. Roy, C.; Lalancette, A.; DeCaumia, B.; Blanchette, D.; Cote, B.; Renaud, M.; Rivard, P. Bio Energy 84; Egneus, H.; Ellegard, A. (Eds.), 1984; Vol. III, 31, Elsevier App. Sc. Publ., London.
6. Knight, J.A.;Gorton C.W.; Stevens, D. J. Bio Energy 84, Egneus, H.; Ellegard, A. (Eds.), 1984; Vol. III, 9, Elsevier App. Sc. Publ., London.
7. Elliott, D.C. International Energy Agency Co-operative Project D-1, Biomass Liquefaction Test Facility, National Energy Administration, Stockholm, Vol. 4.

8. Menard, H.; Belanger, D.;Chauvette, G.; Khorami, J.; Grise, M.; Martel, A.; Potvin, E.; Roy, C.; Langlois, R. in Hasnain, S. (Ed.), Proc. 5th Bioenergy R & D Seminar, Elsevier, New York, 1984, p 418.
9. Nicolaides, G. M. The Chemical Characterization of Pyrolysis Oils, MASc Thesis, Dept. of Chemical Engineering, University of Waterloo, Waterloo, Ontario, Canada, 1984.
10. Shafizadeh, F. J. Anal. Appl. Pyrol., 1982, 3, 283.
11. Funazukuri, T. A Study of Flash Pyrolysis of Cellulose, Ph.D. Thesis, Dept. of Chemical Engineering, University of Waterloo, Waterloo, Ontario, Canada, 1983.
12. Piskorz, J.; Radlein, D.; Scott, D.S. J. Anal. Appl. Pyr., 1986, 9, 121.
13. Radlein, D.;Grinshpun, A.; Piskorz, J.; Scott, D.S. J. Anal. Appl. Pyr., 1987, 12, 39.
14. Golova, O.P. Russian Chemical Reviews, 1975, 44 (8).
15. Marchessault, R. H.; Coulombe, S.; Morikawa, H.; Robert, D. Can. J. Chem., 1982, 60, 2372.
16. Menard, H. Universite de Sherbrooke - private communication.

RECEIVED June 9, 1988

Chapter 17

Product Analysis from Direct Liquefaction of Several High-Moisture Biomass Feedstocks

Douglas C. Elliott, L. John Sealock, Jr., and R. Scott Butner

Pacific Northwest Laboratory[1], P. O. Box 999, Richland, WA 99352

Experimental results are reported for high-pressure liquefaction of high-moisture biomass. The feedstocks included macrocystis kelp, water hyacinths, spent grain from a brewery, grain sorghum field residue and napier grass. The biomass was processed in a batch autoclave as a ten weight percent slurry in water with sodium carbonate catalyst and carbon monoxide gas. Thirty-minute experiments were performed at 350°C with operating pressures ranging from 270 to 340 atmospheres. The oil products were collected by methylene chloride and acetone extractions. Oil yields ranged from 19 to 35 mass percent on a moisture and ash-free basis. The oil products contained from 9.9 to 16.7 percent oxygen with hydrogen to carbon atomic ratios from 1.36 to 1.61. Significant nitrogen content was noted in the oil product from those feedstocks containing nitrogen (kelp, hyacinth, spent grain). Chemical composition analysis by gas chromatography/mass spectrometry demonstrated many similarities between these products and wood-derived oils. The nitrogen components were found to be mainly saturated heterocyclics.

Significant progress has been made over the past fifteen years toward the development of processes for direct production of liquid fuels from biomass. Process research has generally progressed along two lines -- flash pyrolysis and high-pressure processing. Extensive analysis of the liquid products from these two types of processes has demonstrated the significant process-related differences in product composition. However, the effect of feedstock has received a lesser degree of attention.

[1]Operated by Battelle Memorial Institute for the U.S. Department of Energy under Contract DE–AC06–76RLO–1830.

0097–6156/88/0376–0179$06.00/0

Process Research in Direct Liquefaction of Biomass

Two generalized categories of direct liquefaction processes can be identified (1). The first, flash pyrolysis, is characterized by a short residence time in the reactor (~1 second) at relatively high temperature (450-500°C) in order to obtain maximum yield of liquid product. The second, high-pressure processing, is usually performed at lower temperature (300-400°C) and longer residence time (0.2-1.0 hr). A typical operating pressure is 200 atm and often reducing gas and/or a catalyst is included in the process. The differences in processing conditions result in significant differences in product yield and product composition.

Product Analyses. Product analysis in support of the process development research in biomass direct liquefaction began at the rudimentary level of determining solvent-soluble portions of the product. Analysis was soon extended to elemental analyses and proximate analyses, such as ash and moisture. Later, spectrometric analyses were performed followed by detailed chemical analyses used in conjunction with chromatographic separation techniques.

At all stages of development, the significant differences in composition between the products of flash pyrolysis and high-pressure processing have been evident. While polar solvents are most effective for both products, less polar solvents such as methylene chloride and even benzene and toluene have been used as extractants for high-pressure product oils. Comparative analysis has demonstrated the higher oxygen content and higher dissolved water content in the flash pyrolysis oils. Detailed analyses with spectrometric and chromatographic methods have added supporting evidence to these findings.

Variations in Product Due to Feedstock. While process-related differences in product composition have been evident, extensive study of the effect of feedstock on product composition has never been undertaken. Some limited comparative tests can be gleaned from the literature; however, most process research in direct liquefaction of biomass has been performed with woods of various species. Table I provides some of the results available in the literature for non-woody feedstocks. Significant differences in heteroatom content are evident, but only limited chemical analysis is available in most cases.

The researchers at the Pittsburgh Energy Research Center (2-4) steadfastly maintained in their pioneering work on high-pressure liquefaction that their oil products obtained from cellulosic wastes were paraffinic and cycloparaffinic in nature. These products were formed in the presence of sodium carbonate catalyst with an initial overpressure of carbon monoxide at temperatures from 250 to 380°C and pressures from 10.4 to 34.6 MPa with batch residence times of 20 to 60 minutes. They reported the presence of carbonyl and carboxyl functional groups but maintained that there was essentially no aromatic material produced except at higher temperature (then only in very small amounts). These conclusions were based on infrared and mass spectral analysis (2). Later analysis of the sucrose-derived oil (3) included proton nuclear magnetic resonance spectral evaluation but resulted in the same conclusion. Most of the hydrogen was in methylene or methyl groups and about 4 percent

Table I. Product Analyses from Liquefaction Tests with Various Biomasses

Feedstock	Temp. (°C)	Pressure (MPa)	C	H	N (percent)	O	S	H/C
High-Pressure Processes								
(2) newsprint	250	13.8	71.7	7.3	<0.3	~20.6	<0.1	1.21
(2) pine needles & twigs	250	13.8	72.2	8.7	1.05	18.0	0.10	1.43
(2) sewage sludge	250	13.8	77.0	10.7	2.80	8.8	0.64	1.65
(3) cellulose	250	13.8	72.4	7.0	0.004	20.4	0.2	1.15
(3) sucrose	350	27.7	75.2	9.1	--	15.7	--	1.44
(4) municipal refuse	380	34.6	79.8	10.4	3.0	6.8	0.05	1.55
(4) manure	380	31.1	80.4	9.4	3.0	6.9	0.26	1.39
(5) microalgae	400	27.7∞	*81.2	8.6	5.4	3.5	--	1.26
Flash Pyrolysis								
(6) aspen	450	0	53.8	6.7	--	39.3	--	1.48
(7) sewage sludge	450	0	69.4	10.2	5.8	14.5	--	1.75
(8) poplar	500	0	49.8	7.3	0.0	42.8	0.0	1.74
(8) peat	520	0	67.1	9.0	3.4	20.3	0.1	1.59

∞estimate

*the microalgae analysis was calculated from the analysis of product fractions (oil and asphaltene) and the product distribution

was unsaturated but probably olefinic and not aromatic. Some ether linkages were also reported present in the sucrose-derived oil. Mass spectral analysis of the municipal refuse-derived oil (4) identified only two long chain fatty acids with certainty; however, not more than traces of aromatics were determined to be present. The manure-derived oil was found to be largely alicyclic hydrocarbon but contained heterocyclic nitrogen and alkyl phenolics (4). These claims of the saturated hydrocarbon nature of the oil products are at odds with the reported elemental analyses (see Table I). The low hydrogen to carbon ratios dictate that the oil products must contain a large fraction of aromatic or, at least, highly unsaturated compounds.

An algae-derived oil was reported to be principally n-paraffins and olefins with oxygen- and nitrogen-containing straight-chain hydrocarbons (5). Polar compounds were reported to comprise 50-60 percent of the oil. Unfortunately, there is no indication of the type of analysis performed or detailed results of any kind; therefore, it is difficult to evaluate the veracity of these reported results.

The comparison of the peat and wood flash pyrolysis products by Elliott (8) is a good example of the effect of feedstock on product oil composition. The poplar oil typically was composed of phenolic, ketone and furan compounds with a substantial fraction of low molecular weight organic acids. The main components of the peat oil were hydrocarbons, mostly straight chain olefins. Minor quantities of ketones were noted but no acids, aldehydes or furans were identified by mass spectrometry. Phenols were also present in significant quantities.

A significant effort in comparing feedstock effects on product oil composition was reported by Russell et al. (9). Unfortunately, this effort did not include ultimate analysis of the oils for comparison. The report contains qualitative analysis by gas chromatography/mass spectrometry of five product oils derived from cellulose, hops field residue, softwood tree branches, peat, and sewage sludge. Phenols were a major component group for all feedstocks. Ketones and furans were also common. Hydrocarbons, aromatic and otherwise, were also identified primarily in the cellulose and softwood products. Nitrogen-containing products were absent from the cellulose and softwood products but could be found in the peat and sewage sludge-derived oils.

All of the above accounts can be contrasted with the large amount of analytical work on the chemical composition of wood-derived direct liquefaction products which has been reported over the past several years (8,10-18). In all cases the majority of the product oils have been identified as phenolic with only minor amounts of pure hydrocarbon reported.

Liquefaction Experiments with Moist Biomass

At Pacific Northwest Laboratory we have been testing the use of high-moisture biomass (marine and fresh-water biomass, post-harvest field residues and food processing wastes) in a thermochemical conversion system to produce useful fuels. Although the main focus of the work (19) has been gasification (catalytic production of methane) we have also performed a limited number of tests under high-pressure liquefaction conditions.

Feedstock Description. Five high-moisture biomass feedstocks were tested in these liquefaction experiments. They are characterized as follows:

Kelp - The sample used was a freshly harvested macrocystis kelp from Pacific Ocean seabeds off the southern California coast (El Capitan Beach, Santa Barbara Channel). It was packed in ice and flown to our laboratory where it was frozen in a polyethylene bag until used.

Water Hyacinth - Uprooted samples of hyacinths were recovered from the primary treatment lagoon at the Reedy Creek experimental sewage treatment facility near Orlando, Florida. The sample was packed in ice and flown to our laboratory where it was frozen in a polyethylene bag until used.

Spent Grain - The grain sample used was the residue following malting barley and water extraction of the sugars prior to fermentation. The sample was obtained from the Blitz-Weinhard Brewery in Portland, Oregon and was transported to our laboratory where it was frozen until used.

Napier Grass - Napier grass was collected after harvest by University of Florida researchers. The sample was bagged and shipped in a refrigerated container to our laboratory where it was stored in a freezer until used.

Sorghum - Grain sorghum was collected after the harvest of the grain by the University of Florida. The sample containing stems, stalks, and leaves was bagged and shipped in a refrigerated container to our laboratory where it was stored in a freezer until used.

Ultimate analysis, moisture contents and energy contents for the five feedstocks are provided in Table II.

Table II. Analysis of Moist-Biomass Feedstocks

Feedstocks	C	H	N	O	Ash	Moisture (percent)	HHV* (BTU/lb)
	(percent, dried basis)						
Kelp	26.9	4.0	1.2	30.2	38.4	89.0	7150
Water Hyacinth	43.0	5.8	5.6	29.5	15.3	94.9	7730
Spent Grain	48.6	6.8	3.4	35.3	3.4	80.5	9170
Napier Grass	45.2	6.0	<0.1	42.3	5.7	84.4	7870
Sorghum	44.4	5.8	0.4	38.3	7.9	77.0	8040

*HHV = higher heating value of dried biomass

Reactor Conditions. The experiments were performed batchwise in a one-liter, stirred autoclave. Approximately 300 g of moist-biomass was charged to the autoclave in a stainless steel liner as a water slurry containing 5 to 10 weight percent of dry solids. Sodium carbonate was added to the feedstock (approximately 0.1 g/g dry biomass) except in the case of kelp which already contains a high level of alkali as part of its chemical makeup.

The autoclave was then sealed, purged with nitrogen and then pressurized with carbon monoxide (approximately 50 atm). The reactor was heated to 350°C (approximately 30 minutes from 200° to 350°C) and held at that temperature for 30 minutes. The pressure within the autoclave at temperature typically increased from 270 atm to 340 atm over the period of the experiment. At the end of the allotted time cooling water was flushed through the internal cooling coil which brought the reactor temperature down to 200°C within 5 minutes.

Product Recovery and Analysis. After the autoclave had cooled completely, the gas product was vented. The typical gas composition included nearly equal parts of hydrogen and carbon dioxide with a 10-15 percent residual amount of carbon monoxide and minor amounts of hydrocarbons. [These results suggest a strong water-gas shift reaction as catalyzed by the sodium carbonate base (20).] The autoclave was then opened and the liquid product was collected. The autoclave was rinsed with acetone and the resulting wash solution filtered. The liquid product (approximately 250 g) was acidified to pH 2 with dilute HCl and then extracted with methylene chloride (3 x 40 ml).

The soluble and insoluble products were analyzed for elemental content of carbon, hydrogen, nitrogen and oxygen with Perkin-Elmer 240 series instruments. The methylene chloride soluble oil product was also analyzed as a methylene chloride solution on a gas chromatograph equipped with a mass selective detector for qualitative analysis and a similar gas chromatograph equipped with a flame ionization detector for quantitative analysis. Identification of compounds was made by comparison of mass spectra with library listings of known compounds in conjunction with a comparison of chromatograph column retention time with similar known compounds. Quantitative analysis was based on a known amount of internal standard (trans-decahydronaphthalene) with detector response factors determined for various functional group types. Quantitation is estimated at within ± 20 percent. DB-5 capillary columns are used in both chromatographs.

Liquefaction Results and Product Description

Results from the liquefaction experiments with the five moist-biomass feedstocks are given in Table III. The oil yield is based on the combined mass of acetone- and methylenechloride-soluble oils as a percent of the mass of dried feedstock calculated to an ash-free basis. The product oil elemental analysis is the calculated composite analysis for the combined acetone- and methylene chloride-soluble oils.

Table III. Experimental Results for Liquefaction Experiments

Feedstocks	Oil Yield (percent)	C	H	N	O	H/C
		(combined oil analyses)				
Kelp	19.2	76.7	8.9	3.5	9.9	1.38
Water Hyacinth	26.0	76.3	9.9	3.3	10.5	1.54
Spent Grain	34.7	75.2	10.2	3.8	10.8	1.61
Napier Grass	34.4	74.5	8.5	0.4	16.7	1.36
Sorghum	26.6	75.9	8.7	1.7	13.7	1.36

The test results in Table III demonstrate oil product yields for liquefaction of the moist-biomass feedstocks at levels comparable to wood liquefaction. Reported yields for wood liquefaction in aqueous slurries, such as the LBL process (21), have typically been in the 25 to 30 weight percent range. The quality of the moist-biomass liquefaction products fall in a general range which is also similar to reports for wood liquefaction products. However, certain examples of moist-biomass product oils appear to have elemental compositions suggesting higher quality products. Especially interesting are the high hydrogen to carbon ratios for the spent grain and water hyacinth products and the relatively low oxygen contents of the spent grain, water hyacinth and kelp products. A significant difference from wood-derived oils is the high nitrogen content in the oils from spent grain and aquatic biomasses.

Product Analysis Details. The detailed chemical analysis of the five moist-biomass derived oils by gas chromatography and mass spectrometry helps to better define the differences in oil composition. Over 190 different compounds and isomers were identified in the five oils. In order to better understand this large amount of information, the components have been grouped by chemical functionality and these groups are listed in Table IV.

Table IV. Chemical Functional Groups in Moist-Biomass Oil Products*

Compound Group	Kelp	Water Hyacinth	Spent Grain	Napier Grass	Sorghum
esters/aldehydes/ alcohols	0 (1)	0 (0)	0 (1)	3 (4)	0 (0)
cyclic ketones	10 (17)	3 (12)	8 (13)	23 (31)	7 (15)
furans	0 (0)	1 (1)	4 (3)	2 (3)	2 (3)
phenols	32 (17)	35 (30)	30 (18)	45 (36)	34 (25)
methoxy phenols	3 (3)	3 (2)	3 (1)	5 (5)	10 (5)
benzenediols	10 (2)	6 (4)	0 (0)	6 (6)	24 (11)
naphthols	9 (5)	2 (3)	0 (0)	5 (9)	9 (6)
aromatic oxygenates	1 (2)	3 (3)	6 (2)	4 (5)	2 (2)
cyclic hydrocarbons	2 (3)	1 (3)	3 (3)	3 (6)	4 (5)
long-chain hydrocarbons	6 (3)	15 (15)	5 (4)	3 (9)	4 (4)
fatty acids	10 (3)	8 (7)	8 (4)	1 (2)	5 (4)
nitrogen cyclics	16 (11)	16 (11)	19 (7)	0 (0)	0 (0)
amines/amides	3 (1)	6 (3)	15 (7)	0 (0)	0 (0)
percent identified	10.8	18.4	14.3	21.0	18.4

*tabular listing is the mass percent of identified oil components in each compound group; the number in parentheses is the number of individual compounds and isomers in each compound group

The compound groups consist of the following types of compounds:

esters/aldehydes/alcohols - four to six carbon oxygenates
cyclic ketones - five and six carbon rings, many unsaturated, most alkylated
furans - dihydrofuranones, hydroxymethyltetrahydrofuran
phenols - phenol and alkylated (up to five carbons) phenols
methoxyphenols - mono- and dimethoxyphenols and alkylated forms
benzenediols - dihydroxybenzenes and alkylated (up to five carbons) forms
naphthols - naphthols and methylated naphthols
aromatic oxygenates - bismethylguaiacol(?), phenylphenols, benzodioxin(?)
cyclic hydrocarbons - alkylcyclopentenes, alkylbenzenes(?), alkylindans, phenanthrene
long-chain hydrocarbons - C_{14} to C_{27} n-alkanes and olefins
fatty acids - C_{12} to C_{20} saturated and unsaturated acids
nitrogen cyclics - alkylpyrrolidinones, alkylaziridines(?), alkylpyrroles, alkylindoles
amines/amides - C_8 to C_{22} amines/amides(?)

The question marks in the listing above indicate a tentative identification.

Comparison to Earlier Results. These results verify that the carbohydrate structures found in biomass can be converted thermochemically to a mixture of primarily phenolic compounds. Hydrocarbons are not predominant yet they may survive the processing in a significant yield given an appropriate feedstock. Cyclic ketones are the other major component group which can be identified by GC/MS. Low molecular weight oxygenates and furans are minimized by the addition of base to the reaction medium as has been demonstrated by other researchers (22).

The product compositions of the napier grass and sorghum-derived product oils show many similarities to wood-derived oils. In comparing with high-pressure processed oil from Douglas fir the same groups of cyclic ketones, phenols, naphthols, and dihydroxybenzenes dominate. The traces of hydrocarbon in the sorghum and napier grass oils are significantly different from the Douglas fir and are apparently feedstock related. Nitrogen-containing compounds were not found in either the Douglas fir oils or the sorghum or napier grass oils reported here.

The large fraction of nitrogen-containing cyclic compounds is the distinguishing factor between the hyacinth, kelp and spent grain oils when compared to earlier wood oil analyses. Similar compounds were found earlier in peat and sewage sludge oils (9). Our results now extend this trend to high protein feedstocks and green, aquatic plants. It is obvious that a strong correlation exists between nitrogen content of the feedstock and the amount of nitrogen incorporated into the product oil. High-pressure liquefaction even with a reducing gas environment and alkaline catalysis cannot effect a preferential denitrogenation reaction. Substantial amounts of nitrogen are condensed into cyclic systems which remain in the oil product.

Utilization of Oil Products from Moist Biomass. The oil products from these high-moisture biomasses have properties similar to the more widely studied wood-derived oils. The numerous applications of wood-derived oil have been discussed by others (10,22). The moist-biomass oils should be amenable to the same types of applications. In addition, the nitrogen-containing compounds may be useful as chemical commodities. Indoles in particular may be recoverable for use as fragrances or flavors. The hydrocarbon component in the oils may facilitate the direct use of these oils as fuels.

The nitrogen-containing components in some of the moist-biomass oils is a source of concern when considering their use as fuels. Direct utilization of these oils would certainly lead to higher levels of emission of NO_x. The nitrogen-containing components have also been indicated as a source of fuel instability during storage and as cancer-causing agents in various chemical forms. Hydrotreating of these oil products to remove the nitrogen is a possible means of upgrading the products. However, hydrodenitrogenation of the heterocyclic compounds is a difficult and costly procedure compared to hydrodeoxygenation of the phenolics, which would also be accomplished in a hydrotreating type of oil upgrading.

Implications for Future Research. The high-moisture biomass feedstocks can be a source of useful liquid fuel products. The use of the high-moisture biomasses in high-pressure processing will allow their utilization in a thermochemical process without prior drying. Other research on the use of these feedstocks in high-pressure gasification has suggested the feasibility of feeding these materials as a slurry following maceration (19). This same type of feeding should allow direct utilization of high-moisture biomass in high-pressure liquefaction processing. Experimental verification of this type of continuous processing needs to be undertaken.

The use of the nitrogen-containing feedstocks will lead to production of a nitrogen-containing oil product. Direct utilization of this oil product as a fuel will likely require development of appropriate emission control techniques in order to maintain air quality. Alternately, nitrogen-containing components can be removed from the oil product for use as specialty chemicals or by hydrotreating. Further development of hydrotreating technology specific to these oils may be necessary in order to process the heterocyclic nitrogen-containing compounds which require extensive processing in order to effect nitrogen removal.

Acknowledgments

This research was performed as part of a project titled "Low-Temperature Thermochemical Conversion of High-Moisture Biomass Feedstocks" funded through the Biomass Thermochemical Conversion Program Office at Pacific Northwest Laboratory by the U.S. Department of Energy, Biofuels and Municipal Waste Technology Division, S. Friedrich, Program Manager. The authors also acknowledge the help of three coworkers at Pacific Northwest Laboratory, G. G. Neuenschwander, who operated the batch reactor, R. T. Hallen, who performed the chromatography and mass spectrometry, and W. F. Riemath, who performed the elemental analyses.

Literature Cited

1. Beckmen, D.; Elliott, D. C. Can. J. Chem. Eng. 1985, 63, 99-104.
2. Appell, H. R.; Wender, I.; Miller, R. D. Conversion of Urban Refuse to Oil, Bureau of Mines Solid Waste Program, Technical Progress Report-25: Pittsburgh, PA, 1970.
3. Appell, H. R.; Fu, Y. C.; Friedman, S.; Yavorsky, P. M.; Wender, I. Converting Organic Wastes to Oil: A Replenishable Energy Source, Report of Investigations 7560, Pittsburgh Energy Research Center: Pittsburgh, PA, 1971.
4. Appell, H. R.; Fu, Y. C.; Illig, E. G.; Steffgen, F. W.; Miller, R. D. Conversion of Cellulosic Wastes to Oil, Report of Investigations 8013, Pittsburgh Energy Research Center: Pittsburgh, PA, 1975.
5. Chin, L.-Y.; Engel, A. J. Biotech. & Bioeng. Symp. #11, 1981, 171-186.
6. Scott, D. S.; Piskorz, J.; Radlein, D. Ind. & Eng. Chem., Proc. Des. & Dev., 1985, 24, 581-588.
7. Piskorz, J.; Scott, D. S.; Westerberg, I. B. Ind. & Eng. Chem., Proc. Des. & Dev., 1986, 25, 265-270.
8. Elliott, D. C. Analysis and Upgrading of Biomass Liquefaction Products, IEA Co-operative Project D1 Biomass Liquefaction Test Facility project Final Report, Volume 4. National Energy Administration: Stockholm, 1985.

9. Russell, J. A.; Molton, P. M.; Landsman, S. D. "Chemical Comparisons of Liquid Fuel Produced by Thermochemical Liquefaction of Various Biomass Materials" presented at the 3rd Miami International Conference on Alternative Energy Sources, Miami, Florida, PNL-SA-8602, Pacific Northwest Laboratory: Richland, Washington, 1980.
10. Elliott, D. C. Biotech. & Bioeng., Symp #11, 1981, 187-198.
11. Elliott, D. C. In Fundamentals of Thermochemical Biomass Conversion, Elsevier Applied Science Publishers: London, 1985; Chapt. 55, 1003-1018.
12. Karlsson, O., and Björnbom, P. In Fundamentals of Thermochemical Biomass Conversion, Overend, R. P., Milne, T. A., and Mudge, L. K., Eds.; Elsevier Applied Science Publishers: London, 1985; Chapt. 56, pp. 1019-1026.
13. Davis, H. G.; Evans, M. A.; Figueroa, C.; Gansley, R. R.; Schaleger, L. L.; Watt, D. W. In Fundamentals of Thermochemical Biomass Conversion, Overend, R. P., Milne, T. A., and Mudge, L. K., Eds.; Elsevier Applied Science Publishers: London, 1985; Chapt. 57, pp. 1027-1037.
14. Eager, R. L.; Mathews, J. F.; Pepper, J. M.; Zohdi, H. Can. J. Chem, 1981, 59, 2191-2198.
15. Eager, R. L.; Pepper, J. M.; Roy, J. C.; Mathews, J. F. Can. J. Chem., 1983, 61, 2010-2014.
16. Ménard, H., Bélanger, D.; Chauvette, G.; Grisé, M.; Martel, A. Comptes Rendus De L'Atelier de Travail sur la Liquéfaction de la Biomasse, National Research Council of Canada, NRCC23120: Ottawa, 1983; pp. 208-213.
17. Boocock, D. G. B.; Kallury, R. K. M. R.; Tidwell, T. T. Anal. Chem. 55, 1983, 1689-1694.
18. Soltes, E. J.; Elder, T. J. In Organic Chemicals from Biomass, Goldstein, I. S., Ed.; CRC Press Inc.: Boca Raton, Florida, 1981; Chapt. 5, pp. 63-99.
19. Butner, R. S.; Elliott, D. C.; Sealock, L. J., Jr. Proc. of the 1985 Biomass Thermochemical Conversion Contractors' Meeting, PNL-SA-13571, Pacific Northwest Laboratory: Richland, Washington, 1986; pp 193-209.
20. Elliott, D. C.; Hallen, R. T.; Sealock, L. J., Jr. Ind. & Eng. Chem., Prod. Res. & Dev., 1983, 22, 431-435.
21. Figueroa, C.; Schaleger, L. L.; Davis, H. G. In Energy from Biomass & Wastes VI, Klass, D. L., Ed.; Institute of Gas Technology: Chicago, Illinois, 1982; pp 1097-1112.
22. Nelson, D. A.; Molton, P. M.; Russell, J. A., Hallen, R. T. Ind. & Eng. Chem., Prod. Res. & Dev., 1984, 23, 471-475.

RECEIVED March 31, 1988

Chapter 18

An Integrated Spectroscopic Approach to the Chemical Characterization of Pyrolysis Oils

Barbara L. Hoesterey[1], Willem Windig[1], Henk L. C. Meuzelaar[1], Edward M. Eyring[2], David M. Grant[2], and Ronald J. Pugmire[3]

[1]Biomaterials Profiling Center, University of Utah, Salt Lake City, UT 84108
[2]Chemistry Department, University of Utah, Salt Lake City, UT 84112
[3]Fuels Engineering Department, University of Utah, Salt Lake City, UT 84112

The hydrocarbon ("oil") fraction of a coal pyrolysis tar prepared by open column liquid chromatography (LC) was separated into 16 subfractions by a second LC procedure. Low voltage mass spectrometry (MS), infrared spectroscopy (IR), and proton (PMR) as well as carbon-13 nuclear magnetic resonance spectrometry (CMR) were performed on the first 13 subfractions. Computerized multivariate analysis procedures such as factor analysis followed by canonical correlation techniques were used to extract the overlapping information from the analytical data. Subsequent evaluation of the integrated analytical data revealed chemical information which could not have been obtained readily from the individual spectroscopic techniques. The approach described is generally applicable to multisource analytical data on pyrolysis oils and other complex mixtures.

Due to the extremely complex nature of pyrolysis tars obtained from recent or fossil biomass samples, structural and compositional analysis of such tars poses a formidable challenge to analytical chemists. Even when armed with an arsenal of sophisticated analytical techniques, a detailed qualitative analysis requires careful, laborious combination and integration of voluminous chromatographic and spectroscopic data. Complete quantitative analysis of all components of oil is generally not within reach of current analytical methodologies, especially if the tar contains nonvolatile and/or highly polar or reactive components. Although in recent years impressive advances have been made with the physical coupling of two or more chromatographic and/or spectroscopic techniques into so-called "hyphenated" methods, e.g., GC/MS, LC/MS, GC/FTIR, MS/MS, etc., true integration of the analytical data by means of multivariate analysis methods such as canonical correlation

0097–6156/88/0376–0189$06.00/0

analysis is rarely even attempted. Yet, intuitively the potential advantages and benefits of data integration methods are easily understood. With these considerations in mind the authors carried out the present feasibility study of a coal-derived pyrolytic tar using a combination of chromatographic (LC), spectroscopic (MS, IR, PMR, CMR) and chemometrics (factor and canonical correlation analysis) techniques. In order to reduce the complexity of the analytical problem to more manageable proportions, a completely distillable coal tar was selected. Moreover, polar and/or highly reactive components were removed by open column LC. Preliminary results of this integrated analytical approach will be presented here.

Experimental

A pyrolysis tar from a high volatile B bituminous Hiawatha seam coal (Wasatach Plateau field, Utah) was obtained from the fixed bed Wellman Galusha gasifier operated by Black, Sivalls and Bryson in Minneapolis (1). Open column liquid chromatography (LC) on silica gel using four solvents and solvent mixtures of increasing polarity, e.g., hexane; hexane/benzene 8/1; benzene/ether 4/1; and benzene/methanol 1/1, was used to separate the whole tar into broad compound classes as described by McClennen et al. (2). Hydrocarbons constituted more than 50% of the whole tar. The hexane and hexane/benzene eluted fractions constituted complex mixtures, principally composed of hydrocarbons as determined by low voltage mass spectrometry (3). These fractions were combined and further separated by a second LC run. Subfractionation was accomplished as follows: two HPLC pumps and a solvent programmer/mixer were adapted to deliver solvents to the top of a 1.4 by 45 cm open glass column packed with 60-200 mesh silica gel (activated by heating overnight at 175°C); the sample was layered on the column in 10 ml hexane and fractions were eluted from the column with a nonlinear gradient beginning wiht 100% n-hexane (HPLC grade) and ending with 9:1 hexane/benzene over a period of 30 min. Subfraction collection continued for 40 min with the 9:1 hexane/benzene. Sixteen fractions were collected and weighed over a total of 40 minutes. Approximately 1 ml was taken from each sample for low voltage MS analysis. The remaining subfractions were then rotary evaporated and weights of residue were recorded. Elution volumes were calculated from solvent weights. The calculated elution volumes are shown in Figure 1.

Low voltage mass spectra of subfractions 1-15 were obtained on an Extranuclear 5000-1 quadrupole mass spectrometer with Curie-point heating inlet using 1/4 µl glass capillary probe tips as described by McClennen et al. (4). The inlet was heated to 200°C, electron energy was set at 12 eV. Samples were scanned from m/z 20 to m/z 300. All spectra were recorded with an IBM 9000 computer.

FTIR spectra were obtained using a few mg of neat sample on NaCl (salt) disks. The instrument was a Nicolet 7000 series spectrometer, resolution 4 cm^{-1}, 200 scans, operated in the absorbance mode. Samples were scanned from 4000-600 cm^{-1}. Baseline corrected absorbance intensities were recorded for 20-30 peaks in each spectrum which were above 0.01 absorbance units in intensity. No peaks indicating water were seen in the spectra. Infrared data consisted of raw absorbance spectra with wavenumber and intensity values written above each peak. This made the choice of the wavenumber variables difficult especially in the aliphatic stretching region. For example, it was not clear whether peaks with highest intensity at 2924, 2929, and 2918 cm^{-1} in different spectra represented information from the same functional group or not. The various LC subfractions would be expected to show differences based on elution time that represent chemical information. Therefore, peak maxima at slightly different wavenumbers in the same 20 cm^{-1} region were classified as a single variable if the differences occurred more or less randomly along the elution time axis. A distinct trend of wavenumber position vs. time was taken to indicate that more than one variable was present. For example, the peak maximum at 2873 cm^{-1} occurred in the first 6 fractions only, and was replaced by a peak at 2868 cm^{-1} for the remaining fractions 7 to 15. Other spectral regions were straightforward with regard to the choice of variables since sharper peaks with narrower bandwidth were present. All in all, thirty three wavenumber variables were obtained for fractions 1-15 and the data were put into standard format suitable for factor analysis. Prior to the acutal factor analysis, spectra were scaled to 100%, and peak variables in percent were calculated.

Proton NMR spectra of the hydrocarbon subfractions in CD_2Cl_2 with TMS internal standard (80 300 mg sample / 5-10 ml solvent) were taken using a Varian 300 superconducting instrument over the 1-10 ppm region. All spectra were taken under identical conditons so that relative differences in composition could be noted. Integrated peak intensities for eight regions of the spectrum were tabulated for each subfraction, in addition to a table containing an overall view of the number of aliphatic, aromatic and olefinic protons present.

Carbon 13 NMR spectra of the subfractions were also run in CD_2Cl_2 at the same concentrations noted above on the Varian 300 from 0-180 ppm. A 30° tip angle and 5 sec recycle time were used. Peak intensities were measured using integration curves. Spectra were again taken under identical conditions. Twenty three variables were chosen.

Computerized multivariate analysis was carried out using the interactive SIGMA program package developed at the University of Utah Biomaterials Profiling Center which enables careful scaling of spectral intensities, as well as factor, discriminant and canonical correlation analysis (5). Chemical components were numerically extracted using the Variance Diagram technique described by Windig et al. (6).

Results and Discussion

Mass Spectrometry. The first 13 subfractions were chosen for detailed analysis. Mass spectra of the fractions revealed the following trends: Fraction 1 (Figure 2a) shows alkane fragment ions and $C_nH_{2n}^+$ (alkene or cyclo paraffin) molecular ions. (Note the series progression to m/z 254). By fraction 3, $C_nH_{2n-1}^+$ fragment ions begin to dominate over the $C_nH_{2n+1}^+$ fragment ions. Fraction 5 shows fragment ion series at m/z 149, 163, 177, 191 as well as signals at m/z 68, 136, 206. These ion series are frequently seen in resinites found in Utah coal (7) as well as terpenoid components of recent plant matter. Fraction 7 produces a complex spectrum, containing a large number of ion series, including benzenes, indane/tetralins, and dienes or decalins (C_nH_{2n-2}).

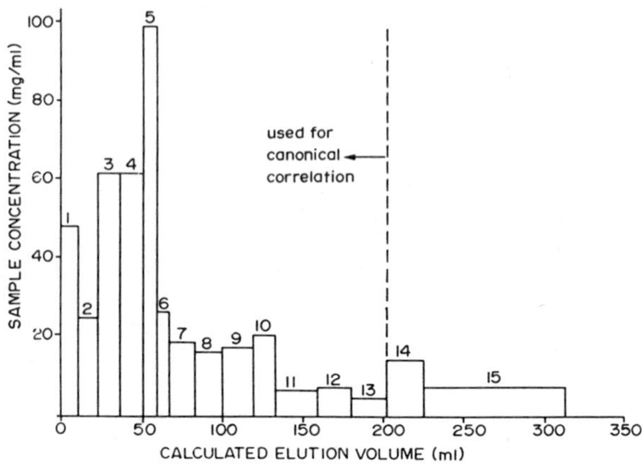

Figure 1. Reconstructed liquid "chromatogram".

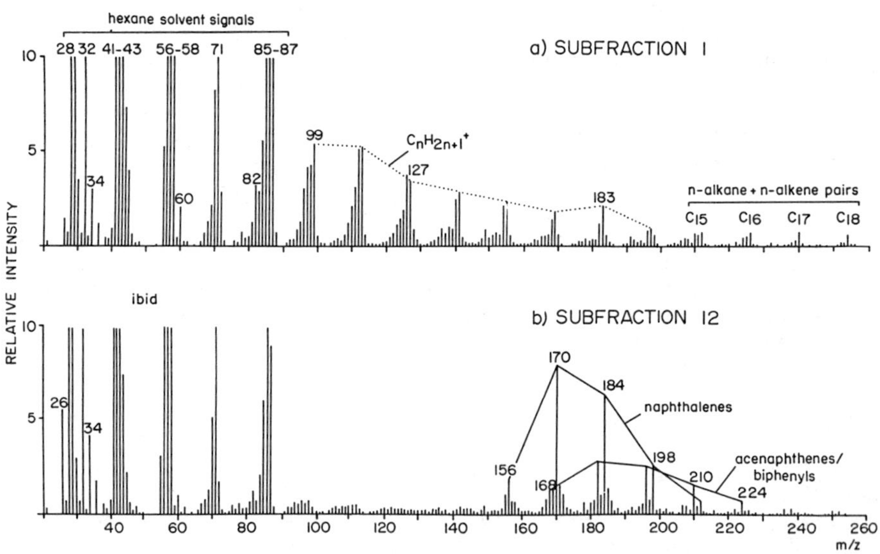

Figure 2. Low voltage mass spectra of (a) subfraction 1 and (b) subfraction 12.

All ion series have tentative identification from the raw spectra. This C_nH_{2n-2} series overlaps the dihydroxybenzenes series at m/z 110, 124, etc., but can be distinguished using LC fractionation (polarity) considerations. C_nH_{2n-2} ions, e.g., decalins, are not present in fraction 8, but naphthalenes are beginning to appear, along with tetralins/indanes. Finally, fractions 10, 11, 12 (Figure 2b) and 13 are dominated by naphthalenes, with acenaphthalenes/ biphenyl ions appearing from fractions 11-13.

Infrared Spectroscopy. Infrared spectra of the fractions (Figure 3) show the following trends: Fractions 1, 2, 3, 4 and 5 show the IR pattern of the "pure" aliphatic compounds. The bands at 2961 and 2929 cm^{-1} indicate the stretchings of methylene and methyl groups and the corresponding bands located at 1456 and 1378 cm^{-1} indicate the bending motion of the CH_3 group. The IR spectra of fractions 7 and 8 start to shows some aromaticity. The weak peak at 1601 cm^{-1} can be assigned to the C = C moiety of a benzene ring and the peaks at 871, 832, 812, and 750 cm^{-1} can be assigned as the out-of-plane bending modes of 1, 2, 3 and 4 adjacent hydrogens, respectively. The CH of the aromatic ring shows up at 3050 and 3012 cm^{-1}.

Nuclear Magnetic Resonance. Table 1 shows the overall aliphaticity/ olefinicity/aromaticity from the carbon and proton NMR data. Both techniques find that fractions 1 and 2 are entirely aliphatic, olefins are present in fractions 3-7, and aromatic compounds begin appearing in fraction 6 (CMR) and fraction 7 (PMR), along with olefin and aliphatic functional groups. Fractions 8-13 contain aromatic moieties, likely with aliphatic substituents. Note that the carbon aliphatic/aromatic ratios decrease among fractions 8 through 13, whereas the proton aliphatic/aromatic ratio's decrease among samples 8-12, with a higher value at fraction 13 than for 12.

Carbon 13 NMR peaks showed the similar trend of increasing aromaticity in terms of increasing elution time. Samples 1 and 2 contained only aliphatic peaks, at 14, 20, 23, 30, 32, 34, 38 and 40 ppm. Three peaks were also found in all subsequent samples. By sample 3, olefinic peaks at 97, 114, 140 ppm were present, and continued to be seen through sample 7. The aromatic peaks 124, 131, 132 ppm were present in sample 6. Sample 7 contained alarge number of aromatic peaks from 122-142 ppm, with the largest peaks at 128 and 142 ppm. Samples 8-13 also contained aromatic peaks from 122-135 ppm. The largest peaks in sample 10, for example, were at 126, 131, 132 and 134 ppm.

Factor Analysis. Several choices had to be made in preparing the data for factor analysis as well as in choosing criteria for selecting the number of factors needed to describe the data space (e.g. eigenvalue > 1.0, ratio adjacent eigenvalues > 2.0, etc.) and the number of factor scores to be used as input into the canonical correlation analysis. These choices may have affected subsequent interpretation of the multivariate spaces and evaluation of the chemometric analysis methods. Table II shows the types of spectral data input into factor analyses of the first 13 subfractions.

Our discussion of the factor analyses presented in Figure 4 will first identify components characteristic of early eluting samples and

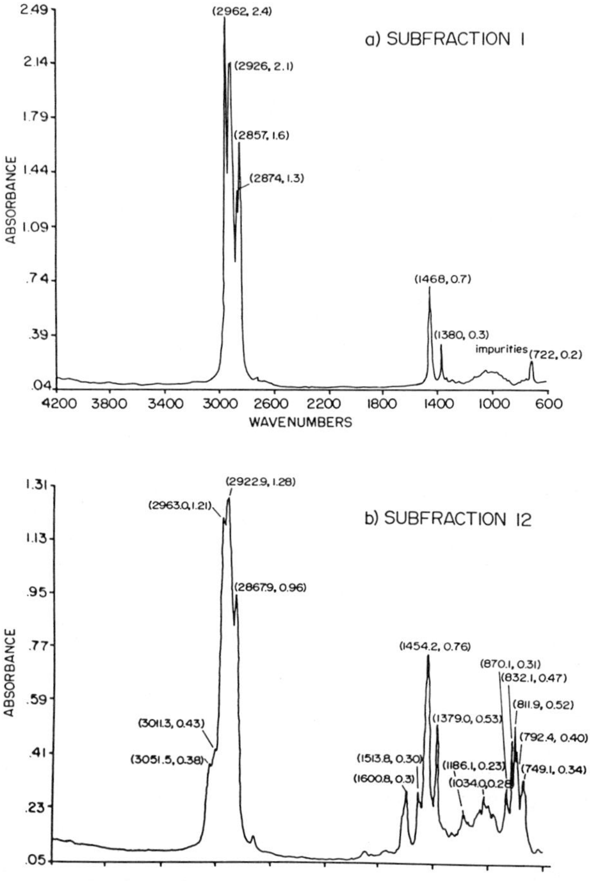

Figure 3. FTIR spectra of (a) subfraction 1 and (b) subfraction 12.

Table I. Integrated Intensities of Aliphatic Olefinic and Aromatic Regions of the NMR Spectrum for LC Subfractions of the Hydrocarbon Fraction of Hiawatha Tar

SUBFRACT.	(Proton NMR Data) ALIPHATIC (.5-4 ppm)	OLEFINIC (4-6 ppm)	AROMATIC (6-9 ppm)	ALIPHATIC H/-AROMATIC H
1	1.0			
2	1.0			
3	0.959	0.041		
4	0.948	0.052		
5	0.962	0.038		
6	0.951	0.049		
7	0.905	0.043	0.052	17.40
8	0.909		0.091	9.99
9	0.827		0.173	4.78
10	0.822		0.178	4.62
11	0.789		0.211	3.74
12	0.786		0.214	3.67
13	0.849		0.151	5.62

SUBFRACT.	(Carbon 13 NMR Data) ALIPHATIC	OLEFINIC	AROMATIC	ALIPHATIC C/-AROMATIC C
1	1.0			
2	1.0			
3	0.98	.02		
4	0.922	.078		
5	0.922	.078		
6	0.855	.063	.052	
7	0.759	.121	.120*	
8	0.596		.404	1.48
9	0.513		.487	1.05
10	0.357		.643	0.56
11	0.333		.667	0.50
12	0.261		.739	0.35
13	0.211		.789	0.27

* integration from 50-150 ppm, olefinic and aromatic arbitraily made equal

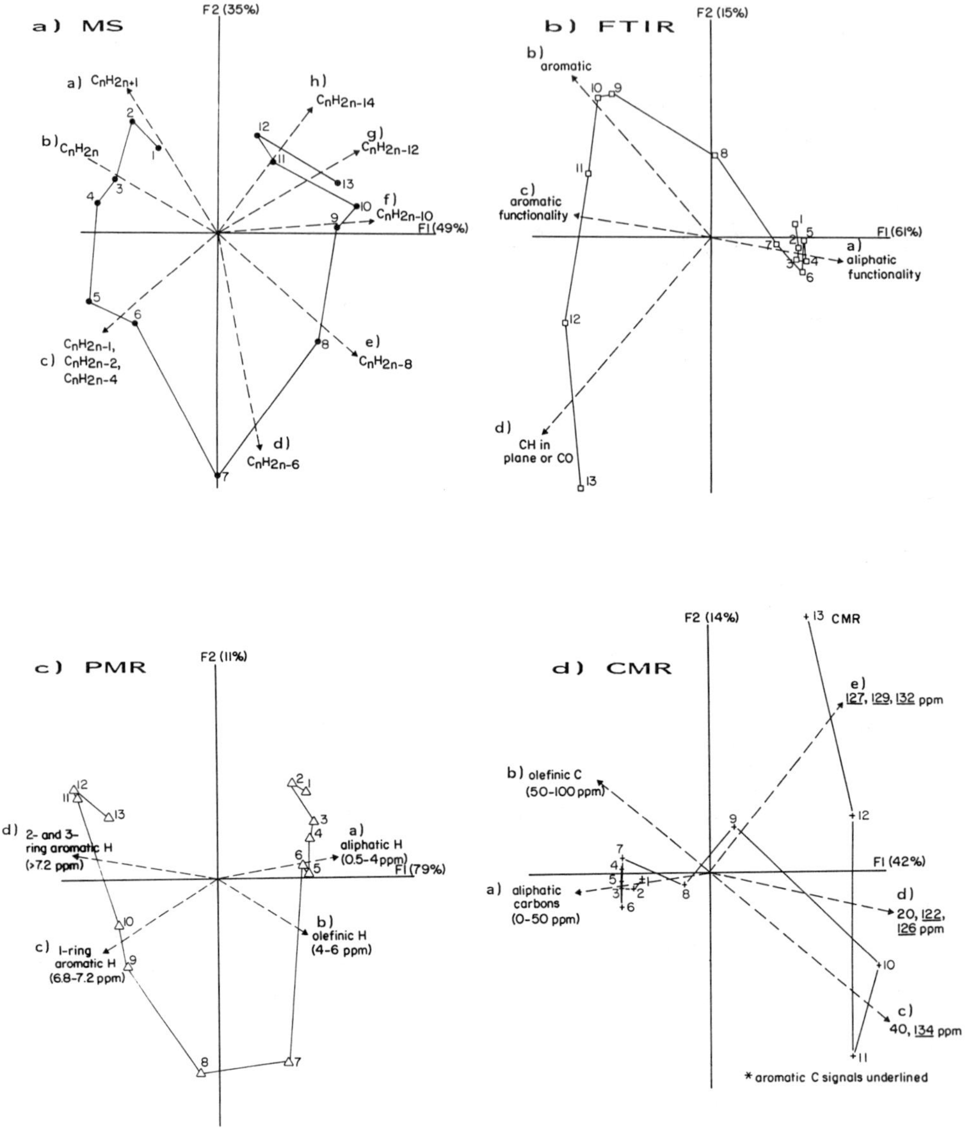

Figure 4. Factor score plots in FI/FII spaces of all four individual data sets.

TABLE II. Details of Individual Factor Analyses

Data	No. Variable	Type	% variance/-variable	Factors with Eigen-values >1.0
Mass Spectrometry	137*	m/z, intensity	0.7	9
Infrared Spectroscopy	28	cm-1, absorbance peak height	3.6	6
Proton NMR	8	ppm regions, integrated area	12.5	1
Carbon 13 NMR	23	ppm, intensity	4.3	6

* 273 nonzero variables were available, the 137 chosen represented those with highest variance.

then move on to later eluting samples. Figure 4a shows a F1 vs. F2 score plot from the MS data. Investigation of the correlated mass peaks loading on factors 1 and 2 (Figure 4a) by means of the variance diagram method revealed 8 components. In Figure 4a components (a) (130°) represents the ion series $C_nH_{2n+1}^{+}$., whereas component (b) (160°) shows $C_nH_{2n}^{+}$ ions from monocyclics or alkenes. A large component (c) (190-240°) contains $C_nH_{2n-1}^{+}$ ions (190°), $C_nH_{2n-2}^{+}$. ion (220°) as well as fragment ions at m/z 149, 163, 177 and 191 characteristic of terpenoid resins or other $C_nH_{2n-4}^{+}$ compounds (240°).

Aromatic compounds such as short (C_n, n=1, 2, 3) alkyl substituted benzenes occur at component (d) (280°C), with longer chain (C_n, n=5, 6, 7) benzenes + tetralins at 320°; component (e). Component (f) at 0° is thought to represent $C_nH_{2n-10}^{+}$ series. Naphthalenes are found at component (g) (30°) and acenaphthene/biphenyl ions are present at component (h) (50°, Figure 5a). Note that the scores in this factor space roughly describe a circle, with the exception of fraction 13, which is found near sample 10. Factor 3 (not shown) distinguishes sample 13 from the others with a component axis containing anthracene/phenanthrene moieties as well as an ion series at m/z 180, 194, 208. Chemical identification of the various ion series is tentative.

Factor analysis was performed on the IR spectra of subfractions 1 to 13 using 28 nonzero wavenumber variables. Five of the orginal 33 variables were unique to spectrum 15 and were not used in the factor analysis of samples 1-13. Figure 4b shows the factor score plots of the IR data on subfractions 1-13 in the F1 vs. F2 factor space. Samples 1-7 are very close together, implying that infrared spectroscopy does not detect much difference between these dominantly aliphatic mixutres in this space. Analysis of the underlying correlation between variables by means of the variance diagram method showed that component (a) (350°) represents methyl and methylene absorptions such as 2870, 2850, 2920, 1460 and 720 cm^{-1}. Component axes (b) (120°) with peak 1516 cm^{-1} and (c) (160°) with 3050, 3015 and 1600 cm^{-1} represented aromatic absorptions. A component

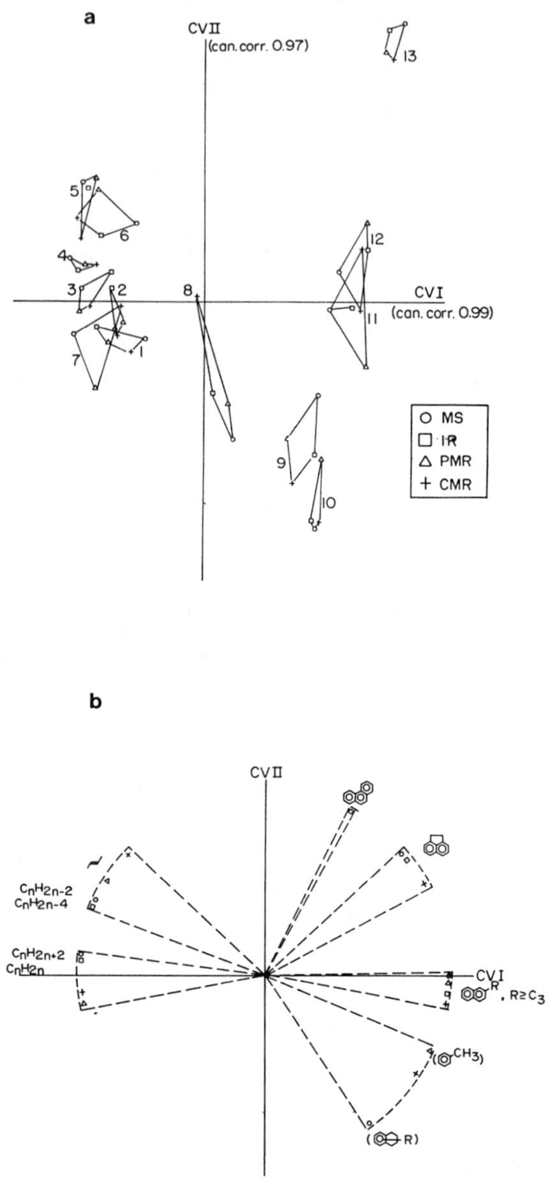

Figure 5. a) Combined score plots of integrated spectroscopic data in "common" CV1/CV2 space.

b) Combined loading plot of integrated spectroscopic data in CV1/CV2 space showing common chemical components.

axis (d) (240°), which points to subfractions 12 and 13, represents peaks 750, 1440, 1030 and 1180 cm^{-1}. Initially, we tentatively assigned 1030 and 1180 cm^{-1} as C-O stretches, but further examination of infrared spectra of aromatic standards showed that these are probably CH in-plane bends, e.g., 1030 cm^{-1} (benzene). An interesting feature of the IR data is the peak at 2868 cm^{-1} which correlates with the aromatic component at 1600, although it is believed to represent a combination of methyl and methylene stretches. Painter et al. (8) also found this behavior in IR spectra of coal macerals. Our data strongly suggest a reinterpretation of this peak assignment.

Several peaks in the F1+ direction of Figure 4b (IR) with low loadings on Factor 1 can be assigned as olefin CH out-of-plane bends. On Factor 3, these peaks showed that fractions 4-6 contained compounds with olefin moieties. These turned out to be important in the combined (canonical variate) space and will be discussed later.

The factor score plot F1 vs. F2 (91% variance) of the PMR data in Figure 4c shows a general distribution of samples forming a circle. The variance diagram of F1 vs. F2 from the proton NMR data shows that the positive F1 axis contains methyl and methylene groups attached to aliphatic (sp^3 hybridized) carbon groups (a) (10°) and olefinic protons (b) (330°). The F1 axis contains aromatic protons, split into two groups. The component axis at 200° (c) represents methyl-substituted benzenes (α CH_3 + 1 ring aromatic), oriented toward fractions 9 and 10. The 170° rotation (d) contains 2-ring and 3-ring aromatics and longer chain aromatic substituents (αCH_2) oriented toward fractions 11-13.

Factor analysis of the CMR data is in Figure 4d. The score plot of F1 vs. F2 (56% of the variance) shows that samples 1-7 appear to be similar in this dimension, oriented along the negative side of F1. Variables in this direction include aliphatic peaks such as at 23, 30, 32 and 38 ppm and (with weak loadings), at 97, 114 and 140 ppm, olefinic carbons. Fraction 8 is somewhat removed from fractions 1-7 but still on F1-, and therefore contains predominantly aliphatic carbons. Fractions 9-13 are widely spread on factor F1+. A component axis at 350° (between samples 10 and 12) represents peaks at 20, 122, 126, 131 and 135 ppm. These aromatic carbons (100-150 ppm) have a coupling constant of about 4 ppm. Assignment of an aromatic structure will be deferred to the canonical correlation discussion. Fractions 10 and 11 have an associated component which includes the peaks at 40 and 134 ppm. Fractions 9 and 13 have an associated component axis with the peaks at 127, 129 and 132 ppm (about 2 ppm apart). All peaks on F1+ (except at 20 and 40 ppm) are aromatic carbons. The 20 and 40 ppm peaks are sp^3 hybridized carbon substituents, probably attached to aromatic rings.

Canonical Correlation Analysis. Our method of canonical variate analysis uses factor scores as input. Since using the minimum number of data reduction factors (eigenvalue > 1.0, 1 in the PMR data) constrained the CV space to too low a dimensionality to take full advantage of all the data, the first 6 factors from each data set were used. It can be seen from the last column of Table II that the PMR data input contained more factors than were significant, and that the MS data used somewhat too few factors (6 out of 9) based on the eigenvalue criterion. Table III shows the canonical variate

TABLE III. TABLE OF CV DIAGNOSTICS

Data File	% Var. in 1st 6 Factors	% Var. Retained in all C.V.'s	% Var. Retained in C.V.'s (corr. > .90)
MS	96	89	71
IR	92	84	81
PMR	100	94	87
CMR	94	81	69

The variance that was not retained in the CV's represents non-correlating variance and was not treated further.

diagnostics. More than 90% of the original factor variance were used from each data set.

Canonical correlation of the factor analysis results from the MS, IR, PMR and CMR data using 6 factors from each data set gave four canonical variate functions with correlation coefficient greater than 0.90. Figure 5 is a score plot of CV1 vs. CV2 for the four data sets. The scores from each fraction analyzed by the four methods are connected by lines. A small polyhedron implies that the methods describe the sample in a similar way in this dimension. The later eluting samples (9-13) appear to group into clusters that are widely separated from one another (e.g., 9 and 10, 11 and 12, 13) whereas early eluting samples (1-7) are close together in this space. Figure 5b shows a consensus picture of the component directions from each method found in this CV space. Correlated variables consistent with an interpretation of aliphatic compounds are clustered around CV1-, near fractions 1-4. Only the MS found an axis of C_nH_{2n} (alkene, cycloparaffin) components in this direction. Fraction 7 appears also in this direction. A component axis of alkyl benzenes (m/z 92, 106...) from the MS data loads weakly in this CV space. From the original factor analyses it can be seen that the mass spectral and PMR data clearly differentiated fraction 7 from the other fractions in the F1 + F2 factor spaces shown. The PMR data showed no unique component associated with fraction 7. This says that the mass spectral picture of fraction 7 is in a sense unique, since confirmation by other data sets is not apparent. A component axis corresponding to olefinic variables (IR, PMR, CMR) appears at 150°, in the direction of fractions 5 and 6. The mass spectral data shows ion series with 2 and 3 units of unsaturation, one or more of these apparently being a double bond. The positive half of CVI reveals three components, each one consistent with an assignment of aromaticity. The PMR and IR (CH in-plane bend modes) show increasing fused ring aromaticity in the ccw direction (300° to 50°). The mass spectral data identified the component at 300° with indane/tetralin, the 0° component as $>C_3$ alkyl substituted naphthalenes and the 50° component as acenaphthene/- biphenyls. The CMR data showed a variable 142 ppm at 300° consistent with an assignment of an indane bridgehead carbon (literature value 143.9 in $CHCl_3$). A component axis at 80° (mass spectral data only) near sample 13 showed peaks characteristic of alkyl anthracene/ phenanthrenes. The original CMR factor analysis showed correlated peaks near sample 13 which fit an interpretation of phenanthrenes better than anthracenes.

The CV3 and CV4 dimensions also showed correlation coefficients of >0.9. The variance in these dimensions ranged from 7% (PMR) to 21% (MS). Since one variable in the PMR data accounts for 12.5% of the variance, no variables were found to load above 0.4 in this dimension, and PMR data will not be further discussed.

Both the CMR and the MS data sets presented a similar space on CV3 and and 4. Samples 7 and 13 were grouped together along CV3+. Variables from the MS with appreciable loadings in this dimension were m/z 92, 106, 134... ("alkyl benzenes") and m/z 146, 160, 174 ("indane/tetralins"). The alkyl benzene components appeared with weak loadings in the CV1 + CV2 space near sample 7 only. The infrared data showed two variables with low loadings that were in common between samples 7 and 13, 2950 and 1450 cm^{-1}. These variables are small portions of the broadest IR peaks and their significance is not clear. Variables from the CMR data, 14 and 34 ppm, implied similar alkyl functionalization, e.g., ethyl. From evidence in the factor analysis of ms data from samples 1 to 15 (not presented here) of fragmentation/pyrolysis dominant in samples 14 and 15 and to a lesser extent 13, and evaporation dominating earlier samples, it is likely that the tetralin/indane ms ion series in samples 7 and 13 arose from different sources. GC/MS data, not completely interpreted yet, determined that sample 13 was dominated by 2 and 3 ring polycyclic aromatics, e.g., higher substituted naphthalenes, biphenyls and larger, whereas sample 7 had benzene and toluene alkyl substituted compounds.

Samples 8 and 9 had associated components from both MS and CMR in CV3 and CV4. Fragment ions m/z 145, 159 and 173 loaded moderately in this space, as did the CMR variable 142 ppm. These are not inconsistent with the interpretation of indane/tetralin found in the CV1 + CV2 space. Sample 10 had associated variables of m/z 142 and 156, C_1 and C_2 alkyl substituted naphthalenes and an "acenaphthene/biphenyl" (m/z 144, 158...) component. The lone variable from CMR, at 131 ppm, might arise from either of these components (2 methyl naphthalene 131.7 ppm or acenaphthene 131.5 ppm $CDCl_3$) considering that we chose variables as virtually $\pm$ 1 ppm.

The only variable loading strongly on CV3 or CV4 from the IR data is 791 cm^{-1}, a CH out-of-plane bend, pointing directly opposite samples 7 and 13, and probably showing some relationship in aromatic substitution patterns.

Conclusions

A consensus interpretation of the major components contained in LC eluted samples from a pyrolysis oil was attempted using MS, IR, PMR and CMR data.

Valuable information was gained by correlating the four analytical techniques. For example, mass spectral peaks of samples containing 2 and 3 units of unsaturation, such as samples 4-6, were shown by PMR, IR and CMR to contain double bonds, whereas mass spectral peaks corresponding to molecules containing one unit of unsaturation such as samples 1 and 2, were apparently cyclic, since only aliphatic functional groups were found by the other methods.

Alkyl benzene components in sample 7 were not represented in the canonical variate space. Mass spectral peaks corresponding to indane/tetralins in samples 8 and 9 were found by CMR to contain indanes, and anthracene/phenanthrene mass spectral molecular ions were found to be predominantly phenanthrene.

Acknowledgments

The expert assistance of Dr. Jim Liang, Mary West and Alireza Pezeshk in obtaining the analytical data, and of Dr. Liang and Dr. Charles Mayne in interpreting the results is gratefully acknowledged. Drs. George R. Hill and Larry L. Anderson are thanked for their stimulating support of this research. The work reported here is part of the research carried out at the University of Utah, under the auspices of the Consortium for Fossil Fuel Liquefaction Science funded under DOE contract RF-4-21033-86-24.

Literature Cited

1. Thimsen, David; et al., Fixed Bed Gasification Research Using U.S. Coals; Vol. 1 Program & Facility Description, USBM HO222001.
2. McClennen, W. H.; Meuzelaar, H. L. C.; Metcalf, G.S.; Hill, G. R. Fuel 1983, 62, 1422.
3. Meuzelaar, H. L. C.; Hoesterey, B. L.; McClennen, W. H.; Hill, G.R. Proc. 10th Ann. EPRI Contractors Conference on Liquid & Solid Fuels, 1985.
4. Tandem Mass Spectrometric Analysis (MS/MS) of Jet Fuels, Part I and II, Air Force Aero Propulsion Laboratory, Wright-Patterson Air Force Base, Ohio, AFWAL-TR-85-2047, 1985.
5. Windig, W.; Meuzelaar, H. L. C. Proc. 34th ASMS Conf. on Mass Spec. All. Topics., 1986, p 64.
6. Windig, W.; Meuzelaar, H. L. C. Anal. Chem. 1984, 56, 2297-2303.
7. Crelling, J. C.; Pugmire, R. J.; Meuzelaar, H. L. C.; McClennen, W. H.; Karas, J. "The Structure and Petrology of Resinite from the Hiawatha "B" Coal Seam." submitted to Energy & Fuel, 1988.
8. Kuehn, D. W.; Davis, A.; Painter, P. C. In ACS Symp. Series 252, Chapter 7, 1984; p 100.

RECEIVED April 14, 1988

Chapter 19

Chemical Characterization of Wood Pyrolysis Oils Obtained in a Vacuum-Pyrolysis Multiple-Hearth Reactor

H. Pakdel and Christian Roy

Chemical Engineering Department, Université Laval, Pavillion Adrien–Pouliot, Sainte–Foy, Quebec F1K 7P4, Canada

A multiple hearth reactor has been used to produce high yield of pyrolysis oil from aspen poplar. The Process Development Unit (P.D.U.) has the capability of achieving a fair fractionation of wood oils by using six heat exchangers (Primary Condensing Unit, P.C.U.) and a series of cooling trap receivers (Secondary Condensing Unit, S.C.U.) at the outlets of the reactor. While the weight average molecular weights ($\overline{M_w}$) of the recovered compounds in the P.C.U. were 342, 528, 572, 393, 233 and 123 from top to bottom of the reactor, the low molecular weight compounds with $\overline{M_w}$ ≈ 100 or below were recovered in the S.C.U., which contained at least 90% of the total water. Silica-gel column chromatography enabled us to fractionate the oil from P.D.U. into fourteen fractions. Aromatic hydrocarbons were collected in Fraction 1 (F1) and F2 followed by elution of moderately polar compounds in F3 to F11 with about 23-35% of P.C.U. oil which can be fully characterized. 8.1% sugar in P.C.U., mainly glucose, was found in F13. Usefulness of ^{1}H-FTNMR and infrared spectroscopy was shown for preliminary characterization of the F12, F13 and F14. Overall 27.89% of the P.C.U. including water and low molecular weight carboxylic acids have been measured and identified so far.

The identification and extraction of valuable chemicals from wood-derived oils is a very important goal for the biomass thermochemical conversion industry (1-2). Pyrolysis oils have been extensively studied and extensive number of compounds have been identified (3-4). However to our knowledge there are only two general methods which have been reported for the fractionation of pyrolysis oils

0097–6156/88/0376–0203$06.00/0

into chemical groups: the solvent extraction method (5) and the adsorption-chromatography method (6). The former technique is rather tedious and quite often the phase separation is difficult due to the emulsion formation. The yield of extraction strongly depends on the solvent volume and extraction repetition number. The adsorption-chromatography method was used for this investigation with further modifications which will be discussed in this paper.

Extensive works conducted by different authors utilizing GC and GC/MS sometimes lead to different results which indicate the difficulties of carrying out accurate detailed analysis of the chemical constituents of pyrolysis oils. Examples of incomplete or even contradictory results can be found in the literature (3-4) and this paper in the analysis of vacuum pyrolysis oils. Other researchers have studied the functional group distribution in pyrolysis oil (8). Although those techniques are long and tedious, they will lead to useful information about wood oil chemistry.

The majority of compounds found in pyrolysis oils are oxygenated with rather similar polarity. Therefore their gas chromatograms, in general, suffer from low resolution and consequently the quantitative analysis will be less accurate. Although this problem may be partially obviated by choosing narrow bore and long capillary columns, but they are very expensive and not very practical. Direct injection of a complex mixture into the gas chromatograph on the other hand tend to deteriorate the column by building-up of non volatile matter in the column inlet leading to gradual decomposition of the column stationary phase. Generally gas chromatography has a limited application and is not meant to be used for very complex and less volatile mixtures. GC/MS is a much more powerful analytical tool but would not be available in a great majority of cases. Besides it is quite costly and requires skilled operators for the interpretation of the results. Therefore development of new methods of separation and fractionation in particular are needed. The primary objective of this work is to develop a separation and fractionation method for better and detailed analysis of pyrolysis oils and to demonstrate fair fractionating capability of the multiple hearth vacuum pyrolysis system. This will eventually enable us to make correlations between the oil properties and pyrolysis operation conditions. Full characterization of the oils will also shed some light on the possible pyrolysis reaction mechanism and upgrading of oils. The secondary objective is to develop methods for the extraction of valuable chemicals such as specialty and rare chemicals which are in increasing demand (9).

Experimental

The wood oil samples which have been characterized in this work have been obtained from pyrolysis of *Populus deltoides* (clone D-38) with no bark in a multiple-hearth vacuum pyrolysis reactor. The Process Development Unit, Fig. 1, (P.D.U.) has been described in detail by one of the co-authors in another paper (1).

The P.D.U. was tested for the production of high yields of oils from wood chips. One objective was to separate the bulk of the aqueous phase from the organic liquid phase by means of

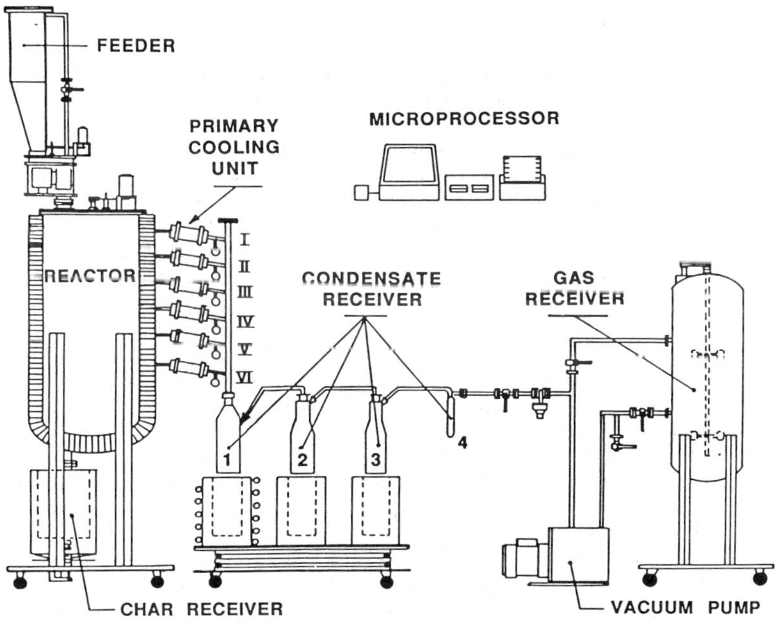

Figure 1. Schematic view of vacuum pyrolysis Process Development Unit (P.D.U.).

fractionation of the oil directly at the outlet of the reactor. This was achieved in the following way. The organic vapor product was removed from the reactor through six outlet manifolds which corresponded to the six heating plates of the reactor, the hearths. The vapors were condensed in two condensing units named Primary and Secondary Condensing Units (P.C.U., S.C.U.). P.C.U. consisted of six heat exchangers in parallel (H-I to H-VI) and S.C.U. consisted of four receivers placed in series (C1 to C4). The pyrolysis oils which were obtained in both condensing units were subjected to sequential elution solvent chromatography. The organic fractions were then analysed as described below. Relationship of all samples and designations are described in Table I.

One gram of the oil sample immediate after pyrolysis was transferred into a glass column with 16 mm i.d. packed with 12.5 g of 60-120 mesh silica-gel in petroleum ether (30-60°C b.p.). Fourteen fractions were collected using different solvents as depicted in Table II. All the solvents were distilled before use and the silica-gel was washed with dichloromethane and dried in air. The oil fractions were dried by rotary evaporator without heat. All the yields are shown in Table II. ^{1}H-FTNMR spectra of 5% solution in DMSO were recorded on XL 200 Varian instrument. Gas chromatographic analyses were performed on a 6000 Varian gas chromatograph with flame ionization detector with two injectors (on column and split). The capillary columns were J & W fused silica: DB5, 30 m X 0.25 mm i.d. and DB1, 30 m X 0.32 mm i.d. The carrier gas was H_e and N_2 as make up gas. The oven temperature was maintained at 50°C for 2 min then programmed to 150°C and 290°C at rates of 4 and 10°C min^{-1} respectively. Water's 840 data and chromatography control station with digital professional 350 computer and LA50 recorder were used as data processor. Various standard mixtures were prepared with the available compounds. Their relative response factors to benzophenone were measured. Silica-gel eluates were added an accurate quantity of benzophenone as internal standard. Their gas chromatograms were compared with the standard mixtures for peaks identification and followed by integrations for their quantifications.

Hydrolysis followed by sugar analysis was carried out according to the silylation technique. The procedure can be found elsewhere (10).

Gel permeation chromatographic (G.P.C.) analysis of the six oils from P.C.U. and one oil from S.C.U. (C1) were performed on ALC/GPC-201 Water's Associate liquid chromatograph equipped with a model R-401 refractometer. Four 30 cm X 7.8 mm i.d. columns packed with 100, 500, 10^3 and 10^4 μ styragel were used in series. The samples were prepared in THF (5%) and 15 μl was injected and eluted with THF. The following standards were used to calibrate the system: polystyrene (M_w = 4000; 2000; 800 and 600), polyethylene glycol (M_w = 450, 300, 200), guaiacol, syringaldehyde and vanilline.

Results and Discussions

Pyrolysis oil, water, char and gas are the wood pyrolysis products. Depending on the liquefaction process, pyrolysis oils composition change significantly. Lack of a standard pyrolysis oil characterization technique has initiated us at the first stage to develop a

Table I. Relationship of the samples studied in the text (See ref. 1 for more details)

Experimental Technique	Nature of Sample	Designation	Remarks
Vacuum pyrolysis	Oil phase	P.C.U.	Primary Condensing Unit
Vacuum pyrolysis	Aqueous phase	S.C.U	Secondary Condensing Unit
Vacuum pyrolysis	Pyrolysis oil	H-I to H-VI	Obtained from the six hearths
SESC, GPC, GC, NMR	PCU oil	F1 to F14	Fourteen fractions of H-I to H-VI oils by SESC
GPC	SCU oil	C1	Receiver 1
IR	Solid residue	hearth-I to hearth-VI	See Fig. 2

SESC = Sequential elution solid chromatography; GPC = Gel permeation chromatography; GC = Gas chromatography; NMR = Nuclear magnetic resonance; IR = Infrared

Table II. Primary Condensing Unit Pyrolytic Oils and the Yield of Various Classes of Compounds Obtained by Silica-Gel Column Chromatographic Analysis (a)

PYROLYSIS OIL SAMPLES FROM	FRACTION 1	FRACTION 2	FRACTION 3-11	FRACTION 12	FRACTION 13	FRACTION 14	TOTAL
H-I	0.11	0.04	26.83	31.06	27.16	1.67	86.87
H-II	0.64	0.23	26.37	32.52	26.28	1.65	87.69
H-III	0.85	0.21	22.99	36.80	26.93	2.34	90.12
H-IV	0.33	0.16	22.32	35.60	30.70	2.24	91.35
H-V	0.45	0.10	24.24	35.35	29.48	2.79	92.41
H-VI	1.44	0.12	33.18	29.09	28.26	3.28	95.37

(a) See the text and table I for description of the classes.
All figures are expressed in weight percentage (as-received oil basis).

Fraction 1: 128 ml with petroleum ether.
Fraction 2-11: 128 ml each with CH_2Cl_2 / Petroleum ether mixture, from 10 to 100% CH_2Cl_2 (10% increments) for F2 to F11 respectively.
Fraction 12: 128 ml with ether.
Fraction 13: 128 ml with water.
Fraction 14: 60 ml with 10% formic acid in methanol.

sub-fractionation technique followed by quantitative gas chromatographic analysis. Interestingly, vacuum pyrolysis yields relatively low percentage of high molecular weight compounds. The high molecular weight compounds with $\overline{M}_w$ > 300 represent approximately 25% of the total vacuum pyrolysis oils. It has been suggested that the high molecular weight less volatile and presumably more polar compounds are produced by incomplete thermal degradation of lignocellulosic materials (8). The following characterization technique enabled us to find app. 5% oligosaccharides with $\overline{M}_w$ > 300. Due to GC limitation, the rest of the high molecular weight compounds can be characterized utilizing GPC, MS-MS, HPLC or combination of those. The work in this line is presently under investigation in our laboratory.

The six oils from the P.C.U. (H-I to H-VI) and an oil sample from the S.C.U. (C1) were analysed by GPC for their molecular weight range distribution. The weight average molecular weights for the H-I to VI were: 342, 528, 572, 393, 233 and 123 respectively. The test for C1 showed weight average molecular weight of about 100. The molecular weight distribution of C1 was as below:

$\overline{M}_w$<100 (40%); 100<$\overline{M}_w$<200 (46.5%); 200<$\overline{M}_w$<300 (10.5%) and 300<$\overline{M}_w$<500 (3%).

The pyrolytic oil in the S.C.U. is about 38.8% of the total oil which consists of 17.9% water and 5.6% carboxylic acids (Pakdel, H.; Roy, C. Biomass, in press). Due to its low average molecular weight however, it is expected that the majority of the S.C.U. oil can be analysed if the interference due to the water can be eliminated.

G.P.C. analysis was carried out for a series of pyrolytic oils obtained in a batch reactor operated at various temperatures similar to the P.D.U. hearth temperatures (11) and the results showed a rather similar weight average molecular weight which indicate a fair selective separation of wood oil constituents at various temperatures in the P.D.U.

The results of silica-gel column fractionation of the H-I to VI oils are shown in Table II. A fairly good reproducibility was obtained for F1 (±0.01%). F2 suffered sometimes with contamination from F3 nevertheless the reproducibility was ± 0.1% obtained with three repetitive fractionations. F3-F12 showed usually about 0-5% loss during the solvent evaporation. F13 and F14 were satisfactory with ± 2 and ± 0.4% reproducibilities respectively. All fractions are designated as F1 to F14 in this part and will be studied separately. F1 to F12 of all condensers were liquids with some differences in their colours and odors. The F13 and F14 were found to be very viscous and nearly solid. They became partially insoluble by redissolving in water and methanol. F1 to F2 were not very soluble in methanol indicating their hydrocarbon nature. F4,5,6 showed some methanol insoluble matter which was identified as high molecular weight carboxylic acids in the range of C_{18} to C_{30} with a maximum at C_{24} (Pakdel, H.; Roy, C. Biomass, in press). These were separated and purified by crystallization in ethanol. Although the high molecular weight carboxylic acids comprise very little percentage of the oil (≈ 0.2%) but their finding is very interesting as only even carbon number acids up to C_{24} have been found earlier in hardwood pyroligneous tars (12). The fractions 7 to 12 were slightly insoluble in dichloromethane and fully soluble in methanol. A preliminary compositional analysis of each fraction was made and more work is in progress. The results are the followings:

Fraction 1. This fraction mainly contains hydrocarbons with an odor typical of a fossil fuel mono-aromatic hydrocarbons. Proton nuclear magnetic resonance spectrum (^{1}H-NMR) was recorded for a sample of H-I.F1 (Hearth-I, Fraction 1) as an example and the results are shown in Table IV. The NMR spectrum showed long alkyl and alkenyl side chains on benzene rings. Recently a series of short side chain alkyl and alkenyl benzenes (up to $C_{11}H_{14}$ & $C_{11}H_{16}$) were reported in wood pyrolysis oil (6). A significant difference was observed in their gas chromatograms of H-I to H-VI. It is interesting to see in Table II that there is an increasing trend in quantity of the hydrocarbons up to the maximum in H-III which falls down in H-IV and again reachs to the maximum at 1.44 in H-VI. Whether further increase of the hearth temperature will increase and produce any more materials has not been tested. However we have made an infrared spectroscopic study of the solid residues which were left behind accidentally in each hearth after the pyrolysis (13). Their infrared spectra are shown in Fig. 2 . Despite of the fact that a bulk of the pyrolytic oils were produced in H-I and II but there are very small changes in hearth-I and hearth-II spectra (see Fig. 2). However significant differences are observed in hearth-IV to VI. The solid residue in hearth-VI which was obtained at 448°C shows only minor bands due to the remaining organic matter mainly lignin or recondensed materials (14). These materials were measured to be approximately 20% of the total residue.

Fraction 2. This fraction was in low abundance in all of the condensers and its quantity was significantly dependent on the hearth temperature. Similar to F1, F2 has also aromatic nature but with slightly high polarity. Interestingly H-V and VI were found to have completely different composition from the rest. Their GC appeared to have moderately resolved peaks.

Fractions 3-11. These fractions have a particular value as their constituents are valuable and many of them can be characterized by gas-chromatography. F3-11 contributed between 22-33% of the hearth oils and 40.5% of the total P.C.U. oil. The total yield of these fractions therefore can be used as quality index to compare different oils. This paper is a small contribution to characterization of these fractions and further work is in progress. Many new compounds have been identified and will be reported elsewhere.

Table III lists some of the compounds which have been identified in these fractions which are mainly mono phenolic types and oxygenated heterocyclic compounds. All the chromatograms of F3 to F10 showed well resolved peaks but F11 suffered from peak broadening which is more likely due to the contamination from F12. Examples of a few chromatograms are presented in Fig. 3 and are compared with a chromatogram of the total oil from H-VI (Fig. 3a). From Fig. 3a, we could only identify a few compounds and the rest

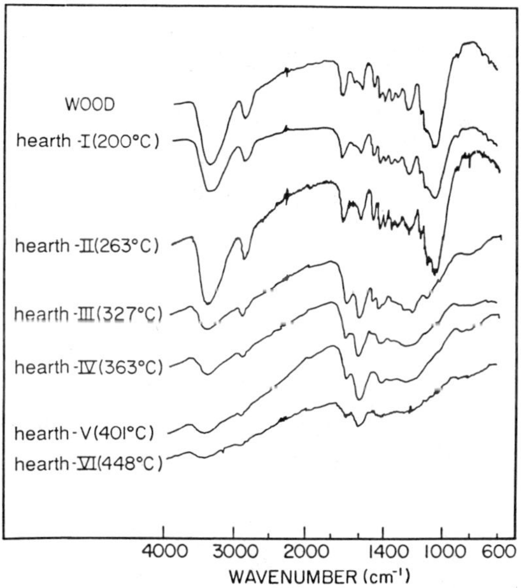

Figure 2. Infrared spectra of wood and solid residues from six hearths (values in parenthesis are hearths temperatures).

Table III. Preliminary Analytical Data Obtained from Primary Condensing Unit (see table I for description of samples)

GC #	COMPOUND	H-I	H-II	H-III	H-IV	H-V	H-VI
---	Water	46.54 (8.24)	50.58 (6.8)	59.45 (6.24)	83.63 (6.54)	85.03 (6.60)	28.52 (5.12)
---	Formic acid	13.89 (2.46)	27.37 (3.68)	38.77 (4.07)	59.33 (4.64)	72.02 (5.59)	11.36 (2.04)
---	Acetic acid	40.50 (7.17)	53.83 (7.91)	62.40 (6.55)	40.15 (3.14)	3.18 (2.42)	9.64 (1.73)
1	Acetol	---	---	---	---	31.82 (2.47)	7.24 (1.3)
2	Furfural	0.28 (0.05)	0.37 (0.38)	0.44 (0.05)	0.89 (0.07)	0.52 (0.04)	0.50 (0.09)
3	Furfuryl alcohol	1.58 (0.28)	2.75 (0.37)	14.3 (1.5)	} 0.52 (0.51)*	4.64 (0.36)	2.45 (0.44)
4	Angelicalactone	0.68 (0.12)	2.90 (0.39)	0.48 (0.05)		---	---
5	Cyclohexanol	---	---	---	0.02 (0.002)	---	---
6	γ-Butyrolactone	---	---	---	---	---	4.79 (0.86)
9	Me-Cyclopenten-1-one	---	---	0.51 (0.05)	---	---	---
10	5-Me-furfural	---	---	1.05 (0.11)	1.92 (0.15)	---	---
11	Phenol	5.2 (0.92)	5.50 (0.74)	9.53 (1.0)	1.41 (0.16)	---	1.28 (0.23)
12	Me-Cyclopentene-ol-one	0.14 (0.02)	0.82 (0.11)	1.81 (0.19)	4.35 (0.34)	1.54 (0.12)	2.73 (0.49)
13	m-Cresol	---	0.26 (0.35)	---	---	---	---
13a	p-Cresol	---	---	---	---	---	0.56 (0.1)
14	Guaiacol	0.02 (0.003)	1.12 (0.15)	0.95 (0.1)	1.79 (0.14)	0.52 (0.04)	4.07 (0.73)
16	3-Hydroxy-2-Me-4-pyrone	0.51 (0.09)	0.59 (0.08)	---	---	0.64 (0.05)	
18	Dimethyl 2,3, phenol	---	---	---	---	---	0.45 (0.08)
19	Dimethyl 2,5, phenol	---	---	---	---	---	0.67 (0.12)
19a	n-propyl-phenol	---	---	---	---	---	8.74 (1.57)

20	Cathecol	1.13 (0.2)	2.01 (0.27)	2.48 (0.26)	4.6 (0.25)	7.73 (0.6)	8.74 (1.57)
22	Hydroquinone	---	---	---	---	0.52 (0.04)	3.34 (0.6)
23	Resorcinol	---	---	---	---	0.52 (0.04)	
24	Syringol	1.81 (0.69)	2.37 (0.5)	10.86 (1.14)	10.23 (0.8)	5.02 (0.39)	8.35 (1.5)
25	Eugenol	3.11 (0.02)	0.45 (0.06)	0.28 (0.03)	0.26 (0.02)	0.26 (0.02)	0.45 (0.08)
27	Isoeugenol	3.22 (0.57)	3.05 (0.41)	6.29 (0.66)	5.37 (0.42)	6.57 (0.51)	12.48 (2.24)
28	Vanilline	---	---	---	---	0.13 (0.01)	---
	Sugars**	17.73 (3.14)	24.25 (3.26)	64.88 (6.81)	87.08 (6.81)	183.33 (14.23)	57.20 (14.05)
	Total	135.44 (23.98)	180.85 (24.43)	273.48 (26.81)	306.76 (23.99)	431.97 (33.53)	173.56 (31.16)

Total in P.C.U. = 1502.06 (27.89%)

* Could not be distinguished
** Total sugars after hydrolysis
All figures in parentheses are expressed in weight percentage (as-received oil basis). Otherwise, numbers are g.

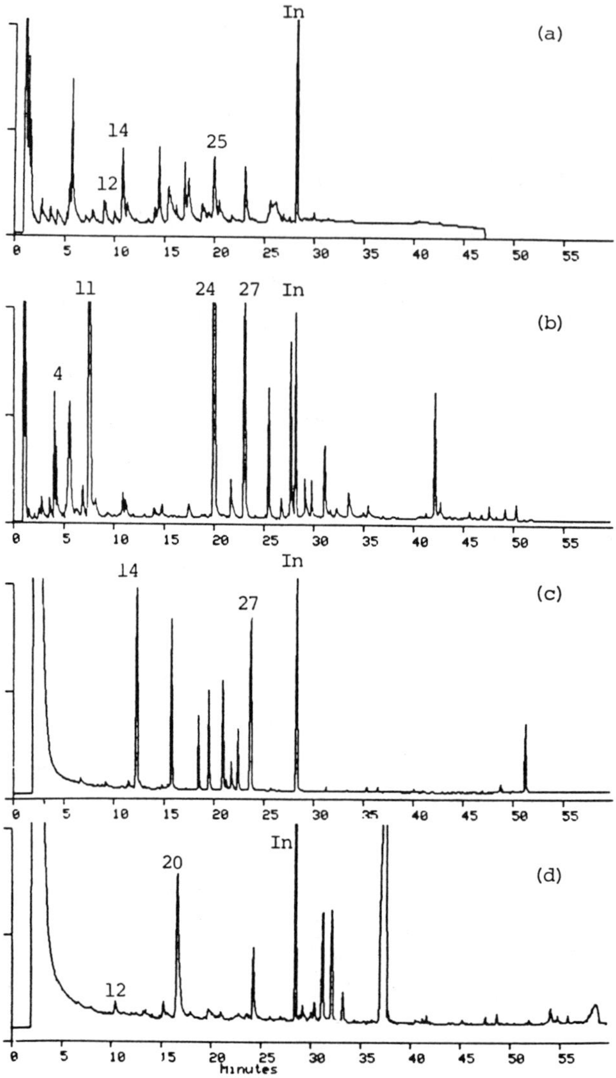

Figure 3. Capillary gas chromatograms of: (a) total H-VI oil; (b) H-I.F1; (c) H-II.F3, and (d) H-III.F9 (see Table I for designation of oils and Table III for identification of peaks), In = Internal standard (bensophenone).

were coeluting and masked with the other non resolved and broad peaks. Consequently, any direct quantitative GC analysis of the total oil will be incomplete and less accurate. The total materials listed in Table III, apart from water, contribute to only 23% of the total dry oil in P.C.U. and they show an interesting distribution of the compounds in the different condensers which can be correlated with their source materials.

Characterization of the low molecular weight carboxylic acids was successfully achieved following benzylation technique developed in our laboratory (Pakdel, H.; Roy, C. Biomass, in press). Formic and acetic acids as decomposition products of cellulose and hemicellulose were the major constituents of pyrolysis oils (15).

Finding of acetol in relatively high abundance in H-V and VI may indicate decomposition of recondensed materials. Since a large number of five carbon atom sugars, the source material for furan derivatives, are associated with hemicellulose especially in hardwood (3) therefore finding of furan derivatives in H-II with maximum abundance gives more support to the selectivity of the separation in the P.D.U. system. Finding of phenol with high proportion in H I to III and some in H-VI is also interesting and supports the suggestion that cellulose is also a source for phenol during wood pyrolysis (3). Since no any significant quantity of phenol was detected in H-V but some in H-VI, the decomposition of recondensed materials in the hearth G is more likely to be the source of phenol in H-VI. n-Propyl-phenol on the other hand has been found in a relatively high percentage in H-VI which indicates lignin as the source of substituted phenols. Although guaiacol, cathecol, eugenol and isoeugenol are spreaded in all condensers with an increasing trend toward H-VI therefore it may be true that their production starts as soon as lignin starts to degrade, presumably at 250°C or even below. Earlier these compounds have been found in the tarry residue from lignin pyrolysis (16). P-Cresol which we have identified in H-VI has been recognized to be a degradation product of the lignin (17) at temperature around 440°C.

Fraction 12. Under the gas chromatographic conditions used, we were unable to observe any well resolved peaks. Typically the chromatogram showed lots of unresolved broad peaks indicating its higher polarity than the previous fractions. Its pH value was also significantly lower than the other fractions. Hydrogen distribution of F12 obtained by ^{1}H-NMR spectrum is shown in Table IV and compared with the total oil. Comparison of these two spectra shows that the F12 has higher hydroxyl group than the initial oil.

Fraction 13. This fraction contains the highest percentage of the high polar compounds, oligosaccharides in particular. Two techniques, ^{1}H-FTNMR spectroscopy and hydrolysis were applied to further study of these fractions which are briefly discussed in this paper as below.

NMR spectra were recorded for the oils from the hearths and their F13s and comparisons were made. Fig. 4a and 4b represent the NMR spectra of H-VI and its F13 as an example. Both spectra were recorded in DMSO. All the samples were freely soluble in DMSO. The peak assignments are shown in Table IV. The hydroxyl group

Table IV. % Hydrogen Distribution in Vacuum Pyrolysis Oils from ^{1}H-NMR

Assignment	Sample[a]				
	H-I	H-I.F1	H-I.F12	H-VI	H-VI.F13
Aliphatic	10.0	-	4.0	12.0	2.0
α $C\underline{H}_2$ to aromatic ring or cyclic $C\underline{H}_2$	31.0	18.0	2.0	38.5	14.0
β $C\underline{H}_2$ to aromatic ring	-	19.5	-	-	
γ $C\underline{H}_3$, γ $C\underline{H}_2$, γ $C\underline{H}$ or further away from ring	-	56.0	-	-	
Olefinic	-	2.0	-	-	
Cyclic hydroxyl and methoxy	30.0	-	47.0	29.5	60.0
Aliphatic hydroxyl, olefinic	7.0	-	17.0	4.0	15.0
Phenolic	4.0	-	3.0	3.0	2.5
Aromatic	16.0	4.5	8.0	12.0	5.5
Aldehydic	2.0	-	1.0	1.0	1.0

a: See Table I for description of samples

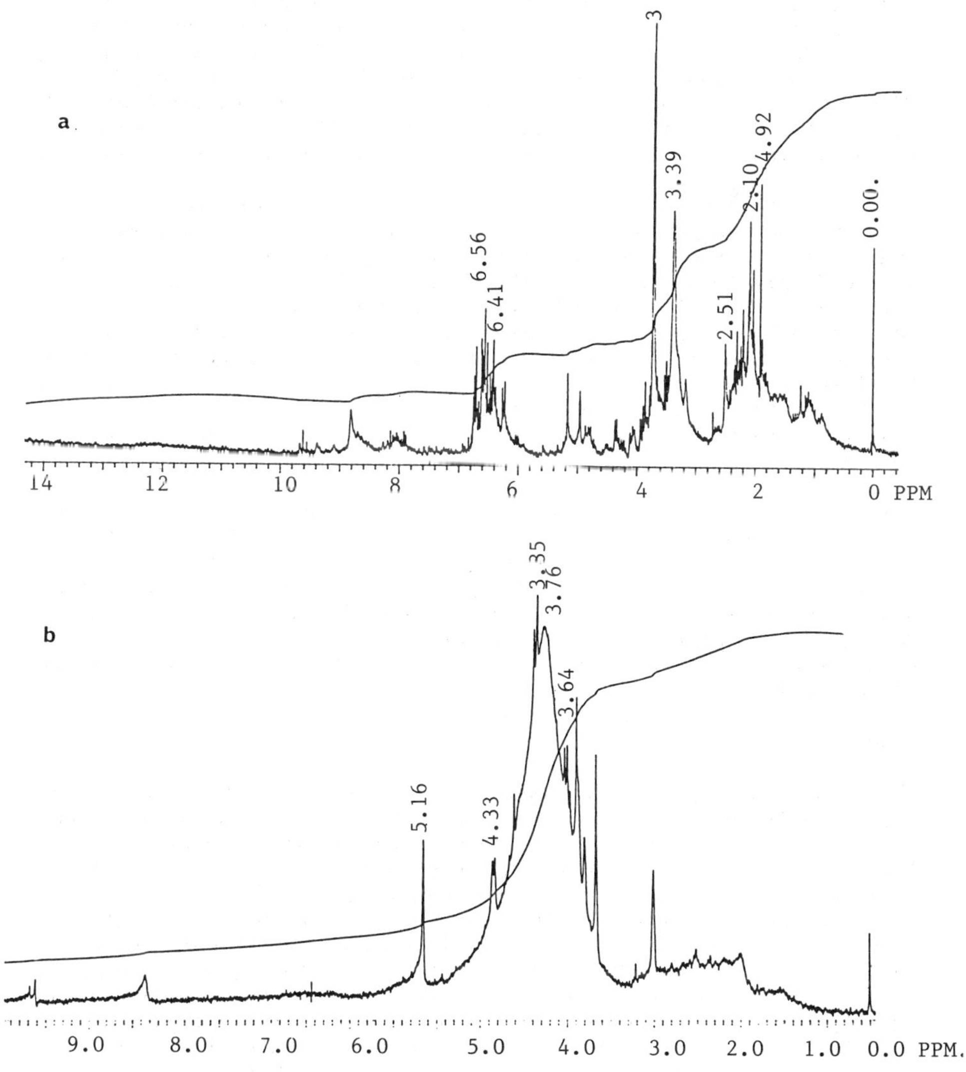

Figure 4. ^{1}H-FTNMR spectra of wood pyrolysis oils: (a) H-VI; (b) H-VI.F13 (see Table 1 for designation of oils).

assignments were also confirmed by addition of a few drops of deuterium oxide before recording the spectrum. Consequently we observed that approximately 50% of the resonance bands in the range of 3-4.2 ppm are due to the sugars hydroxyl groups. Similarly approximately 30% of the resonance bands in the range of 4.2-5.5 ppm are also due to the hydroxyl groups. Their spectra have some similarities with the typical sugar NMR spectrum. From their NMR spectra we observed an increasing trend to the high hydroxyl content from F13 of H-I to VI.

The hydrolysis experiments were carried out to measure the oligosaccharide content of the oils. Primarily analysis was made for the total sugar content of the oils following silylation technique. We found only levoglucosan with 0.3, 0.4, 0.7, 1.7, 3.4, and 4.2% in H-I to H-VI respectively. Hydrolysis of the total oils were carried out and their sugar content were measured. α and β Glucose and xylose were found the most dominant sugars in hydrolysates. Their total sugar content were 3.14, 3.26, 6.81, 6.81, 14.23 and 14.05% for H-I to VI respectively. Similar experiments were carried out on F13 of all hearths. The results indicated a fair sugar recoveries in F13s.

Fraction 14. This fraction comprised a small portion of the oils and a steady increase was observed in their quantities from H-I to VI. They may have presumably polymeric structure. Their infrared spectra showed a weak and broad hydroxyl stretching vibration band and very weak C=O and C-O absorption bands at 1700 and 1600 wavenumbers respectively. C-H stretching and vibration bands were also observed at very low intensities at 2920 and 1520 wavenumbers respectively.

Conclusion

The sequential elution chromatographic technique has been found particularly helpful in separating whole oil produced by vacuum pyrolysis of wood into chemically distinguishable fractions. More than 30% of the P.C.U. oil that eluted first (F1 to F11) could be analysed by GC and GC-MS with less unambiguity.

Now that sensitivity problems are being overcome, infrared spectroscopy and wet chemistry will be overtaken in some instances by high field ^{1}H-NMR, a non destructive technique, and MS-MS spectroscopy and spectrometric techniques for characterization of the other fractions namely F12, 13 and 14 in this investigation.

Fourier transform ^{13}C-NMR which has not been utilized in this investigation is likely to make a large contribution also in the near future to study these fractions.

Since variation in the distribution of the separated fractions and their composition in different hearths were in good agreement with their generally accepted source materials, the P.D.U. is considered to be capable of selective separation of the oils.

Even though they are only present in relatively low concentration, the compounds such as methyl cyclopentene-ol-one, 3-hydroxy-2-methyl-4-pyrone, isoeugnol, for example, can serve to characterize various oils and monitor the pyrolysis liquefaction procedure. They can also be separated and extracted as fine chemicals.

Acknowledgments

This study was supported by the Bioenergy Development Program (Energy, Mines and Resources Canada) and Petro-Sun International, Inc. (Boucherville, Québec).

Literature Cited

1. Lemieux, R.; Roy, C.; de Caumia, B; Blanchette, D. Production Analysis and Upgrading of Oils from Biomass; ACS Symposium: Denver, Col. April 5-10, 1987; pp 12-20.
2. Perdrieux, S. Le Bois National, 1983, 2 juillet, 23-26.
3. Soltes, Ed. J.; Elder, T.J. In Organic Chemicals from Biomass; Goldstein, I.S., ed; CRC Press: Boca Raton, 1981; p 63.
4. Elliott, D.C. Analysis and Upgrading of Biomass Liquefaction Products. Final Report, Biomass Liquefaction. Tests Facility, IEA, Co-Operative Project D-1, Jan. 10, 1984.
5. Boocock, D.G.B.; Sherman, K.M. Can. J. Chem. Eng., 1985, 63, 627-633.
6. Hubert, G.D.; Eames, M.A.; Figueroa, C.; Gansley, R.R.; Schaleger, L.L.; Watt, D.W. In Fundamentals of Thermochemical Biomass Conversion; Overend, R.P.; Milne, T.A.; Mudge, L.K., eds; Elsevier Applied Science Publication: London, New York 1984; p 1027.
7. Ménard, H.; Bélanger, D.; Chauvette, G.; Gaboury, A.; Khorami, M.; Grisé, M.; Martel, A.; Potvin, E.; Roy, C.; Langlois, R. In Fifth Canadian Bioenergy R & D Seminar; Hasnain, S. ed.; Elsevier Applied Science Publisher: New York, 1984; p. 418.
8. Nicolaides, G.M. M.Sc. Thesis. University of Waterloo, Waterloo, 1984.
9. Singh, B.B. World Congress III of Chemical Engineering; Tokyo-Japan, 1986, Vol. I: p 23.
10. Ménard, H.; Grisé, M.; Martel, A.; Roy, C.; Bélanger, D. In Fifth Canadian Bioenergy R & D Seminar. Hasnain, S., ed.; Elsevier Applied Sciences Publisher: New York, 1984; p 440.
11. Brouillard, D. M.Sc.A. Thesis, Université de Sherbrooke, Sherbrooke, 1986.
12. Goos, A.W. In Wood Chemistry; Wise, L.E.; Jahn, E.C., eds; Reinhold: New-York, 1952; p 826.
13. Roy, C.; Lemieux, R.; Pakdel, H.; de Caumia, B.; Blanchette, D. World Congress III of Chemical Engineering; Tokyo, Japan, 1986, Vol. 1: p 621.
14. Ahmed, A.; Grandmaison, J.L.; Kaliaguine, S. J. Wood Chem. Technol. 1986, 6, 219-248.
15. Shafizadeh, F.; Furneaux, R.H.; Cochran, T.G.; Scholl, J.P.; Sakai, Y. J. Appl. Polym. Sci. 1979, 23, 3525-39.
16. Allan, G.G.; Mattila, T. In Lignin Their Occurence, Formation, Structure and Reaction; Sarkanen, K.V.; Ludwig, C.H., eds; Wiley: New York, 1971; Chapter 14.
17. Fletcher, T.L.; Harris, E.E. Tappi, 1952, 35, 536.

RECEIVED March 31, 1988

Chapter 20

Chemical Influence of the Oils Obtained by Hydropyrolysis of Wood

Influence of Heating Rate, Temperature, and Catalyst

N. Soyer, F. Hyvrard, C. Bruneau, and A. Brault

Laboratoire de Chimie Organique et de l'Environnement, Ecole Nationale Supérieure de Chimie de Rennes, Avenue du Général Leclerc, 35700 Rennes–Beaulieu, France

The hydropyrolysis of poplar wood has been studied in autoclaves, under an initial pressure of an inert gas, with and without iron powder used as catalyst, in the 250-350°C range. A comparative study based on the yields and chromatographic analysis, (HPSEC, GC/MS) of the oils has been carried out. The results of slow liquefactions, where 340°C was reached within 90 minutes and those of rapid liquefactions where the same temperature was obtained within 3 minutes, indicated that if a high reaction temperature was required for good oil yields, the influence of heating rate was also an important parameter. The temperatures located in the 250-350°C range were found to be critical for the liquefaction yields. A very fast heating, decreasing the residence time at those intermediate temperatures, led to yields in oil soluble in CH_2Cl_2 greater than 40 % or to yields in acetone solubles greater than 50 %. The presence of iron was absolutely necessary in the case of slow liquefaction but this additive increased the production of oil only slightly in the case of rapid liquefaction.

The thermochemical liquefaction of wood in water to produce fuel or chemical intermediates has been studied intensively over the last decade (1). The earlier works used either sodium carbonate as soluble catalyst and carbon monoxide as reducing gas (2) (3), or nickel catalyst (4) or palladium on activated charcoal (5) in the presence of hydrogen. Then it has been shown that the presence of a reducing gas was not necessary if iron powder was used as additive, with moderate heating rates (6). When the wood suspended in water was rapidly heated to 350°C, and then quenched, no catalyst was necessary and a yield in acetone solubles as high as 50 wt % was obtained (7) (8).

For a better understanding of the liquefaction reaction in water, especially during the heating up period, we investigated a comparative study of different parameters as follows : the influence

0097–6156/88/0376–0220$06.00/0

of two different heating rates (about 4°C/min. and 110°C/min.) up to a temperature located in the 250-350°C range, the influence of the residence time at the reaction temperature, and the influence of the addition of iron. This study was based on the yields in the liquefaction products, and the chemical characterization of the oils produced.

Experimental

The corresponding works were carried out with two different autoclaves.

The slow hydropyrolysis experiments were performed in a 1L stirred autoclave Engineers AFP 1005, with 63 g of dry wood sawdust (particule size less than 4 mm), 300 g of water, 9 g of iron powder (Merck 3800 ; 150 μm) corresponding to 14 weight percent on dry wood and pressurised to 4 MPa with an inert gas. Helium was chosen for a better chromatographic determination of the gases formed, and it was verified that the same results were obtained in the presence of nitrogen. With such equipment, the temperature of 340°C was reached within about 90 minutes.

The fast hydropyrolysis experiments were carried out in a 35 mL autoclave with 2.3 g of dry wood sawdust, 10 g of water, 0.3 g of iron powder under an initial pressure of inert gas of 4 MPa. The temperature was raised to 340°C within about 3 minutes by using a fluidized sand bath.

In both cases, the common points were : a similar initial pressure, weight ratios (water/wood) near 4.5, identical relative amounts of iron (∿ 14 % on dry wood). The products soluble in dichloromethane were called oils, those only soluble in acetone, not fluid at room temperature, were named tars. The solid residue consisted of insoluble compounds : char and iron residue. The gases evolved during the reaction were mainly carbon dioxide (at least 90 %) and carbon monoxide. The aqueous phase also contained organic compounds, in about 8 % yield in a slow liquefaction at 340°C, consisting mainly of carboxylic acids and alcohols, but they were not studied further.

Influence of the Heating Rate and the Final Temperature

The results of the experiments stopped at 250, 300 and 340°C are gathered in Table I.

Table I. Effect of Heating Rate and Temperature

Temperature (°C)		250		300		340	
Heat-up time (min)		1.4	50	2	65	3	90
wt %	(Oil + Tar)	4.5	44.6	38.6	47.7	50.0	48.8
	Oil	1.7	20.8	28.6	40.3	44.3	42.6
	O in Oil	ND	27.2	27.2	22.7	26.3	21.0
	CO_2	0.4	8.4	3.0	11.0	6.8	17.0
	Char	83.7	17.3	28.2	4.2	0.2	2.6
Standard deviation (Oil %)						1.4	0.9

(14 % Fe ; no residence time at the final temperature)

During the heating period up to 340°C, the transformation

of the wood components is always less for a given temperature, when the heating is fast. The higher contents of oxygen in the oils and the lower production of carbon dioxide are relevant to this evolution. However the rapid hydropyrolysis at 340°C leads to interesting results : the conversion of wood is almost complete and the yield in oil is around 45 % on dry wood. A slight improvement in the oil yield (46.6 %) is obtained when the temperature is raised to 350°C, but this is not very significant. No recondensation reactions occur during the heating period of fast liquefactions carried out in the presence of iron : the solid residue decreases, the amount of tar also decreases between 280 and 350°C, and the oil yield increases regularly.

A comparison of the different sizes of the molecules present in the oils can be performed by high performance size exclusion chromatography (HPSEC). On Figure 1, we can see the curves obtained with the oils produced at 340°C according to the two heating rates. In this comparative study, the calibration is arbitrarily based on n-alkanes. On the standard curve, the location of some phenolic compounds and cyclopentanone has been reported. The oil produced by a fast process is relatively less degraded, in that it contains more heavy molecules ; the thermal cracking reactions are of little importance during heating, and the oil produced is close to that obtained during a slow liquefaction at 250°C.

We also studied the production of 12 typical compounds during the heating period from 270 to 340°C by means of gas phase chromatography with capillary columns using 3-ethylphenol as the internal standard. The data are summarized in Table II.

Table II. Yields in Typical Products (mg/g wood)

Temperature (°C)	270	300	320	340	350	340
Heat-up time (min)	1.6	2	2.6	3	3.1	90
Cyclopentanone	-	-	-	tr	0.15	0.9
2-methylcyclopentanone	-	-	-	tr	tr	0.6
3-hydroxy-2-butanone	0.3	0.4	0.5	1.2	0.7	-
Hydroxyacetone	0.8	1.6	1.4	2.1	0.4	-
2-methyl-2-cyclopentenone	-	0.3	0.5	1.3	2.5	ND
Furfural	5.5	4.6	4.9	6.1	2.8	-
3-methyl-2-cyclopentenone	-	tr	tr	0.3	0.4	ND
Methylfurfural	0.4	0.2	0.7	1.0	0.6	-
2-hydroxy-3-methyl 2-cyclopentenone	-	0.2	0.6	1.1	0.8	-
Guaiacol	-	-	tr	0.4	0.8	4.1
Phenol	1.1	3.5	7.1	8.8	8.6	8.2
Syringol	-	-	tr	1.6	3.2	10.0

(14 % Fe ; no residence time at the final temperature)

The results obtained for the fast hydropyrolysis show that the first light compounds formed at low temperatures (270-300°C) seem to result from the degradation of cellulose. We can notice three hydroxyketones (hydroxybutanone, hydroxyacetone and hydroxymethyl cyclopentenone), furfural and methylfurfural. Phenol is also formed. During the heating, the yields in those former compounds increase and new compounds such as cyclic ketones are also detected together with phenolic products resulting from the degradation of lignin (especially guaiacol and syringol). The amount of phenol is almost constant above 320°C. When the temperature is raised to 350°C, the decrease in the yields in the former products indicates that they are unstable intermediates in our hydropyrolysis conditions. Moreover, the production of cyclic ketones and phenolic compounds increases, which shows a more and more efficient degradation of heavy molecules. During slow hydropyrolysis, the same evolution is observed up to 340°C, but for a given temperature, that degradation is always more advanced than in the case of fast hydropyrolysis. So, the hydroxyketones are no longer detected at 300°C, and at 340°C, the yields in cyclic ketones, guaiacol and syringol are higher.

Influence of a Residence Time at 340°C

The results of Table III indicate that in the presence of iron, and after a reaction time of 30 minutes at 340°C, the yields in acetone solubles (oil + tar) are the same for both heating rates. However, the percentage of oil is greater in the case of rapid hydropyrolysis, while slow hydropyrolysis only slightly promotes the condensations into char and the production of carbon dioxide.

Table III. Effect of Residence Time at 340°C and Influence of Iron

	% Fe	14		0			
Heat-up time to 340°C (min)		90	3	90	76	3	3
Residence time at 340°C (min)		30	30	30	0	0	30
wt %	(Oil + Tar)	45.1	45.2	22.2	32.5	50.0	34.1
	Oil	36.8	40.9	13.0	23.7	36.9	29.6
	O in Oil	19.6	20.9	22.0	ND	25.1	22.3
	CO_2	18.6	13.5	11.1	8.6	5.2	10.4
	Char	3.5	<0.2	26.7	20.2	1.6	12.4

The oils resulting from the two processes also have very similar characteristics : their oxygen contents are very close and their NMR and IR spectra appear to be similar. Nevertheless, as shown in Figure 2, the HPSEC analyses show that the oil obtained with rapid heating contains more heavy compounds.

Thermogravimetric studies of the oils under a nitrogen atmosphere show that the two kinds of oil have the same thermal behavior. As shown in Figure 3, around 400°C, the thermal degradation begins and light hydrocarbons are formed, and at 750°C the whole oil sample has vanished, while the oil obtained by rapid hydropyrolysis, without any residence time at 340°C, gives a solid black residue that is not degraded (about 20 % of the initial sample).

As can be seen in Table IV, during the first 10 minutes at 340°C the hydroxyketones and furfurals disappear, while the amounts of cyclic ketones, guaiacol and syringol clearly increase. Then the amount of these compounds remain almost constant. This chemical stability is also found in the (oil + tar) yields for the residence times 15 and 30 minutes at 340°C (around 45.5 %). The amounts of phenolic compounds in the oils obtained after 30 minutes at 340°C seem independant of the heating rate.

So if the liquefaction is carried out in the presence of iron and if a reaction time at 340°C is held, the hydropyrolysis of wood leads to almost identical results whatever the heating rate may be.

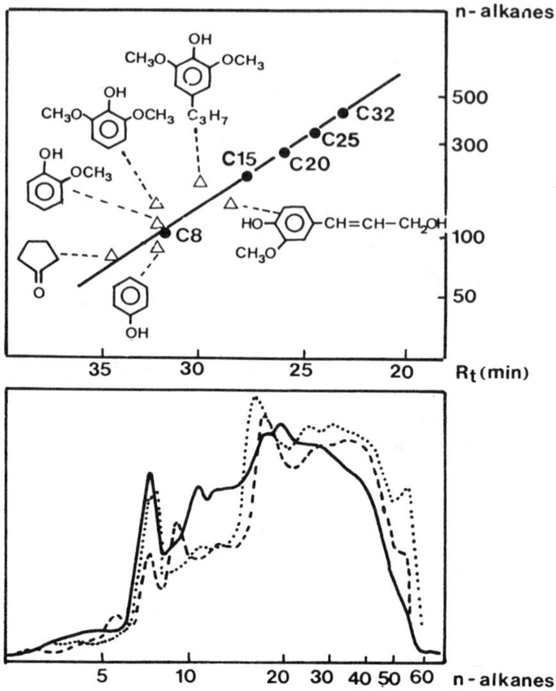

Figure 1. HPSEC of the Oils : Calibration Curve and Effect of Heating Rate (—— 90 min to 340°C ; - - - 3 min to 340°C ; ... 50 min to 250°C ; 14 % Fe) ; Column : Ultrastyragel 100 Å ; 2 x 30 cm ; Eluent : THF (0.5 $mL.min^{-1}$) ; Sample : 100 μ L ; (20 $mg.mL^{-1}$) ; RI detection

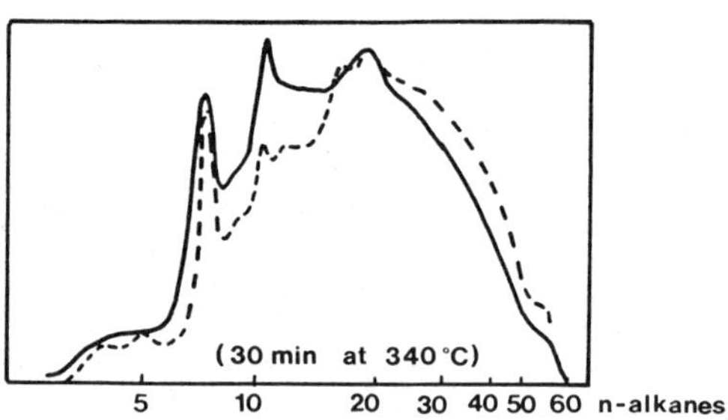

Figure 2. Effect of Heating Rate (—— 90 min to 340°C ; - - - 3 min to 340°C ; 14 % Fe)

It is noteworthy that the amounts of cyclic ketones and phenolic compounds which appeared to be final compounds in the liquefaction, give an indication of the progress of the reaction.

Influence of Iron

If the liquefactions carried out in the presence of iron with a reaction time of 30 minutes at 340°C lead to comparable results, whatever the heating rate may be, it is no longer the case when iron is not added. The results summarized in Table III show clearly that a slow hydropyrolysis without iron gives a lower oil yield and an important production of char.

Therefore, in the absence of iron, the heating rate has a great influence on the yields in liquid products at the expense of char. As shown in Table III, a low heating rate (76 min. to 340°C), implying a residence time of 30 minutes between 250 and 340°C, leads to a conversion of 20.2 % of wood into char, when with an equivalent residence time at 340°C after a rapid hydropyrolysis, the production of char is only 12.4 % . Moreover, a residence time of 30 minutes at 300°C without iron, leads to the formation of a black powder (17.2 %) that can be considered as an irreversible condensation product. Consequently, it appears that the residence time in the 250-300°C range is critical for the direction of the conversion of wood either into liquid compounds or on the contrary towards char, but then the liquids obtained at higher temperature are less sensible to recondensation into solids.

Table IV. Yields in Typical Products (mg/g wood) as a Function of the Residence Time at 340°C

% Fe	14					0
Heat-up time to 340°C (min)	3				90	3
Residence time at 340°C (min)	0	10	15	30	30	0
Cyclopentanone	tr	0.8	0.5	0.8	1.7	-
2-methylcyclopentanone	tr	0.4	-	0.5	1.3	-
3-hydroxy-2-butanone	1.2	0.5	-	-	-	0.5
Hydroxyacetone	2.1	tr	-	-	-	1.9
2-methyl-2-cyclopentenone	1.3	6.8	5.0	5.3	ND	0.5
Furfural	6.1	-	-	-	-	10.4
3-methyl-2-cyclopentenone	0.3	1.5	1.3	1.7	ND	-
Methylfurfural	1.0	-	-	-	-	1.3
2-hydroxy-3-methyl 2-cyclopentenone	1.1	-	-	-	-	-
Guaiacol	0.4	2.7	3.2	3.5	3.45	-
Phenol	8.8	9.3	9.9	9.9	9.0	9.2
Syringol	3.2	7.7	6.5	7.9	6.5	-

The absence of iron doesn't greatly change the distribution of the molecular weights of the oil resulting from rapid hydropyrolysis with a reaction time at 340°C, as can be seen on Figure 4. On the contrary, at 300°C, the presence of iron also leads to a higher proportion of heavy compounds, but as it also increases the yield in oil (28.6 % instead of 21 % without iron, at 300°C), we can deduce that iron improves the conversion into oil of heavy fragments coming from the thermal degradation of the different polymers of the initial wood.

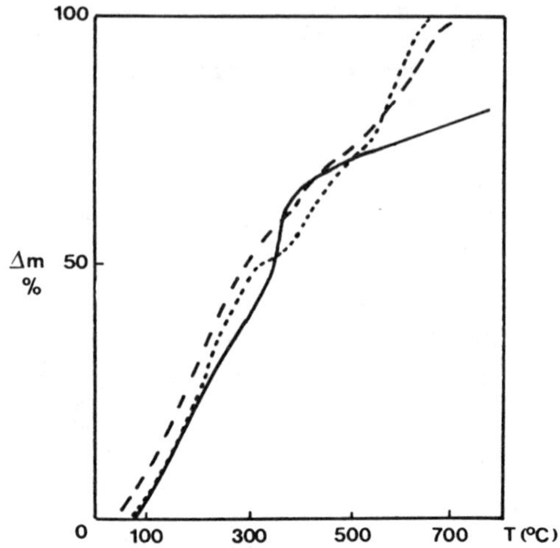

Figure 3. Thermogravimetry of Oils (Rapid Hydropyrolysis : - - - 30 min at 340°C ; —— 0 min at 340°C) (Slow Hydropyrolysis : ... 30 min at 340°C)

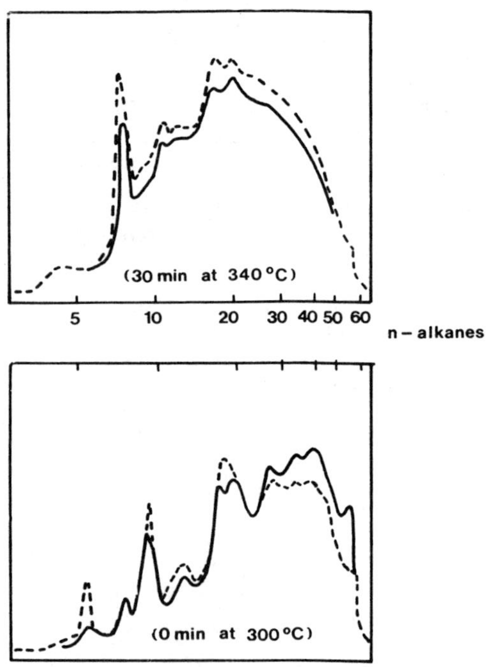

Figure 4. Effect of Iron in Rapid Hydropyrolysis (—— 14 % Fe ; - - - 0 % Fe)

As can be seen in Table IV, no cyclic ketones, no guaiacol and no syringol are detected during rapid hydropyrolysis carried out without iron, but relatively important amounts of furfural and phenol are formed and the presence of hydroxyketones is observed.

The oxidation of iron into Fe_3O_4 that takes place during slow liquefaction (6) and progressively during the residence time at 340°C in the case of rapid hydropyrolysis, would result from further reactions with some products of liquefaction.

Conclusion

All these experimental works provide information on the progress of the liquefaction reaction. It seems that in the first step, at a low temperature (around 250°C) the molecules released by the destruction of the vegetal polymers are very unstable and can follow two competitive directions : either irreversible condensation reactions into insoluble char, or conversion into relatively stable liquids. It appears that in the absence of iron, the second direction is of little importance at below 300°C and the condensation reactions are favoured. A fast heating to temperatures above to 300°C and consequently a very short time at the intermediate temperatures promotes the evolution of the primary products into stable oils. Iron would act as a catalyst at a low temperature, inhibiting the condensation reactions at the beginning of the degradation. Moreover, the compounds resulting from liquefaction in the presence of iron are more stable than those obtained without catalyst and are less modified during sustained heating at 340°C.

Acknowledgments

This study was supported by the "Centre National du Machinisme Agricole, du Génie Rural, des Eaux et Forêts" (CEMAGREF d'Antony).

Legend of Symbols

ND : no determination ; tr : trace

Literature Cited

1. Moffatt, J.M. ; Overend, R.P. Biomass 1985, 7, 99-123.
2. Appell, H.R. ; Fu, Y.C. ; Friedman, S. ; Yavorsky, P.M. ; Wender, I. US Bureau of Mines Technical Report of Investigation n° 7560, 1971.
3. Appell, H.R. ; Fu, Y.C. ; Illig, E.G. ; Steffen, F.W. ; Miller, R.D. US Bureau of Mines Technical Report of Investigation n° 8013, 1975.
4. Boocock, D.G.B. ; MacKay, D. ; McPherson, M. ; Nadeau, S.J. ; Thurier, R. Can. J. Chem. Eng. 1979, 57, 98-101.
5. Meier, D. ; Larimer, D.R. ; Faix, O. Fuel 1986, 65, 910-915.
6. Bestué - Labazuy, C. ; Soyer, N. ; Bruneau, C. ; Brault, A. Can. J. Chem. Eng. 1985, 63, 634-638.
7. Beckman, D. ; Boocock, D.G.B. Can. J. Chem. Eng. 1983, 61, 80-86.
8. Boocock, D.G.B. ; Poretta, F. J. Wood Chem. and Technol. 1986, 6, 127-144.

RECEIVED March 31, 1988

Chapter 21

Catalytic Hydrotreating of Biomass-Derived Oils

Eddie G. Baker and Douglas C. Elliott

Pacific Northwest Laboratory[1], P. O. Box 999, Richland, WA 99352

Pacific Northwest Laboratory (PNL) is investigating the catalytic upgrading of biomass-derived oils to liquid hydrocarbon fuels. Tests have been conducted in a 1-liter, continuous-feed, fixed-bed catalytic reactor at 250-450°C and 2,000 psig. Catalytic upgrading is envisioned as the second stage in a two-stage process to produce hydrocarbon fuels from biomass. Direct thermochemical liquefaction of biomass produces a phenolic oil containing very few hydrocarbons. Hydrotreating these oils can produce nearly pure hydrocarbons, primarily aromatics and naphthenes which would be useful in gasoline blending.

Two types of biomass-derived oils have been studied at PNL. The first type of oil is produced by high pressure liquefaction at relatively long residence times. Oils identified as TR7 and TR12 in Table I were produced by this type of process at the Albany, Oregon Biomass Liquefaction Experimental Facility. These highly viscous oils consist primarily of substituted phenols and naphthols (1-2). The other type of oil is produced by low-pressure, flash pyrolysis at somewhat higher temperature and very short residence times. These oils are highly oxygenated and contain a large fraction of dissolved water (3). Because of the soluble water, these oils have a much lower viscosity (4). The flash pyrolysis oil produced at Georgia Tech is typical of this type of oil. The fourth oil shown in Table I was made at PNL by pretreating the Georgia Tech pyrolysis oil to produce an oil more similar to the high-pressure oils. Details of the pretreating step are given by Elliott and Baker (5).

[1]Operated by Battelle Memorial Institute for the U.S. Department of Energy under Contract DE–AC06–76RLO–1830.

0097–6156/88/0376–0228$06.00/0

TABLE I. Feedstock Oils for Hydrotreating Tests

	TR7		TR12		Georgia Tech Pyrolysis Oil		Treated Georgia Tech Pyrolysis Oil	
	As Fed	Dry	As Fed	Dry	As Fed	Dry	As Fed	Dry
Elemental Analysis, wt %								
Carbon	74.8	77.5	72.6	76.5	39.5	55.8	61.6	71.6
Hydrogen	8.0	7.9	8.0	7.8	7.5	6.1	7.6	7.1
Oxygen	16.6	14.1	16.3	12.5	52.6	37.9	30.8	21.1
Nitrogen	< 0.1	< 0.1	< 0.1	< 0.1	< 0.1	< 0.1	< 0.1	< 0.1
Ash	0.5	0.5	3.0	3.0	0.2	0.3	0.0	0.0
Moisture	3.5	0.0	5.1	0.0	29.0	0.0	14.1	0.0
Density, g/ml @ 55°C	1.10	--	1.09	--	1.23	--	1.14[a]	
Viscosity, cps, @ 60°C	3,000	--	17,000	--	10[c]	--	14,200	
Carbon Residue, wt%[b]	13.5	13.9	26.9	28.3	--	27-31[c]	--	

a) at 20°C
b) TGA simulated Conradson carbon
c) Viscosity and carbon residue were measured for other similar pyrolysis oils

Hydrotreating Biomass-Derived Oils

Given that biomass can be converted to a liquid product that is primarily phenolic, then oxygen removal and molecular weight reduction are necessary to produce usable hydrocarbon fuels. Upgrading biomass-derived oils differs from processing petroleum fractions or coal liquids because of the importance of deoxygenation. This topic has received only limited attention in the literature (6-11). Single ring phenolics and cyclic ketones present in biomass-derived oils can be upgraded to gasoline boiling range hydrocarbons by deoxygenation. Hydrogenation of the aromatic structure is not desirable if high octane gasoline is the intended product. Batch reactor tests with model compounds showed sulfided cobalt molybdenum (CoMo) catalysts to be the best choice for deoxygenation without saturating the resulting aromatics (7). Polycyclic compounds such as naphthol must be deoxygenated and cracked to make gasoline boiling range hydrocarbons. Hydrocracking of polycyclic aromatics requires saturation of one of the rings prior to cracking (12). Nickel molybdenum (NiMo) catalysts are more effective for ring saturation than CoMo catalysts. The heavy fraction of biomass-derived oils is not as well characterized and the reaction mechanism for upgrading is unknown. Use of an acidic support (such as a zeolite) compared with alumina may be beneficial for upgrading the high molecular weight fraction (13).

Experimental

The reactor system used for this study is a fixed-bed reactor operated in an upflow mode. It is shown in Figure 1. Operation in the downflow mode (trickle-bed) plugged the outlet line of the reactor with coke-like material and tests in this mode were discontinued. The oil feedstock, preheated to 40-80°C, is pumped by a high-pressure metering pump. Hydrogen from a high-pressure cylinder is metered through a high-pressure rotameter into the oil feed line before entering the reactor vessel. The reactor is 7.5 cm I.D. by 25 cm and holds approximately 700 ml of catalyst.

A two phase flow pattern exists in the reactor. Gas and volatile products move through the reactor quite rapidly. Unconverted, non-volatile material does not leave the reactor until it reaches the top of the liquid level and overflows into the product line. Pressure in the system is maintained by a Grove back-pressure regulator. Liquid product is recovered in a condenser/separator and the off gas is metered and analyzed before it is vented.

Catalysts used in the most recent hydrotreating tests are shown in Table II. The Harshaw catalysts are conventional extruded CoMo and NiMo hydrotreating catalysts. The Haldor Topsoe catalysts are a composite system using low activity rings in the bottom of the bed to prevent plugging from carbon and metals, and high activity extrudates in the top of the bed. The Katalco and Shell catalysts are commercially available hydrotreating catalysts. The last three catalysts are hydrocracking catalysts which incorporate the metal hydrogenation component on an acidic support for cracking.

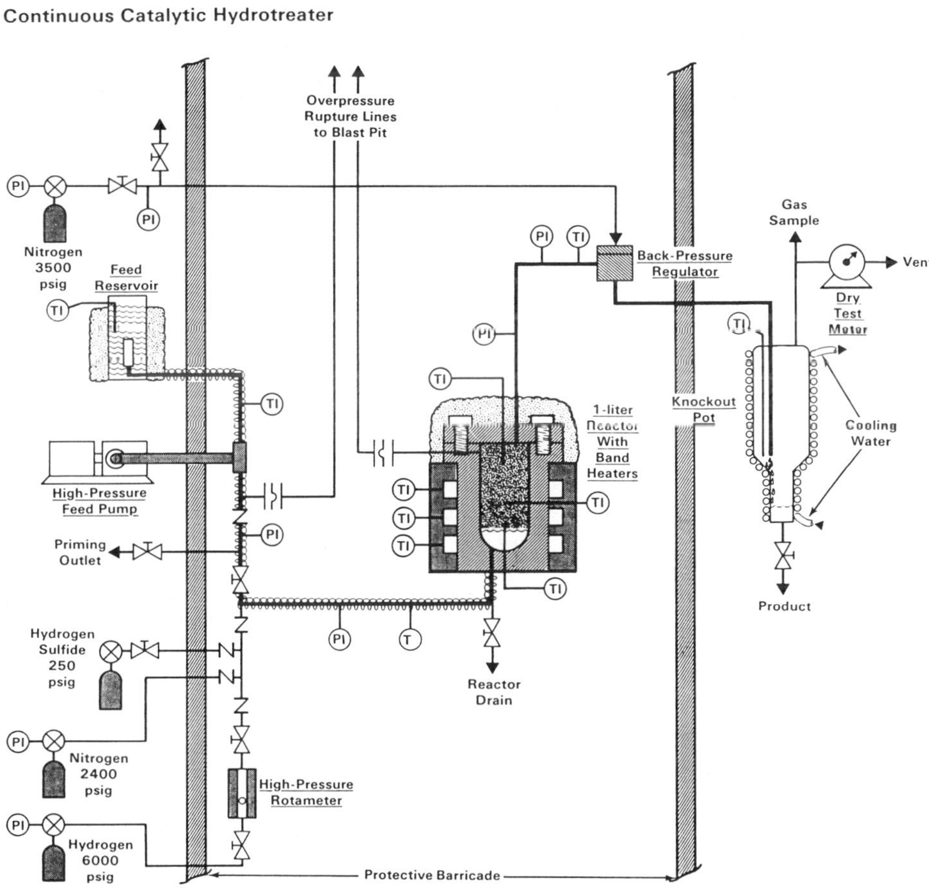

Figure 1. Flow Schematic of 1-liter Continuous Hydrotreater

TABLE II. Catalysts Used for Hydrotreating/Hydrocracking Tests

Supplier	Catalyst ID	Active Metals	Weight Percent	Support	Form[a]
Harshaw	HT 400	CoO MoO_3	3 15	Al_2O_3[c]	1/8-in E
Harshaw	HT 500	NiO MoO	3.5 15.5	Al_2O_3	1/8-in E
Haldor Topsoe[b]	TK 710	CoO MoO_3	2 6	Al_2O_3	3/16-in R
Haldor Topsoe[b]	TK 750	CoO MoO_3	2.3 10.0	Al_2O_3	1/16-in E
Haldor Topsoe[b]	TK 770	CoO MoO_3	3.4 14.0	Al_2O_3	1/16-in E
Katalco	KAT4000	CoO MoO_3	3.5 14.0	Al_2O_3	1/32-in E
Shell	S411	NiO MoO_3	2.67 14.48	Al_2O_3[c]	1/20-in T
PNL/Linde	CoMo/Y	CoO MoO_3	3.5 13.9	Y-zeolite/ Al_2O_3	1/16-in E
PNL/Grace	CoMo/SiAl	CoO MoO_3	3 13	13% Al_2O_3 in SiO_2	3/16-in T
Amoco	NiMo/Y	NiO MoO_3	3.5 18	Y-Zeolite/ Al_2O_3[c]	1/16-in E

a) E = extrudate R = ring T = tablet, size given is O.D.
b) All three catalysts used in a layered bed.
c) Includes phosphorus oxide.

Results and Discussion

Most of the test work has been done with two product oils from the Albany, Oregon Biomass Liquefaction Experimental Facility. Both oils were produced in an alkali-catalyzed, reducing gas environment with long residence times and high pressure (3000 psig). The TR12 represents a PERC-type recycle oil slurry process, and TR7 represents the LBL-type aqueous slurry single pass process. Detailed process descriptions (14) can be found elsewhere.

Only a limited number of tests have been conducted with pyrolysis oils. They behave much differently than the high-pressure oils and will be discussed separately.

Tests with Cobalt-Molybdenum Catalysts

Because of their effectiveness for hydrodeoxygenation, we have used primarily CoMo catalysts in our studies. Table III shows results obtained with the TR12 oil and the Haldor Topsoe composite catalyst system at about 400°C, 2,000 psig and three different space velocities. Typically, the liquid product yield from the TR12 oil is about 0.9 l/l of oil fed. At the low space velocity (0.11 h^{-1}) the oil is 96% deoxygenated and is about one-third high quality aromatic gasoline (C_5 - 225°C). At even lower space velocities (~0.05 h^{-1}) a liquid product containing about 60% gasoline and almost no oxygen can be produced. At higher space velocities (up to 0.44 h^{-1}) deoxygenation is still high, nearly 80%, but hydrogen consumption decreases 50% or more resulting in a lower H/C ratio, higher density, and lower gasoline yield. The theoretical hydrogen requirement to deoxygenate TR12 is about 200 l/l of oil. This indicates that at the low space velocity 350 l H_2/l oil is being used for hydrogenation, hydrocracking and other reactions. At the highest space velocity only about 50 l H_2/l oil is being used by these other reactions.

TABLE III. Results of Hydrotreating TR12 Oil with Haldor Topsoe Composite Catalyst

Run No.	HT-34	HT-34	HT-34
Temperature, °C	397	395	403
Pressure, psig	2,020	2,015	2,030
Space Velocity, LHSV, h^{-1}	.11	.30	.44
Hydrogen Consumption, l/l oil fed	548	296	212
Product Yield, l/l oil fed	.92	.88	.94
Deoxygenation, wt%	96	87	79
Product Inspections			
Oxygen, wt%	0.8	2.5	3.8
H/C ratio, mole/mole	1.5	1.3	1.3
Density, kg/l	0.91	1.0	1.03
Yield C_5 - 225°C, LV%	37	24	11

Table IV shows results from hydrotreating TR7 with Harshaw $CoMo/Al_2O_3$ catalyst. Results obtained with the Harshaw catalyst and TR12 oil were similar to those obtained with the Haldor Topsoe composite catalyst. This similarity indicates that the differences between Tables III and IV are due primarily to the oil. At a space velocity of 0.1 h^{-1}, the TR7 oil is completely deoxygenated and almost entirely converted to gasoline boiling range hydrocarbons. At similar processing conditions the oil produced from TR7 is higher quality than the oil obtained from TR12 as shown in Figures 2 and 3. Analysis of TR7 and TR12 oils has shown that TR7 is composed primarily of single ring phenolics (1) which when deoxygenated become gasoline boiling range aromatics. The TR12 oil is composed primarily of double ring phenolics which require additional cracking and hydrogenation to produce light distillates.

TABLE IV. Results of Hydrotreating TR7 Oil with Harshaw $CoMo/Al_2O_3$*

Run. No.	HT-15	HT-14	HT-14
Temperature, °C	398	394	389
Pressure, psig	2,003	2,021	2,026
Space Velocity, LHSV, h^{-1}	0.10	0.30	0.55
Hydrogen Consumption, l/l oil fed	616	435	202
Product Yield, l/l oil fed	0.99	1.0	0.88
Deoxygenation, wt%	~100	94	88
Product Inspections			
Oxygen	0.0	1.1	2.6
H/C ratio, mole/mole	1.65	1.41	1.32
Density, kg/l	0.84	0.91	0.96
Yield C_5-225°C, LV%	>87	60	28

* adapted from reference (11)

Results to date indicate 400°C is about the optimum temperature for hydrotreating biomass-derived oils. At 350°C a much poorer quality oil is produced. At 450°C the product quality improves somewhat compared with 400°C but the yield is reduced due to increased gas production. Only a limited number of tests have been run at pressures other than 2000 psig. These results showed significant loss of product quality at 1500 psig and no appreciable improvement at 2300 psig compared to results at 2000 psig.

Results With Other Catalysts

Our catalyst development effort is aimed at better hydrocracking capability. Hydrodeoxygenation is achieved readily with the CoMo catalyst in a sulfided form. Better distillate yield with minimal aromatic saturation is our current goal. To this end, we have been

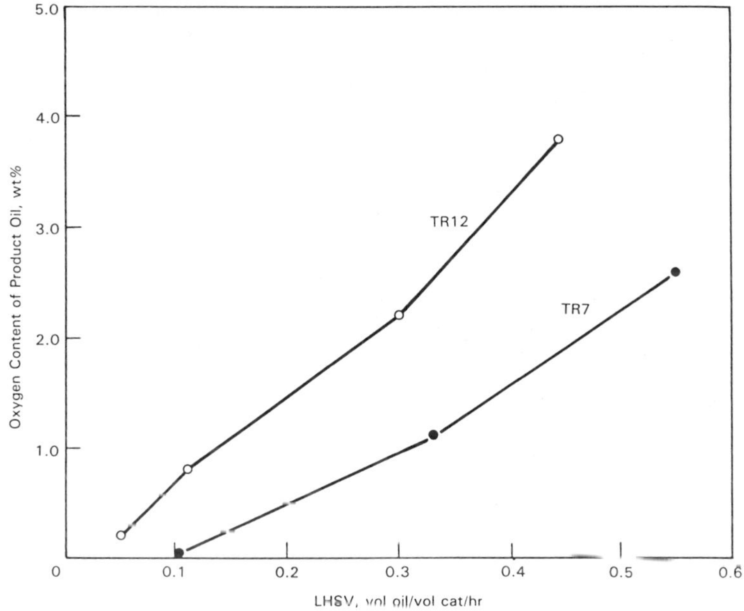

Figure 2. Oxygen Content of Hydrotreated Oils (400°C, 2000 psig)

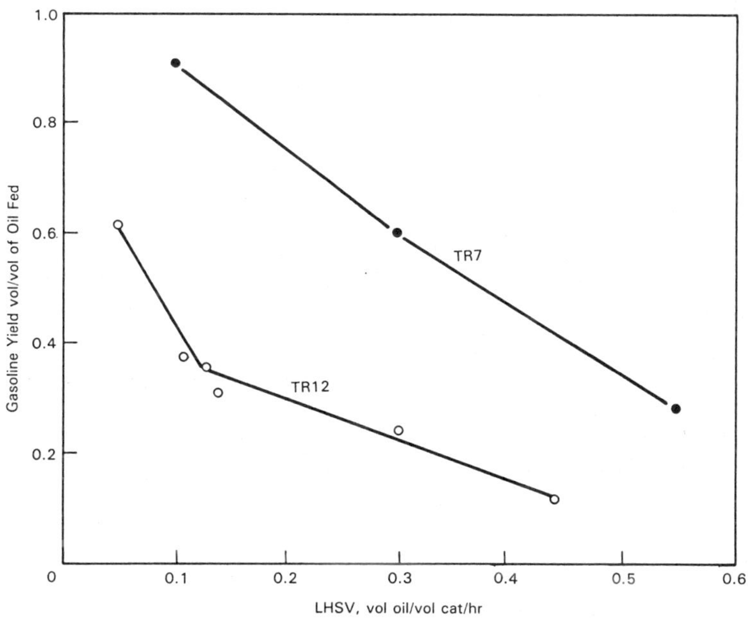

Figure 3. Yield of Gasoline Boiling Range (C_5 - 225°C) Oil (400°C, 2000 psig)

in contact with various catalyst manufacturers to learn what suggestions for wood-oil processing might be made based on present knowledge of petroleum processing. Several hydrocracking catalyst samples have been obtained for testing. These include the Shell S-411, Katalco 4000, and Amoco NiMo/Y catalysts shown in Table II. In addition we made two hydrocracking catalysts at PNL by impregnating cobalt and molybdenum salts on a Y-zeolite and a silica-alumina cracking catalyst.

Comparison of the results obtained to date with different hydrotreating and hydrocracking catalysts shows fairly consistent results with all catalysts. As seen in Figure 4 and 5, all the catalytic results appear as a single set of data with little variation among the different catalysts. Only the zeolite-based catalysts shown in Figure 5 improved results compared with the conventional CoMo hydrotreating catalysts. Note that the difference between Figure 4 and Figure 5 is due to feedstock.

Results with Pyrolysis Oils

When the Georgia Tech pyrolysis oil was hydrotreated with a sulfided CoMo catalyst at conditions similar to those used with TR7 and TR12 the runs had to be terminated due to severe coking in the bed. The temperature had to be reduced to 250°-270°C to prevent coking. The properties of the oil produced at these low temperatures are shown in Table I under the heading of treated Georgia Tech pyrolysis oil. This oil was further hydrotreated at 350°C and 2,000 psig with a sulfided CoMo catalyst, and the results were similar to those obtained with TR7 and TR12. This is the basis for a proposed two stage upgrading process for biomass pyrolysis oils (5). We are currently studying the processing of pyrolysis oils in a single step by varying the temperature in the catalyst bed from 270°C at the inlet to 400°C at the outlet.

Catalyst Deactivation

To evaluate catalyst deactivation a 48-hour test run was completed with TR12 oil and the Haldor Topsoe catalyst. Figure 6 shows the trend of deoxygenation and hydrogen consumption at an LHSV of 0.1 h^{-1}. Hydrogen consumption and the H/C mole ratio (not shown) fell rapidly in the early stages of the test and then leveled off. Deoxygenation fell throughout the test.

Two causes of deactivation have been postulated. The initial deactivation is likely due to coking of the catalyst which we have shown in earlier tests occurs primarily in the first ten hours (11). The longer term deactivation is probably due to buildup of metals, primarily sodium, from the oil. The TR12 oil contains about 3% ash, mostly residual sodium catalyst from the liquefaction process.

Conversion of the catalyst from the sulfided form to the oxide form could also cause a loss of activity. We have not seen any evidence that this occurs in the short term. Analysis of the spent catalyst indicates it is still sulfided and sulfur has not been detected in the off-gas or the product oils.

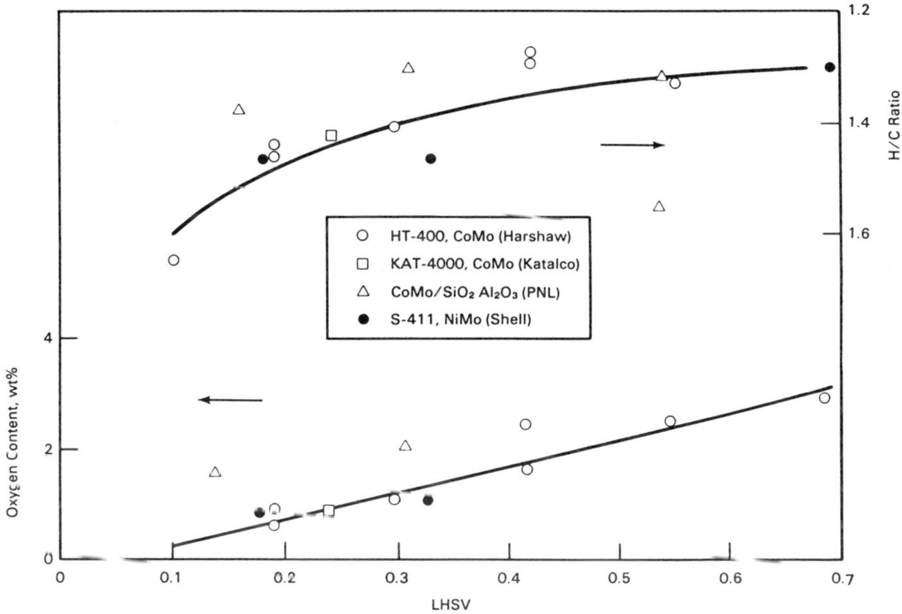

Figure 4. Effect of Catalysts on Quality of Hydrotreated Oil from TR7 (400°C, 2000 psig)

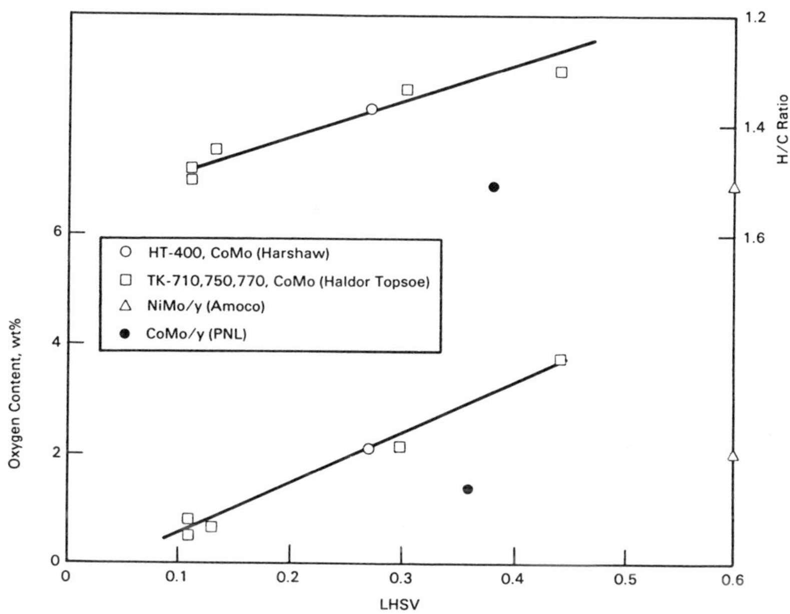

Figure 5. Effect of Catalysts on Quality of Hydrotreated Oil from TR12 (400°C, 2000 psig)

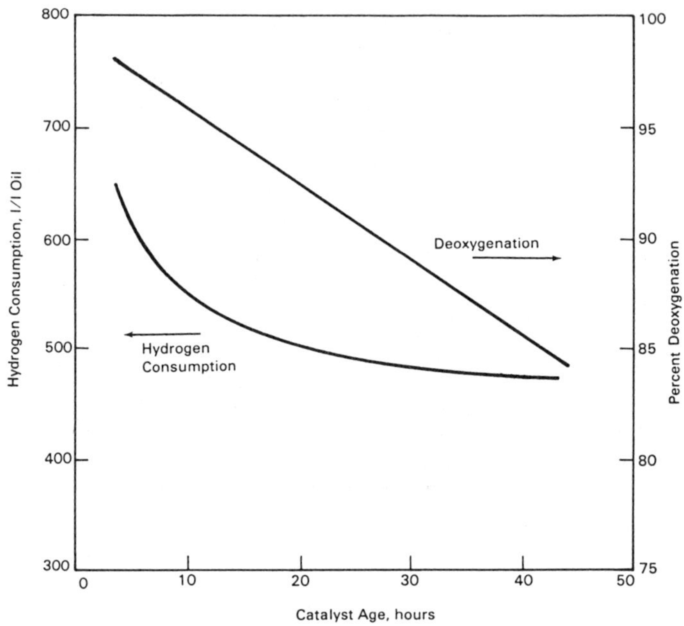

Figure 6. Catalyst Deactivation with Haldor Topsoe Catalyst and TR12 Oil (400°C, 2000 psig)

Conclusions

A variety of biomass-derived oils have been upgraded by catalytic hydrotreating in a 1-liter reactor system. Specific conclusions from our studies are as follows:

- High yields of high-quality gasoline (C_5 - 225°C boiling range) can be produced from biomass-derived oils; however, low space velocities (long residence times) are required. At high space velocities a low oxygen, highly aromatic crude oil is produced.

- Cracking and hydrogenation of the higher molecular weight components are the rate limiting steps in upgrading biomass-derived oils. Catalyst development should be directed at these reactions. Process development in biomass liquefaction should be directed at minimizing the production of multiple ring compounds.

- The TR7 oil is superior to TR12 and both are much superior to pyrolysis oils as feedstocks for conventional catalytic hydrotreating to produce hydrocarbon fuels.

- Pyrolysis oils can be upgraded by catalytic hydrotreating, however, a low-temperature catalytic pretreatment step is required. Studies are underway to combine the pretreating and hydrotreating in a single step.

- Residual sodium catalyst needs to be removed from liquefaction products to prevent rapid catalyst fouling.

- Coke formation on the catalysts is evident in all cases (as expected) but appears to be more extensive with the TR12. Extended time operation may lead to higher coking levels, but this may also be related to metal poisoning effects.

Acknowledgments

The authors acknowledge the dedicated efforts of W. F. Riemath and G. G. Neuenschwander of PNL who assisted in reactor operation and product analysis. The work was funded by the Biomass Thermochemical Conversion Program office at PNL under the direction of G. F. Schiefelbein, D. J. Stevens, and M. A. Gerber. We would also like to acknowledge the support of S. Friedrich of the Biofuels and Municipal Waste Technology Division of the U. S. Department of Energy.

Literature Cited

1. Davis, H., G. et al. "The Products of Direct Liquefaction of Biomass," in Fundamentals of Thermochemical Biomass Conversion, R. P. Overend, T. A. Milne, and L. K. Mudge, Eds; Elsevier Applied Science Publishers, New York, 1985; p 1027.
2. Elliott, D. C. "Analysis and Comparison of Products from Wood Liquefaction," in Fundamentals of Thermochemical Biomass

Conversion, R. P. Overend, T. A. Milne, and L. K. Mudge, Eds. Elsevier Applied Science Publishers, New York, 1985; p 1003.
3. Scott, D. S. and J. Piskorz. Can. Jour. Chem. Eng. 1984, 62, 404-412.
4. Beckman, D. and D. C. Elliott. Can. J. Chem. Eng. 1985, 63, 99.
5. Elliott, D. C. and E. G. Baker. "Hydrotreating Biomass Liquids to Produce Hydrocarbon Fuels," in Energy from Biomass and Wastes X, IGT, Chicago, 1987; p 765-784.
6. Li, C. L., Z. R. Yu, and B. C. Gates. Ind. Eng. Chem. Proc. Des. Dev. 1985, 24, 92.
7. Elliott, D. C. Amer. Chem. Soc. Div. Petr. Chem. Prepts. 1983, 28, 3, 667.
8. Elliott, D. C. and E. G. Baker. "Upgrading Biomass Liquefaction Products Through Hydrodeoxygenation," in Biotechnology and Bioengineering, Symposium No. 14, John Wiley and Sons, New York, 1984; p 159.
9. Soltes, E. J. and S. C. K. Lin. "Hydroprocessing of Biomass Tars for Liquid Fuel Engines," in Progress in Biomass Conversion, Volume 5, Academic Press, New York, 1984; p 1.
10. Furimsky, E. Catal. Rev-Sci. Eng. 1983, 25, 3, 421.
11. Elliott, D. C. and E. G. Baker. "Hydrodeoxygenation of Wood-Derived Liquids to Produce Hydrocarbon Fuels." Presented at the 20th IECEC, Miami, Beach, 1985; SAE Paper No. 859096.
12. Langlois, G. E. and R. F. Sullivan. Adv. Chem. Ser. 1970. No. 97, 38.
13. Haynes, H. W., J. F. Parcher, and N. E. Helmer. Ind. Eng. Chem. Process Des. Dev. 1983, 22, 401-409.
14. Thigpen, P. L. and W. L. Berry. In Energy from Biomass and Wastes VI; Klass, D. L., Ed.; Institute of Gas Technology; Chicago, 1982; p 1057.

RECEIVED December 30, 1987

Chapter 22

Chemical Modeling of Lignin

A Monte Carlo Simulation of Its Structure and Catalytic Liquefaction

Peter M. Train[1,3] and Michael T. Klein[2,4]

[1]Department of Chemical Engineering, University of Delaware, Newark, DE 19716
[2]Center for Catalytic Science Technology, University of Delaware, Newark, DE 19716

A stochastic model of lignin structure and liquefaction was developed. Lignin was viewed as an ensemble of phenolic ring-comprising oligomers. Monte Carlo simulation of lignin structure, via the placement of random numbers on cumulative probability functions for the substitutes on each ring position, provided $N_p \geq 1000$ starting lignin oligomers. Subsequent Monte Carlo simulation of lignin liquefaction, via the comparison of separate random numbers with transition (reaction) probabilities, chronicled the yields of gas, light-liquid, tar and char product fractions. The effects of catalyst decay and effectiveness were probed.

Chemical modelling is the quantitative use of model compound information in the analysis of a real reaction system. Its premise is that "intrinsic" chemistry can be defined as that shared by a model compound and the moiety within the real reaction system the model is meant to mimic. Extrinsic factors, such as kinetic coupling (1), molecular weight-related diffusion limitations (2), and chemical inhomogeneities, render the chemistry observed of a real system different than the intrinsic. The challenge of chemical modelling is to address these extrinsic factors and relate the model compound results to the reactions of real systems. Herein we use lignin liquefaction by thermal and catalytic methods as a vehicle with which to develop a stochastic methodology for the mathematical description of a real feedstock in terms of its model compounds.

Such stochastic modelling was advanced by Klein and Virk (3) as a probabilistic, model compound-based prediction of lignin pyrolysis. Lignin structure was not considered explicitly. Their approach was extended by Petrocelli (4) to include Kraft lignins and catalysis. Squire and coworkers (5) introduced the Monte Carlo computational technique as a means of following and predicting coal pyrolysis routes. Recently, McDermott (6) used model compound reaction pathways and kinetics to determine Markov Chain states and transition probabilities, respectively, in a rigorous, kinetics-oriented Monte Carlo simulation of the reactions of a linear polymer. Herein we extend the Monte Carlo

[3]Current address: Amoco Oil Company, Naperville, IL 60566
[4]Address correspondence to this author.

0097–6156/88/0376–0241$06.75/0

methodology to include lignin structure and some of the extrinsic challenges of chemical modelling.

The simulation model development is divided into three sections. The first discusses the probabilistic modelling of lignin structure, and the use of probability distribution functions to generate representative lignin moieties. The second section details the depolymerization of lignin using stochastic kinetics. The final portion describes the combination of these elements into a Monte Carlo simulation and also presents representative predictions

Stochastic Description of Lignin Structure

Lignin is a phenolic copolymer derived from woody plants wherein its abundance is roughly 30% by weight. Kraft lignins are residues of industrial pulping processes, whereas a milled wood (MW) lignin is a laboratory preparation aimed at isolation of a material that resembles its native form. Lignin structure and properties are dependent on both the parent wood type and method of isolation, which can cause drastic changes in bond types and molecular weight. There exists little agreement on absolute lignin molecular weights or molecular weight distributions (MWD), although the cumulative MWD illustrated in Figure 1 for kraft lignin, obtained from Westvaco Company technical literature, and the weight average molecular weight of 11,000 (degree of polymerization ≈ 60) for MW lignin are reasonable (7,8). These were used herein.

The chemical composition of kraft and MW lignin described by Marton (9) and Freudenberg (10), respectively, provided both the basis for the selection of model compounds for experimental study as well as the existence probabilities of various chemical moieties in the initial lignin polymers. The latter were the initial conditions of the simulation model. The Marton and Freudenberg models provide reasonable lignin structural information that is easily decoupled from the model. Our aim is the construction of a simulation that can accept any structural information as input, and the Marton and Freudenberg structures were thus convenient vehicles. Certainly results from modern NMR and mass spectral methods can be incorporated easily.

Inspection of the Marton and Freudenberg structures revealed both lignins to be an ensemble of single-ring phenolics containing six substituents, two of which, on average, doubling as interunit linkages. The identities of substituents on the six positions of an aromatic ring were therefore required to define a lignin monomer unit. Two of these, P1 and P2 in Table I, were oxygen-containing substituents and linkages, whereas two others, H1 and H2, were either hydride substituents or an interunit link. The remaining two positions were always hydride substituted and are not considered further here.

Mathematically, lignin was the simultaneous occurrence of a set of phenolic units that were in turn defined by the simultaneous juxtaposition of H1, H2, P1, and P2 substituents. The identities and frequency distributions of these substituents, as discerned through scrutiny of the Marton and Freudenberg structures, are illustrated in Table I, which illuminates the differences between kraft and milled wood lignins.

The H1 and H2 substitutents were ethers, methylene, ethylene, and diaryl bridges or hydride substituents. Thus these positions were either involved in an interunit link or hydride substituted in the present model. Greater proportions of the thermally labile ether linkages were in the MWL while the kraft lignin contained greater proportions of more stable methylene, ethylene, diaryl and diaryl linkages. The P1 and P2 substituents could be an ether, such as that listed above for the H positions, or a terminal methoxyl, hydroxyl or hydride substituent.

The requirement that an interunit linkage bridge one unit to another placed conditions on allowed combinations. For example, ethylene and methylene linkages were always between the H2 position of one ring and the H1 position of the next. Likewise, ß-ether linkages that originated from either of the P-positions on one unit terminated at the H1 position of the next. Diaryl ether moieties bridged the P1 and/or P2 positions of two units, whereas diaryl linkages connected two H2 positions.

The information in Table I was reapportioned into the normalized cumulative probability distribution functions (PDF) shown in Figures 2 and 3 for kraft and MW lignin, respectively. The P and H substitutents were reorganized as either P-position terminal substituents or interunit linkages to facilitate probabilistic construction of representative lignin oligomers, as follows.

Model Polymer Generation. The procedure is illustrated in Figure 4. First, the degree of polymerization of a lignin oligomer was determined by comparing a random number to the MWD illustrated in Figure 1. Aside from end effects, treated separately, specification of a monomeric unit required specification of one interunit link and two terminal substituents. The linkage between monomeric units n and $n + 1$ was determined by selecting, at random, a linkage from the PDF of Figure 2a for kraft lignin or Figure 3a for a MW lignin. The linkage could be an ether or a hydrocarbon. We consider the former occurrence first.

When the linkage was determined to be an ether originating at a P-position, a second random number was drawn to determine its ring position. In kraft lignins, 60% of the ether links originated at P1 and 40% originated at P2; all ether linkages were modelled to originate at P2 in MWL. Any substituent on the remaining P-position was determined by comparison of another random number to the cumulative probability distribution function for terminal substitutents, Figure 2b or 3b, for kraft and Milled Wood lignins, respectively. Since all H-positions were either hydrocarbon links or hydride substitutents, the latter substituent was placed on any remaining H-positions for this instance where the interunit link was an ether.

When the linkage was determined to be a hydrocarbon originating at an H-position, the remaining P-position substituent(s) was(were) determined through comparison of a separate random number with the PDF of Figure 2b (or 3b). Any remaining H-position was then assigned a hydride substituents.

By way of example, consider the generation of the three-unit section of a linear lignin polymer illustrated in Figure 5; this also illustrates the specification of end units. Suppose the first step, i.e., the random determination of the linkage between units 1 and 2, produces an ethylene linkage. This will bridge the H2 position of unit 1 and the H1 position of unit 2. The remaining H-position on unit 1 will be hydride substituted, and the remaining P positions on unit 1 would be determined at random after first renormalizing the PDF for P1 and P2 terminal substituents (Figures 2b and 3b) so as to remove the choices leading to linkages: the end unit of a linear polymer can only be linked to one other unit. Suppose, then, that the random tests of the now-conditional PDF led to the completed unit 1 illustrated in Figure 5.

The linkage between units 2 and 3 would be determined next. The illustrated ß-ether was specified as originating from the P2 position of unit 2. Then the P1 position substituent PDF was renormalized by removing the linkage choices. Comparing random numbers with this conditional PDF would then specify the P substituent on unit 2, as, perhaps, that depicted in Figure 5. The H2 position would be hydride substituted.

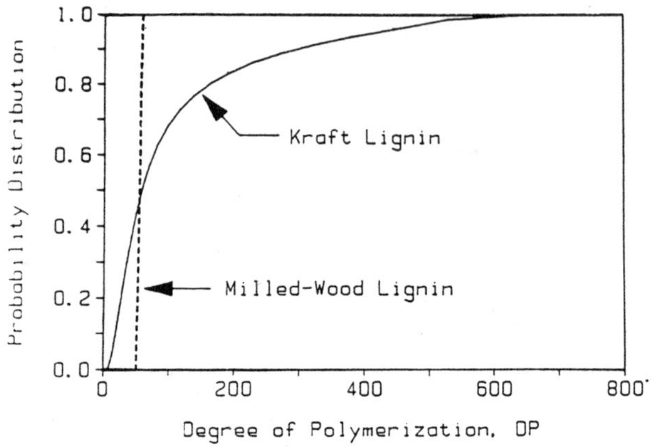

Figure 1. Cumulative Probability Distribution Functions for Lignin Molecular Weight.

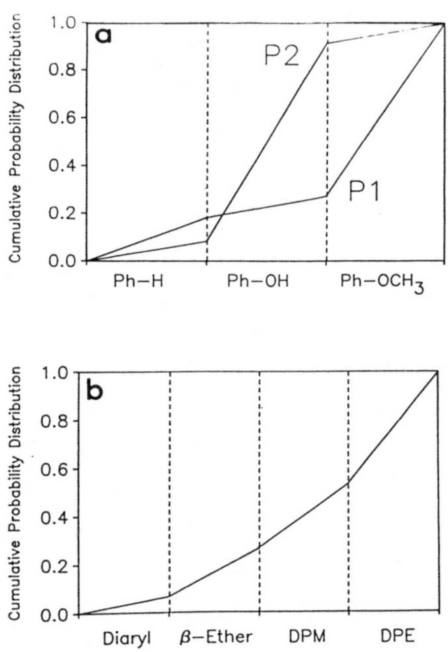

Figure 2. Cumulative Probability Distribution Functions for Kraft Lignin Substituents.
a. Terminal P-positions.
b. Interunit Linkages.

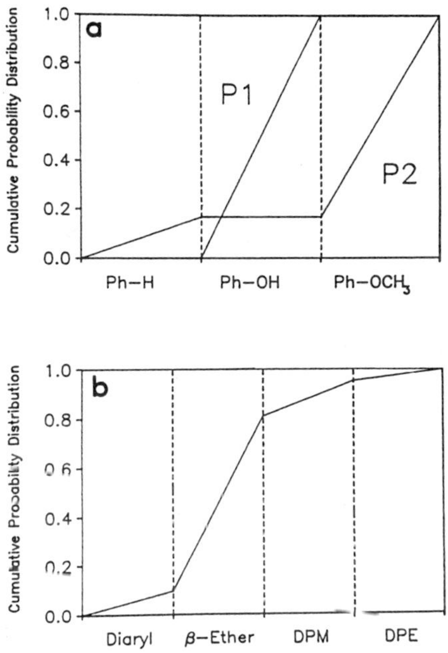

Figure 3. Cumulative Probability Distribution Functions for Milled Wood Lignin Substituents.
a. Terminal P-positions.
b. Interunit Linkages.

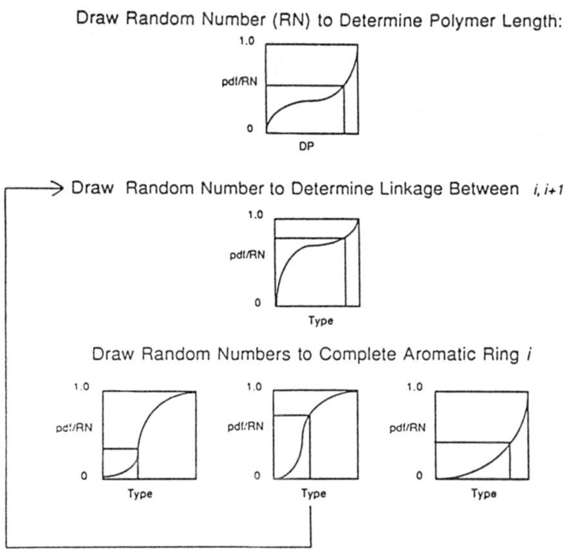

Figure 4. Logic of Construction of Model Lignin Oligomers.

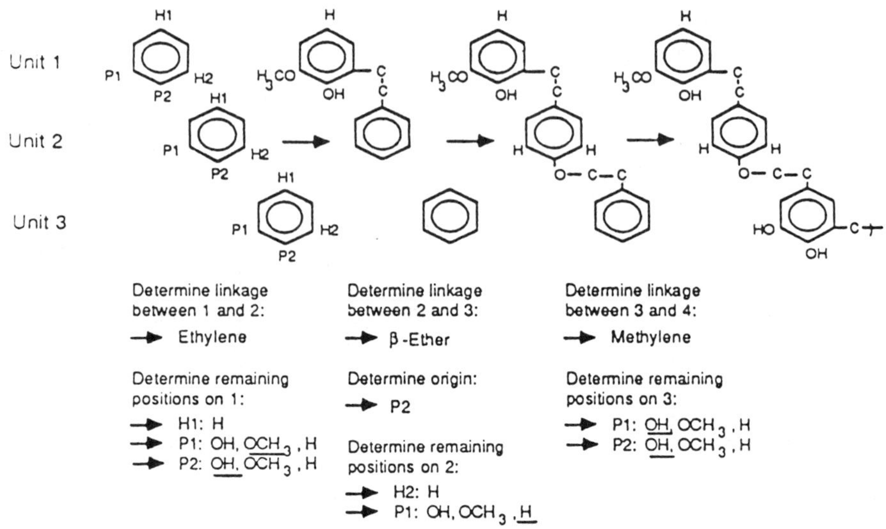

Figure 5. Example of Construction of a Model Lignin Oligomer.

	H1, P1, H2, P2	Kraft Lignin		Milled-Wood Lignin	
		1	2	1	2
H	Ph-O-C-C-Ph	0.18	0.0	0.77	0.0
	Ph-C-C-Ph	0.53	0.31	0.05	0.15
	Ph-C-Ph	0.24	0.25	0.18	0.10
	Ph-Ph	0.0	0.13	0.0	0.20
	Ph-H	0.06	0.31	0.0	0.55
P	Ph-O-R	0.21	0.14	0.0	0.68
	Ph-OCH_3	0.57	0.07	0.83	0.0
	Ph-OH	0.07	0.71	0.0	0.32
	Ph-H	0.14	0.07	0.17	0.0

Table I. Kraft and milled wood lignin bond types.

The substituents on the final unit would be determined randomly as well. Possible choices determined from the PDF of Figures 2 and 3 are shown in Figure 5.

This method was used to construct a large set $N_p \geq 1000$ of lignin oligomers whose average conformed to the PDF of Figures 1, 2, and 3. Monte Carlo simulation of the reaction of each allowed mathematical description of lignin liquefaction.

Lignin Depolymerization

The simulation of lignin liquefaction combined a stochastic interpretation of depolymerization kinetics with models for catalyst deactivation and polymer diffusion. The stochastic model was based on discrete mathematics, which allowed the transformations of a system between its discrete states to be chronicled by comparing random numbers to transition probabilities. The transition probability was dependent on both the time interval of reaction and a global reaction rate constant. McDermott's (6) analysis of the random reaction trajectory of the linear polymer shown in Figure 6 permits illustration.

Initially, the 11 monomer units depicted (ø) were connected by 10 reactive bonds. These bonds could be either intact or cleaved at all subsequent times. The simulated depolymerization proceeded by allowing fixed time intervals, Δt, to pass in series. After the passage of Δt, each of the bonds was tested for reaction by comparing the value of a random number against the associated transition probability. If the transition probability was greater than the random number, the reaction occurred. Otherwise, the bond remained intact. The result of testing each bond then defined the new state of the polymer. This procedure is analogous to a first-order Markov chain through time, in which each discrete polymer state is dependent soley on the immediately previous state. Averaging the results of N_s = 10,000 Markov chains provided a result of desired accuracy.

Application of these ideas to lignin liquefaction required definition of allowable lignin states and the probabilities associated with transitions from one state to another. The random construction just considered provided the initial (t = 0) states. The model compound reaction pathways provided the set of allowable states for $t > 0$, and the model compound kinetics and selectivities provided the transition probabilities.

McDermott (6) *et al.* developed the rigorous expressions shown in Table II for the transition probabilities of bonds reacting according to the prototype reaction sequences of: $A \rightarrow B$, first order; $A \rightarrow B$ and $A \rightarrow C$, first order; $A \rightarrow B \rightarrow C$, first order; $A \rightarrow B$, second order, etc. For example, the transition probability for the irreversible first-order reaction of A to B has the general form:

$$P = 1 - e^{-k_{AB}\Delta t} \qquad (1)$$

where P is the transition probability, k_{AB} is derived from the deterministic (model compound) rate constant and other extrinsic factors, and Δt is the length of time allowed in the test for a transition. In the present simulation the global rate constant k_{AB} had the form $k_{AB} = \eta\phi\omega_c k_c + k_t$ to account for catalyst and effectiveness and thermal-to-catalytic ratios, as developed below.

The results summarized in Table II were used herein as the basis for the dependence of the transition probability on the global rate constant. The details of

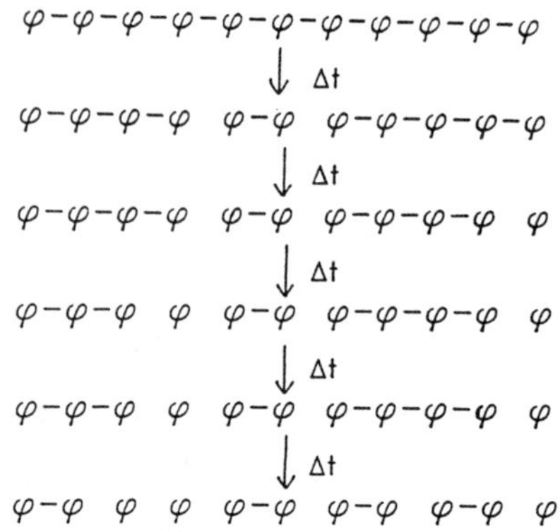

Figure 6. Random Trajectory of the Reaction of a Linear Polymer.

Transition probabilities
A ⟶ B, First-order, Irreversible $P_{AB} = 1 - \exp(-k_{AB}\,\Delta t)$
A ⟶ B and A ⟶ C, First-order, Irreversible Probability of reaction of A: $P_{AB} = 1 - \exp(-(k_{AB} + k_{AC})\,\Delta t)$ Selectivity to products: $S_{AB} = \frac{k_{AB}}{k_{AB} + k_{BC}}$; $S_{AC} = 1 - S_{AB}$
A ⟶ B ⟶ C, First-order $P_{AB} = 1 - \exp(-k_{AB}\,\Delta t)$; $P_{BC} = 1 - \exp(-k_{BC}\Delta t)$ $P_{AC} = -\frac{k_{BC}}{k_{AB} - k_{BC}} P_{AB} + \frac{k_{AB}}{k_{AB} - k_{BC}} P_{BC}$
$A \xrightarrow{k_{0/1}} B_1 \xrightarrow{k_{1/2}} B_2 \xrightarrow{k_{2/3}} \ldots B_{N-1} \xrightarrow{k_{N-1/N}} B_N$, **First-order** $P_i = \sum_{i=1}^{N} (-1)^{N-i} \frac{\prod_{j=1}^{N} k_{j-1/j}}{k_{i-1/i} \prod_{j=1}^{i-1} (k_{j-1/j} - k_{i-1/i}) \prod_{j=i+1}^{N} (k_{i-1/i} - k_{j-1/j})} P_{i-1,i}$ $P_{i-1,i} = 1 - \exp(-k_{i-1/i}\,\Delta t)$
A ⇄ B, First-order, Reversible: conceptually similar to a series A → B → A → B...
2 A ⟶ B, Second-order, Irreversible The transition probability is updated after each time step: $P_{AB}(t) = 1 - \exp(-k'_{AB}(t)\,\Delta t)$ $k'_{AB}(t) = (1 - P_{AB}(t-1))\,k'_{AB}(t-1)$; $k'_{AB}(0) = k_{AB}\,A_0$

Table II. Transition probabilities for Monte Carlo simulation.

the use of these probabilities and the issues of accuracy and precision have been described elsewhere (6).

As indicated above, lignin moieties were subject to both thermal and catalytic reactions. The catalytic rate constant was modified by multipliers to account for catalyst aging and effectiveness (11).

The literature suggests that the large quantities of water, methanol and char formed during the thermal and catalytic liquefaction of lignins can adversely affect the lifetime of the catalyst (12). One separate (13) experimental probe of the decay of during the HDO of o-hydroxydiphenylmethane showed that the exponential decay (11) model for catalyst deactivation provided the best fit to the data. This is shown in Equation 2, and was used in the present model to help fix ideas. Clearly more appropriate activity models could be used without challenge of concept.

$$\phi = e^{-\alpha C_c} \tag{2}$$

The molecular weight differences between lignin and its model compounds also complicate the use of model compound kinetics in a predictive simulation. The mobility of a high-molecular weight polymer would be much less than that of smaller model substrates (14). As for catalyst decay, a simple model was used to probe transport issues. For a first order, irreversible reaction in an isothermal, spherical catalyst pellet with equimolar counterdiffusion, the catalyst effectiveness factor and Thiele modulus provide the relevant information as

$$\eta = \frac{1}{\phi} \; \frac{3\phi \coth 3\phi - 1}{3\phi} \tag{3}$$

$$\phi = \frac{r_p}{3} \left(\frac{k_c \rho_b}{D_e} \right)^{1/2} \tag{4}$$

For relatively short polymer chains (DP <200), Rouse's model (15) for coiling polymers in dilute solutions, where $D_e \approx 1/M$, dictated that the Thiele modulus be proportional to $DP^{\frac{1}{2}}$. The entangled difffusion of longer polymer chains (DP>200) was modelled using deGennes (16) reptation theory, where $D_e \approx 1/M^2$. Under these conditions, the Thiele modulus was proportional to DP.

The diffusion coefficient in Eq. 4 was therefore dependent on DP. Numerical values were estimated as follows. For short chains, where the diffusivity was inversely proportional to DP, i.e., $D_e = a/DP$, the constant "a" was determined from the Wilke-Chang (17) model compound diffusivity where DP = 1. For longer polymers, where $D_e = b/DP^2$, the scaling coefficient "b" was obtained by matching with a/ DP at DP = 200.

With models for catalyst decay and effectiveness now in hand, the simulation of lignin liquefaction could be achieved given the initial lignin structure (as described earlier) and model compound reaction pathways and kinetics, both thermal and catalytic. Construction of a random polymer, as outlined earlier, began the simulation. This structural information combined with the simulated process conditions to allow calculation of the reaction rate constants, selectivities and associated transition probabilities. The largest rate constant then specified the upper limit of the reaction time step size.

For each time step two numbers were drawn to determine both the starting location and the direction of testing intact linkages along the polymer chain. This removed bias induced from testing linkages from end to end in the same direction. The polymer chains were mathematically circular in that the testing of one end unit was followed by a test of the other. Two additional random numbers were chosen to determine the initial location and travel direction for testing for the reactions of the substituents on each monomer unit.

Each intact interunit linkage along the polymer chain was tested in turn for reaction by comparing a drawn random number to the transition probability corresponding to the linkage type. If the probability was greater than the random number drawn, the bond cleaved, and the bond status was updated accordingly. In general several parallel reactions were modelled. The transition probability $P = 1 - \exp(-\Sigma k_{AB}\Delta t)$ was therefore used frequently. Given the occurrence of a reaction, then, its products were determined by comparing product selectivities to a second random number.

All other non-linkage substituents on a given monomer unit were tested for reaction analogously. The substituent disappearance transition probability was compared to the first-selected random number, and when reaction did occur, its products were determined comparing a second random number to the transition selectivities. Interunit linkages and other non-linking substituents of aromatic rings were tested alternately along the entire model polymer chain.

To illustrate the procedure, consider the reaction depicted in Figure 7 of the three-unit model polymer generated earlier. After calculation of rate constants (see below), selectivities and transition probabilities determined from Table II set the upper limit of Δt. Four random numbers were then selected independently to determine the starting location and direction of travel for testing phenolic moieties and interunit linkages. Assume that this dictated that linkage testing for the first time step was to begin at unit 2 and proceed to units with higher indices, and, moreover, that testing of aromatic rings was to begin at position 3 and proceed circularly through lower-numbered positions.

The ß-ether linking units 2 and 3 was thus tested first. Since unit 3 contains ortho hydroxyl substituents, the linkage most closely resembled guaiacylglycol- ß -guaiacyl ether (GGE) (18). Assume that comparison of the transition probability for GGE with a random number indicated cleavage. Then the status of the bond was updated to reflect cleavage to two fragments, the first changing the P2 position of unit 2 to a hydroxyl substituent and the second changing the H1 position of unit 3 to an ethyl substituent.

The ring of unit 3 was tested next. Assume that comparison of a random number to the catechol transition probability in this instance determined that reaction did not occur. Testing of the phenolic moiety commenced at unit 2.

Recall that cleavage of the 2-3 link transformed the phenolic moiety of unit 2 to phenol. Assume that the next-drawn random number was less than the transition probability for reaction of phenol. Since the only possible reaction is dehydroxylation, the aryl hydroxyl would be replaced by a hydride substituent, and the formation of water would be recorded.

Testing continued at the linkages between units 1 and 2. This bridging moiety most closely resembles diphenylethane (DPE). In general, cleavage was possible at three different locations: between the unit 1 aryl carbon and linkage alkyl carbon; between the two alkyl carbons; and between the linkage alkyl and unit 2 aryl carbon. Therefore, one random number was drawn to determine if any reaction occurred, and a second random number provided the products through its comparison to the selectivities along the different pathways. Cleavage between the 2 alkyl carbons, for example, transformed the H2 position of unit 1 and H1 position of unit 2 to methyl substitutents.

The final trial during the first Δt was at the phenolic unit 1, which resembled guaiacol. Assume that the reaction occurred upon comparing the reaction probability with the drawn random number. Then another random number was compared to the reaction selectivities to determine the reaction products. For the guaiacols, there were three possibilities for reaction: demethoxylation to a phenol; demethylation to a catechol; and reaction to char. The calculated selectivities of the three pathways must sum to 1, so each was assigned a segment of a number line ranging from 0 to 1. The segment on the number line at which the random number fell determined the reaction pathway. If the reaction was to char, the phenolic unit would be updated such that further testing was not possible. In this way, the interunit was still active for reaction but the phenolic positions were not.

After the entire chain was tested for reaction, the size of each oligomer was determined along with the quantities of char and light products such as carbon monoxide, methane and water. The time was then incremented by Δt and the process was repeated at the new reaction time. When the desired time was reached, one Markov chain had been completed.

New model polymers were randomly generated and reacted until the predetermined Monte Carlo sample size was reached. The quantities of all products at each time were added to the sum recorded from all previous polymers. Afterward, the weight yield in g/g_{lignin} of each product of interest was calculated at each time step by multiplying its cumulative product sum by the ratio of its molecular weight and the cumulative molecular weight of all the model polymers reacted. The results were organized as product yield versus isothermal batch reaction time.

The database of model compound reaction pathways and kinetics and pathways gathered from the sources summarized in Table III was used to determine the allowable transitions and calculate associated transition probabilities. For illustrative purposes, Figure 8 presents a network of thermal and catalytic reaction pathways depicting the transitions of the lignin phenolic (P) moieties. Only the least-substituted analog of each phenolic class (veratroles, guaiacols, catechols, anisoles, phenols, and hydrocarbons) is presented in Figure 8, but it is to be understood that the hydrocarbon positions H1 and H2 in the actual lignin and model lignin polymer would be occupied typically by the substituent and linkage types discussed previously.

Model Predictions and Discussion

The liquefaction model allowed comparison of the thermal and catalytic liquefaction of kraft and milled-wood lignins by accepting the information in either Figures 1 and 2 or Figures 1 and 3 as input distribution functions.

Kraft Lignin. Simulations of isothermal, batch pyrolysis and isothermal catalytic liquefaction of kraft lignin were at 380°C. The transition probabilities for the catalytic liquefaction were for reaction at 2250 psig over a presulfided catalyst in proportions of 2.47 g_{lignin}/g_c.

Results are described as the temporal variation of tar, light product and residue fraction classes. Within the single-ring portions of the tar fraction, the guaiacols' yield is the sum of all monomeric products that contain ortho hydroxyl and methoxyl substitutents. Catechols comprise the single-ring products that contain ortho hydroxyl substituents. Monomeric products with a single methoxyl substituent were classified as anisoles, those with monohydroxyls as phenols, and those without oxygen-containing substituents as hydrocarbons.

Reactant	Primary Products	Thermal Parameters $\log_{10}A$ (s^{-1})	Thermal Parameters E* (kcal/mol)	Source	Catalytic Parameters $\log_{10}A$ ($1/g_c s$)	Catalytic Parameters E* (kcal/mol)	Source
Veratrole	Guaiacol	11.4	48	[19]	4.8	21.7	[13]
	Anisole	10.6	47.4	[19]	x	x	
	Phenol	8.7	41.6	[19]	x	x	
Guaiacol	Catechol	12.6	49.3	[19]	5.2	25.1	[13]
	Phenol	11.7	48.2	[19]	6.1	28.3	[24]
	Char	8.4	42.7	[19]	1.8	15.9	[24]
Catechol	Phenol	x	x		6.8	29.3	[24]
	Char	4.9	25.2	[20]	-0.5	10.6	[24]
Anisole	Phenol	9.1	42.2	[19]	8.0	29.7	[25]
	Benzene	17.2	69.7	[19]	x	x	
Phenol	Benzene				-1.9	6.7	[24]
DPM	Tol + Benz	12.7	66.0	[21]	2.9	21.18	[13]
OHD	Ph + Tol	9.6	43.4	[19]	9.2	36.1	[24]
	o-Cr + Benz	x	x		7.8	35.2	[24]
DPE	2 Tol	10.2	45.3	[21]	†	†	
	EtBz + Benz	8.8	44.5	[21]			
Phenylphenol	Ph + Benz	x	x		-2	8.2	[24]
Biphenol	2 Phenol	x	x			7.26[1]	[24]
PPE	Ph + EtBz	11.1	45.0	[22]			
GGE	Guaiacol + AV	5	20.8	[23]			
Phenyl Ether	Ph + Benz	14.8	72.1	[19]	8.7	35.4	[24]
4PP	2 Phenol				8.7	32.2	[13]

DPM = Diphenylmethane | Tol = Toluene
Benz = Benzene | OHD = 2-hydroxydiphenylmethane
Ph = Phenol | o-Cr = o-Cresol
DPE = Diphenylethane | EtBz = EthylBenzene
PPE = Phenethyl Phenyl Ether | GGE = Guaiacylglycol-β-guaiacyl Ether
AV = Acetovanilone | 4PP = 4-Phenoxyphenol

† $-\log_{10}k_{330} = 7.26$

Deactivation Parameters [13]	α, (mol O/g_c)
Single-Ring Phenolics	589.2
Carbon Linkages	273.5
Ether Linkages	20.0

Table III. Model compound reaction pathways and kinetics

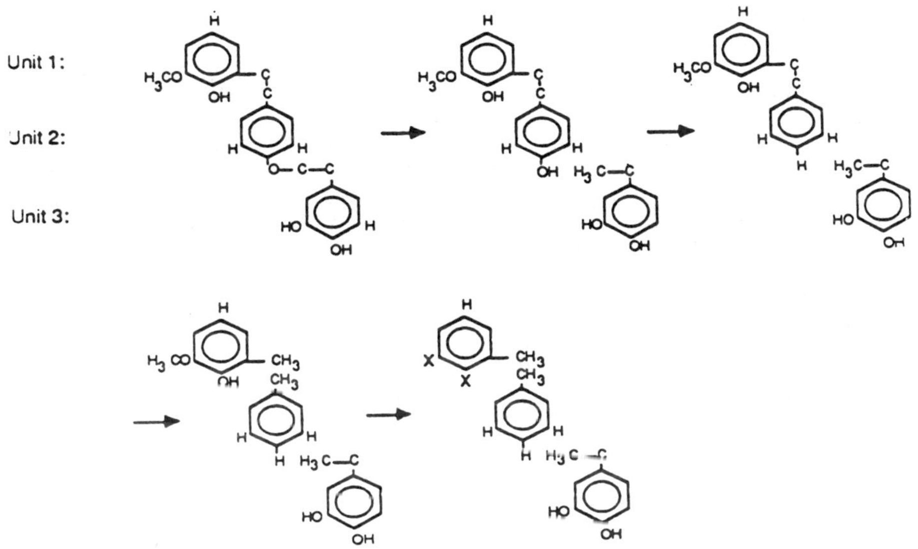

Figure 7. Sample Reaction Trajectory of a Model Lignin Oligomer.

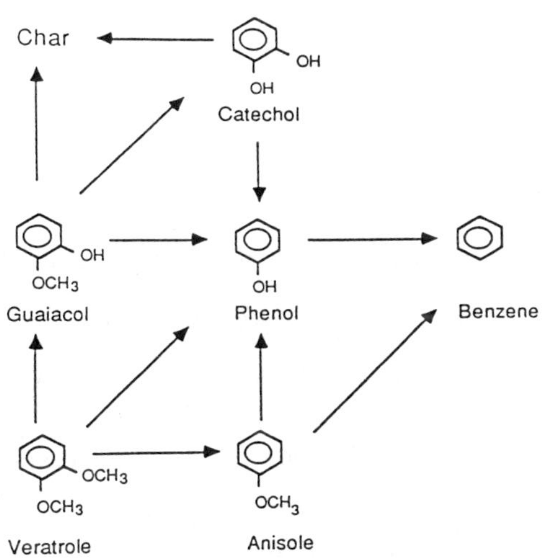

Figure 8. Operative P-position Reaction Pathways.

Pyrolysis. Figure 9 shows the predicted yields of tar fraction products from kraft lignin pyrolysis at 380°C. The total yield of single-ring (monomeric) products increased quickly to 0.04 by 5 min and gradually rose to 0.10 after 120 min. The initially rapid depolymerization corresponded to cleavage of the thermally labile ß-ether and ethylene linkages. The ultimate yield of two-ring products reached 0.042 by 120 min. Yields of three-ring products passed through a maximum of 0.028 at 30 min, and decreased slightly to 0.022 after 120 min.

Figure 9b illustrates the temporal variation of the monomeric products within the tar fraction. Guaiacols and catechols achieved maximum yields of 0.018 and 0.024, respectively at 60 min. Yields to both decreased through secondary decomposition, and phenols became the most abundant single-ring products after about 65 min. The phenols' yield reached 0.045 by 120 min, where simulated yields of anisoles and hydrocarbon products reached 0.007 and 0.002 respectively.

Thus, roughly 66% of the single-ring product fraction from pyrolysis consisted of compounds with two oxygen-containing substituents at 60 min. The balance was composed of monophenolics, anisoles and hydrocarbons, which in total were found in yields nearly equal to that of the total dioxygenates after 120 min.

The yields of gases and light liquids rose steadily with pyrolysis time. After 120 min, the yield of methane reached 0.005, while that of carbon monoxide approached 0.003. Each evolved exclusively from reactions of methoxyl substituents. The lack of appreciable product water underscored the thermal stability of hydroxyl substituents.

Residue was the material unaccountable for in the light product and tar fraction. At low times this could be viewed as "unreacted" lignin. The simulated residue yield decreased to 0.093 by 5-7 min, during which time the rate of monomer formation was greatest. The residue yield decreased gradually to 0.85 by 120 min.

Catalytic Liquefaction. The predicted yields of products in the tar fraction are illustrated in Figure 10. Yields of one-, two-, and three-ring products are depicted in Figure10a and increased with time to 0.24, 0.09, and 0.04, respectively by 30 min. The rate of formation of each was most rapid in the first five minutes.

Figure 10b shows that products with ortho oxygen-containing substitutents accounted for only a small fraction of the total single-ring product yield. The yields of guaiacols and catechols reached 0.02 and 0.04 by 4 min, and decreased with time to trace levels. Monophenolics were the most abundant products, and reached a total yield of 0.19 after 30 min. The yield of single-ring hydrocarbons rose steadily to 0.03.

Water, CO, and methane were the major light products and reached yields of 0.0058, 0.0038 and 0.0048 respectively, after 30 min. Both methane and carbon monoxide were primary products whereas water, evolved by ring dehydroxylation, was largely a secondary product from the reactions of phenolic units.

The residue yield decreased to 0.65 by 30 min, which represents higher conversion than attained in neat pyrolysis because of the catalytic cleavage of thermally stable diphenyl alkanes and ethers.

Figure 11 summarizes the effects of catalyst deactivation and internal transport limitations. Simulations run in which both were either included or neglected provided limiting cases. Intrinsic kinetics resulted in the highest yields of single-ring products, where allowing for catalyst deactivation and diffusional limitations resulted in the lowest single-ring product yield.

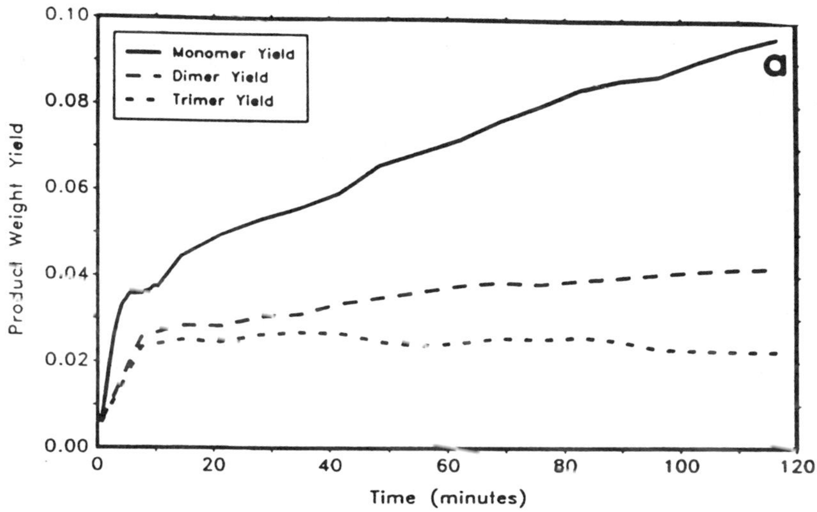

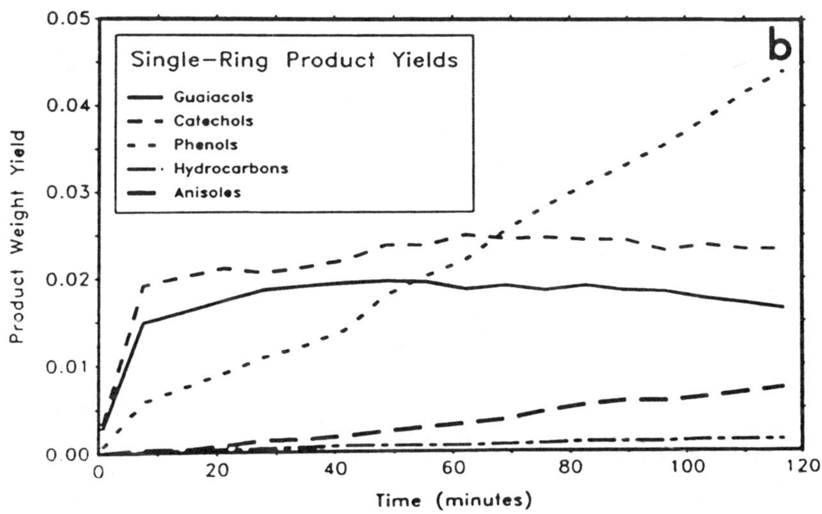

Figure 9. Predicted Tar Fraction Product Yields of Kraft Lignin Pyrolysis at 380°C.
a. Oligomers' Totals.
b. Monomers' Totals

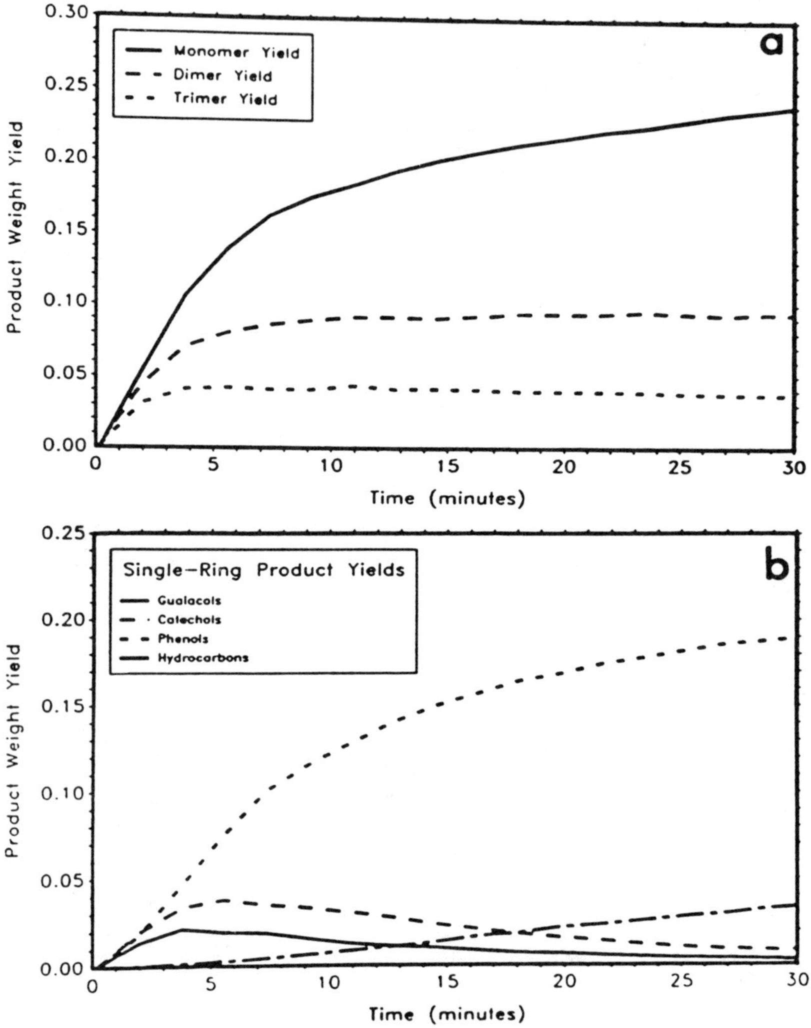

Figure 10. Predicted Tar Fraction Product Yields of Kraft Lignin Liquefaction at 380°C.
a. Oligomers' Totals.
b. Monomers' Totals.

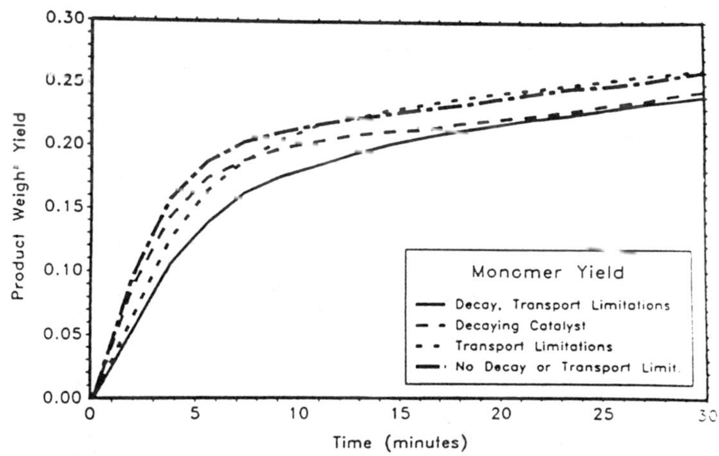

Figure 11. Influence of Catalyst Decay and Effectiveness on Monomers' Yield.

The effects of polymer diffusion and catalyst deactivation were also considered separately. The remaining curves of Figure 11 illustrate. The diffusional limitation was largest initially, where polymer molecular weight was high, and single-ring products evolved at lower rates initially than in the limiting case of no internal transport limitations. Moreover, inspection of the monomer yield after 30 min showed that transport had little effect on the ultimate evolution of single-ring products. Finally, the remaining curve in Figure 11 illustrates that the effect of catalyst deactivation was largest at longer reaction times, and lower ultimate yields of single-ring products were achieved.

Summary of Kraft Lignin Simulations. The foregoing simulations reveal the dramatic influence of the catalyst on the yields of single-ring products; the residue yield correspondingly decreased by roughly 50% when a catalyst was added. The identities of the major products in the single-ring product fraction were also affected. Whereas products with two oxygen-containing substituents (guaiacols and catechols) accounted for nearly 50% of the monomeric products evolved from pyrolysis, they accounted for only 5% of the products resulting from catalytic liquefaction. This was attended by an increase in the yields of phenols and hydrocarbons upon addition of the catalyst.

Milled-Wood Lignin. Simulation of the thermal and catalytic liquefaction of milled-wood lignin was at 380°C and, for catalysis, 2250 psig H and a loading of $CoMo/\gamma\text{-}Al_2O_3$ of 2.47 g_{lignin}/g_{cat}.

Pyrolysis. The temporal variation of the yields of predicted products are illustrated in Figure 12. The total yield of single-ring products rapidly reached 0.43 by 10 min, after which time their reaction to char caused their yields to decrease to 0.32 by 120 min. The yields of two-and three-ring products reached 0.13 and 0.05, respectively, by 10 min, and thereafter decreased to 0.07 and 0.03 by 120 min.

Within the single-ring product fraction, the guaiacols' yield shown in Figure 12b reached 0.33 by about 10 min. Their secondary decomposition to catechols, phenols and char led to a rapid decrease in their yield to a value of 0.07 by 120 min. The yields of catechols and phenols reached about 0.08 and 0.16, respectively, by about 120 min. The catechols underwent secondary decomposition to phenols and char; neither hydrocarbons nor anisoles formed. Thus, the tar fraction was composed primarily of compounds with two oxygen-containing substituents.

Methane and CO were the only light products since phenolic hydroxyls were thermally stable. The yield of the former reached 0.010, whereas the maximum CO yield was 0.0064. After rapid decreasing to 0.43, as the reactive other linkages cleaved, the residue yield increased to 0.6 by 120 min while the single-ring product yield decreased. This was a result of the secondary reactions of the guaiacyl- and dihydroxyl-containing compounds to char.

Catalytic Liquefaction. Results of the simulated catalytic liquefaction of milled-wood lignin are illustrated in Figure 13. The total yield of single-ring products in Figure 13a increased with time to 0.57 after 30 min. Dimer and trimer yields attained maxima of 0.13 and 0.05, respectively, at 2 min, after which time they decreased to 0.08 and 0.01 by 30 min.

The variations with time of the yields of the single-ring products are illustrated in Figure 13b. Yields of guaiacols reached 0.13 by 2 min, and decreased to trace levels by 30 min. The formation of catechols was not as rapid, as product yields reached 0.12 after 4 min. The yield of catechol decreased to zero through secondary reactions to phenols. Their yield rose to 0.43 by 15 min,

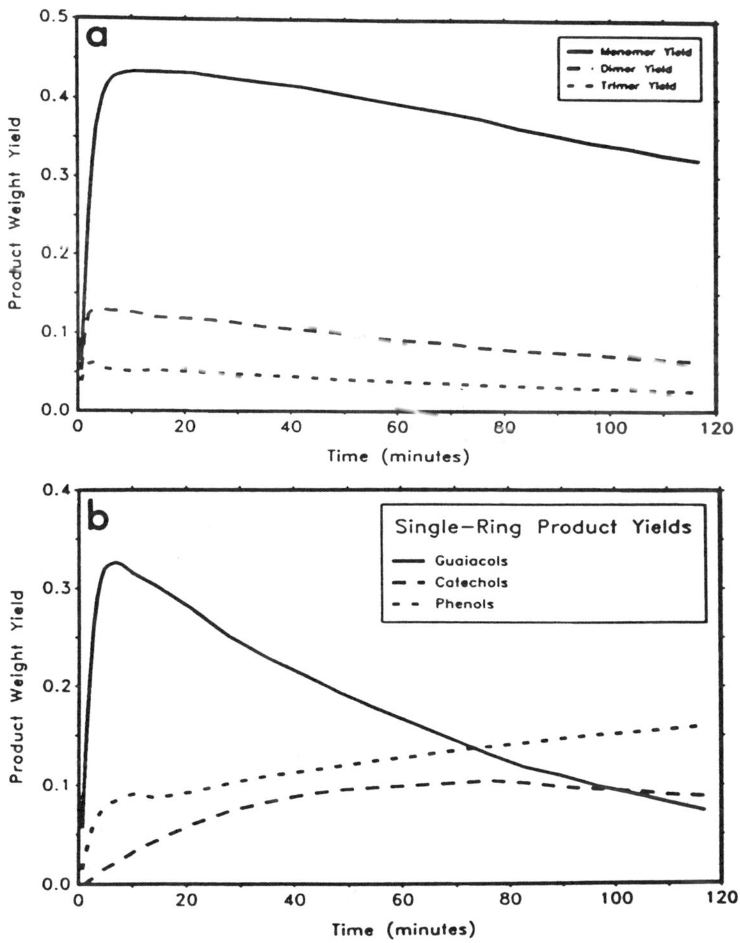

Figure 12. Predicted Tar Fraction Product Yields of Milled-Wood Lignin Pyrolysis at 380°C.
a. Oligomers' Totals.
b. Monomers' Totals.

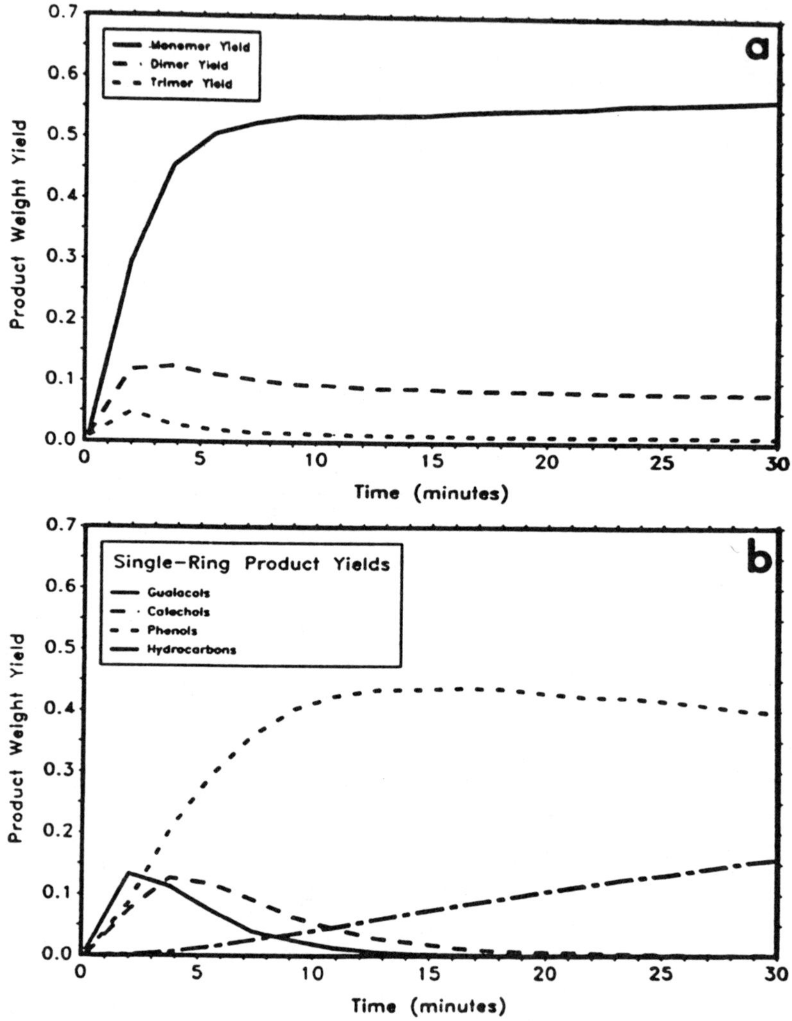

Figure 13. Predicted Tar Fraction Product Yields of Milled-Wood Lignin Liquefaction at 380°C.
a. Oligomers' Totals.
b. Monomers' Totals.

and their secondary decomposition was to hydrocarbons, which accounted for 0.16 of the tar fraction at 30 min.

The light products were methane, CO, and H_2O. Both methane and carbon monoxide were primary products, and their respective yields reached 0.007 and 0.011 by 15 min. A secondary product, water, attained yields of 0.015 by 30 min. The yield of water was high because HDO reactions of catechols were facile over the catalyst. The residue fraction decreased to 0.35 by 5 min.

The effects of catalyst decay and transport were qualitatively similar to those discussed for the simulation of kraft lignin. Intrinsic kinetics resulted in the highest yields of monomeric products, and the presence of both deactivation and internal transport limitations resulted in the lowest.

In summary, the addition of the catalyst increased the yield of the single-ring product fraction. The sum of the yields of single-ring products formed from thermolysis reached a maximum value of 0.43 before decreasing to 0.36 by 120 min. In the catalytic simulation, secondary reactions to stable single-ring products were competitive with reactions forming char. This is because HDO of phenolics with two oxygen-containing substitutents, which, thermally, char readily, was facile over the catalyst. Thus the yield of single ring products resulting from the simulation of catalytic liquefaction increased rapidly to 0.53, and thereafter rose gradually to 0.61 after 120 min.

Summary

A stochastic formalism for using the experimental results of model systems in the analysis of the reaction of complex macromolecular substrates has been developed. The approach is flexible and renders lignin structure and its implications explicit; the Monte Carlo model can easily accommodate a variety of molecular weight distributions, lignin types and processing strategies. In addition, the effects of catalyst deactivation and transport limitations in catalyst pores were modelled easily by coupling a diffusion model to species' molecular weights. Evaluation of the approach is forthcoming in a detailed comparison of predictions vs. pyrolysis and catalytic liquefaction experiments with actual lignins.

Acknowledgments

This work was supported by the NSF (Grant No. CPE 840 4452). We are grateful for useful discussions with Dr. John B. McDermott.

Notation

k_{AB}	pseudo first-order rate constant, s-1
η	effectiveness factor
ø	catalyst activity function
ω_o	catalyst loading, g_c/cm^3
k_c	catalytic rate constant, cm^3/g_cs
k_t	thermal rate constant, s^{-1}
ϕ	Thiele Modulus
r_p	pellet radius, cm

ρ_c catalyst density, g/cm^3

D_e effective diffusivity, cm^2/s

P probability

DP degree of polymerization

Δt time interval

α deactivation rate constant, g_c/mol O unrecovered

C_c mol oxygen unrecovered per unit mass of catalyst, mol O/g_c

Literature Cited

[1] Boudart, Michel. *Kinetics of Chemical Processes*; Prentice-Hall, Englewood Cliffs, NJ , 1968.

[2] Spry, J.C. and Sawyer, W.H., Paper presented at 68th Annual A.I.Ch.E. Meeting; Los Angeles, CA (November 1975)

[3] Klein, M.T. and Virk, P.S. MIT Press Energy Lab Series, 1981.5.

[4] Petrocelli, F.P. and Klein, M.T. *Chem. Eng. Commun.*, **30**(6), 343, 1984.

[4b] Petrocelli, F.P. and Klein, M.T. *Fuel Science and Technology International*, **5**(3), 291-327, 1987.

[5] Squire, K.R., Solomon, P.R., Di Taranton, M.B., and Caragelo, R.M. *ACS Division of Fuel Chemistry Preprints* **30**(1):386, 1985.

[6] McDermott, J.B., Libanati, C., LaMarca, C., and Klein, M.T. A Stochastic Model of Depolymerization Kinetics 1. Development Approach. (in press)

[7] Pellinen, J. and Salkinoja-Salonen, M. *J. Chromotagr.* **328**:299, 1985.

[8] Soundararajan,T N. and Wayman, M. *J. Polymer Sci., Part C* **30**(1):386, 1985.

[9] Marton, J. *Lignins: Occurrence. Formation. Structure and Reactions.* Wiley, New York, 1971.

[10] Freudenberg, K. *Science* **148**:595, 1965.

[11] Froment, G.F. and Bischoff, K.B. *Chemical Reactor Analysis and Design.* Wiley, New York, 1979

[12] Vogelzang, M.W., Li, C-L., Schuit, G.C.A., Gates, B.C. and Petrakis, L. *Journal of Catalysis* **84**:170,1983

[13] Train, P.M. *Stochastic and Chemical Modelling of Lignin Liquefaction.* Ph.D. Thesis, University of Delaware, 1987.

[14] Baltus, R.E. and Anderson, J.L. *Chem. Eng. Sci.*, **38**(12), 1959-1969, 1983.

[15] Rouse, P.E. *Journal of Chemical Physics* **21**(7):1272, 1953.

[16] de Gennes, P.-G. *Journal of Chemical Physics* **55**(2), 1971.

[17] Sherwood, T.K., Pigford, R.L. and Wilke, C.R. *Mass Transfer.* McGraw-Hill, New York, 1975.

[18] McDermott, J.B., Klein, M.T. and Obst, J.R. *Ind. Eng. Chem. Process Des. Dev.*, **25**(4), 1986.

[19] Klein, M.T. *Model Pathways in Lignin Thermolysis..* Ph.D. Thesis, Massachusetts Institute of Technology, 1981.

[20] Jegers, H.E. *Thermolysis of Lignin..* Masters thesis, University of Delaware, 1982.

[21] Petrocelli, F.P. and Klein, M.T. *Macromolecules* **17**:161, 1984.

[22] Klein, M.T. and Virk, P.S. *Ind. Eng. Chem. Fundam.* **22**:35, 1983.

[23] McDermott, J.B. *Chemical and Stochastic Modelling of Complex Reactions: A Lignin Depolymerization Example*. Ph.D. Thesis. University of Delaware. 1986
[24] Petrocelli, F.P. *Chemical Modelling of the Thermal and Catalytic Depolymerization of Lignin*. Ph.D. Thesis, University of Delaware, 1985.
[25] Hurff, S.J. and Klein, M.T. *Ind. Eng. Chem. Fundam.* **22**:426, 1983.

RECEIVED April 14, 1988

Chapter 23

Biomass to Gasoline

Upgrading Pyrolysis Vapors to Aromatic Gasoline with Zeolite Catalysis at Atmospheric Pressure

James Diebold and John Scahill

Chemical Conversion Research Branch, Solar Energy Research Institute, 1617 Cole Boulevard, Golden, CO 80401

The primary pyrolysis vapors generated by the fast pyrolysis of biomass at atmospheric pressures consist initially of low-molecular-weight compounds, but which polymerize upon condensation. Prior to condensation, these primary vapors have been found to be very reactive with ZSM-5 catalyst to produce methyl benzenes boiling in the gasoline range. This gasoline is predicted to have very high blending octane numbers. By-products are coke, carbon oxides, water, naphthalenes, ethylene, propylene, and some phenols. The effect of different by-products on the theoretical gasoline yield is examined. Preliminary results, generated with a reactor having a fixed bed of 100 g of catalyst, are examined for the continuous feeding of never-condensed primary vapors and compared to feeding methanol in the same reactor. The conversion of primary pyrolysis vapors made from biomass is a relatively new research and development area which is showing early promise. The extent to which the product slate can be manipulated by process variables will impact heavily on the viability of this process.

The conversion of biomass materials to high octane gasoline has been actively pursued for many years. Historically, methanol was made in very low yields by the destructive distillation of hardwoods. More recently, the manufacture of methanol has been by the reaction of synthesis gas over catalysts at high pressures. In theory, any carbon source can be used for this catalytic generation of methanol, but in practice, biomass has not been advantageous relative to coal or natural gas. Other approaches to making liquid fuel from biomass have involved the fermentation of biomass to ethanol in a rather slow process. The conversion of biomass to alcohols is technically feasible, but the utilization of the alcohols as transportation fuels will require modifications to the

0097–6156/88/0376–0264$06.00/0

distribution systems and to the individual automobiles. The high-pressure liquefaction of biomass to oxygenated liquids followed by high-pressure catalytic hydrogenation to form hydrocarbons is one approach to convert biomass to liquid fuels. (1) However, in the last decade, Mobil has developed the use of a zeolite catalyst for the conversion of methanol to gasoline at atmospheric pressures. (2) This process has recently been commercialized and is now in operation in New Zealand. (3) The zeolite catalyst used in the Mobil process is a medium pore zeolite, which has shape selectivity to restrict the products from methanol to methylated benzenes, isoparaffins, and olefins, while preventing the formation of coke in the catalyst pores. (4) This catalyst is known as ZSM-5, and its commercial use is controlled by Mobil.

The reactivity of high-molecular-weight vegetable oils with ZSM-5 was reported in 1979 (5) and, in fact, ZSM-5 catalyst is very reactive toward most small oxygenated species to convert them to methylated benzenes and other products. (6-7) Although alcohols appear to be some of the best feedstocks for ZSM-5 catalysis due to their low coking tendencies, the petroleum industry has long made use of zeolite catalysts for the cracking of very heavy hydrocarbons to produce gasoline and about 5 to 15 weight percent coke. This suggests that the formation of coke and the need for frequent catalyst regeneration will heavily impact the reactor design, but that significant coke formation can be part of a viable commercial process. The thrusts of our early efforts were to briefly examine the effect of possible stoichiometries and to generate preliminary experimental results from passing primary pyrolysis vapors, made by the fast pyrolysis of sawdust at atmospheric pressure in a vortex reactor, over a fixed bed of ZSM-5 catalyst. The production of these oxygenated vapors is addressed in another chapter of this book.

Stoichiometry Considerations

Although the hydrocarbon products have an unusual feedstock independence, the chemistry involved with different feedstocks over the ZSM-5 catalyst varies considerably with the functionality of the oxygen in the feedstock. As seen in Table I, hydroxyl and methoxy groups tend to reject oxygen in the form of water; aryl ethers reject a nearly equal amount of oxygen in water and carbon monoxide; carbonyl and formate groups reject oxygen largely as carbon monoxide, and carboxyl groups reject oxygen mostly as carbon dioxide and water. With the model compounds listed in Table I, many of these trends may also be a function of reaction conditions as well as reactants. A model compound study coupled with a process variable study is in progress at SERI with the free-jet, molecular beam/mass spectrometer (FJMBMS) (see chapter in this book by Evans and Milne).

The method of oxygen rejection which occurs over the catalyst has a very important impact on the potential yield of hydrocarbons, especially for a feedstock like biomass which has a relatively low hydrogen content. Although the products formed from a few compounds reacted with ZSM-5 are known for certain conditions, methods to manipulate the by-product slate are essentially unreported.

TABLE I. Reported Distribution of Oxygen in Inorganic By-Products with Various Feedstocks over ZSM-5 Catalyst

Compound	Oxygen Radical in Reactant*	% of Oxygen in By-Products			
		H_2O	CO	CO_2	Ref.
methanol	H, M	100	--	--	(8)
dimethyl ether	E, M	100	--	--	(8)
guaiacol	H, M	96	3	1	(10)
glycerol	H	92	7.5	0.5	(9)
xylenol	H	93	6	1	(10)
eugenol	H, M	89	9	2	(10)
anisole	M	88	12	tr	(10)
2,4 dimethyl phenol	H	87	12	1	(10)
o-cresol	H	80	17	3	(10)
starch	H, E	78	20	2	(7)
isoeugenol	H, M	77	19	4	(10)
glucose	C1, H, E	75	20	5	(7)
dimethoxymethane	M, E	73	6	21	(8)
xylose	C1, H, E	60	35	5	(7)
sucrose	H, E	56	36	8	(7)
n-butyl formate	C2	54	46	0	(8)
diphenyl ether	E	46	46	8	(10)
furfural	C1, E	14-22	75-84	2.5-3.0	(9,7)
methyl acetate	C2	54	10	36	(11)
acetic acid	C2	50	4	46	(11)

*H = hydroxy; M = methoxy; E = C-O-C; C1 = carbonyl; C2 = carboxy

However, the desirability to reject oxygen as carbon oxides becomes quite obvious by examining potential product slates which are possible from the stoichiometry of the reacting primary pyrolysis vapors, $CH_{1.2}O_{0.49}$. The best hydrocarbon yields would be attained with oxygen rejection as carbon dioxide, with the excess hydrogen used to reject oxygen as water. If the liquid hydrocarbon product corresponds to xylene, C_8H_{10}, rather than to more hydrogen-rich hydrocarbons such as olefins, C_nH_{2n}, then the potential hydrocarbon yield will be higher from a hydrogen-poor feedstock like wood pyrolysis vapors. If carbon monoxide is the assumed carbon oxide, more carbon is needed to reject the oxygen, which decreases the potential hydrocarbon yields. If the by-product gases are a mixture of carbon oxides, methane, olefins, etc., then the gasoline yields would be lowered due to the noncondensible hydrocarbons in the gases. If the olefins in the by-product gases were recycled through the catalyst bed to extinction, the gasoline yields should increase not only due to the conversion of the olefins to aromatics, but also due to the release of hydrogen in the aromatization process which could help counter coke forming (dehydrogenation) reactions. Coke formation is typically an aromatization reaction to produce polycyclic aromatic hydrocarbons containing residual hydrogen, so a reasonable coking reaction to assume is one which produces water and coke but no gasoline at all. In summary, the more desired reactions produce carbon oxides and water as by-products. Undesired reactions produce noncondensible hydrocarbons and/or coke and water.(12)

Experimental

The primary pyrolysis vapors, used as feedstock, were produced by fast pyrolysis in a vortex reactor from coarse softwood sawdust, as discussed in another chapter of this book. After the vapors left the vortex reactor system, they were allowed to cool to the desired catalytic reaction temperature, as they passed through a tubular transfer line to the catalytic reactor. The transfer line was located inside of a series of six tubular furnaces, which allowed the vapor stream to equilibrate to the desired temperature. The residence time of the vapor in the transfer line was about one-half second prior to reaching the 2.5-cm diameter catalytic reactor, shown in cross-sectional view in Figure 1. The catalytic reactor used with the pyrolysis vapors had a 30-cm-long fixed bed of 100 g of ZSM-5 containing catalyst (MCSG-2), which was located in the middle of the sixth furnace section. Only a small slipstream of the pyrolysis vapors passed through the catalytic reactor. The catalyst was in the form of 1.4-mm diameter extrudate and supplied by Mobil Research and Development Corp. in a cooperative agreement with the Solar Energy Research Institute. Although the catalyst composition is proprietary to Mobil, the ability of the catalyst to maintain good activity in the presence of high-temperature-steam environments implies a fairly high silicon to aluminum ratio in the ZSM-5 crystal lattice. The temperature of the catalyst bed was measured with an axial thermocouple inside a 6-mm thermowell. The temperature profile of the bed was determined by moving the axial thermocouple within the thermowell. A sintered stainless steel filter rated at 5 micrometers was used to remove char fines from the pyrolysis vapors. The products were collected in water-cooled condensers. The pressure in the reactor was slightly above the local atmospheric pressure at about 95 kPa.

Analysis of the organic condensates was with a 5-micrometer, wide bore capillary column having a length of 60 m. The capillary column was coated with one micrometer of cross-linked methyl silicones. With helium as the carrier gas, the temperature profile started at 0°C for 4 minutes, followed by a temperature ramp of 8°C/min until a temperature of 260°C was reached. Detection of the eluted organics was by flame ionization detection (FID). Identification of the major peaks was by reference materials, whereas the minor peaks were identified by a combination of the FJMBMS at SERI and a Hewlett-Packard 5890 gas chromatograph/5970 Mass Selective Detector (GC/MSD) using a 1.5-m high-performance cross-linked methyl silicone capillary column. The noncondensible gases were analyzed with a Carle GC designed for refinery gas analysis, which used thermal conductivity detection (TCD) and was calibrated with a gravimetrically prepared reference mixture. Electronic grade methanol (99.9% pure) was used for comparison to the softwood feedstock.

Experimental Runs

Methanol. To verify the activity of the catalyst and to gain experience in the operation of the catalyst system, methanol was metered into a preheater tube located inside of the transfer line heated to 500°C. This preheating proved to be too severe and the products which emerged from the catalytic reactor were dominated by hydrogen and carbon monoxide in a 2:1 ratio, as shown in Table II. This would be expected from thermal decomposition of the methanol prior to reaching the catalyst. This experiment was repeated using a preheating temperature ramp to just reach 400°C at the entrance to the catalytic reactor. The catalytic reactor was held at a nominal 400°C prior to the addition of the methanol at a space velocity (WHSV) of 0.9 g methanol per gram of catalyst per hour. The noncondensible gas composition is shown in Table II and was rich in hydrogen, olefins (ethylene and propylene), and alkanes (methane, propane, and isobutane) The gaseous olefins could be used to alkylate the reactive isobutane to result in a highly branched-chain gasoline fraction. (13) The liquid product contained relatively little alkanes or olefins and was dominated by methylated benzenes, such as toluene, xylenes, and trimethyl benzenes. Relatively very small amounts of naphthalenes were seen. The temperature profile of the catalytic reactor immediately prior to and also during this experiment is shown in Figure 2. The location of the temperature-profile maximum was quite stable, which indicates that the catalyst was not significantly deactivated during the short time of the experiment. This low rate of catalyst deactivation is consistent with published data. (14) The ratio of gasoline to water in the condensates suggests that the gasoline yield was only about one-third of the potential due to the formation of the noncondensible hydrocarbons. This product slate is

TABLE II Calculated Molar Compositions of Net Product Gases Produced During Cracking of Primary Pyrolysis Vapors or Methanol (ZSM-5 containing catalyst - Mobil's MCSG-2)

Feed	Softwood Pyrolysis Vapor ($CH_{1.2}O_{0.49}$)			Methanol (CH_4O)	
Reactor	Thermal (Run 55) (650°C)	Catalytic (Run 76-C) (400°C)	Catalytic (Run 88-C) (485°C)	Thermal (Run 74-C)* (500°C)	Catalytic (Run 75-C) (400°C)
H_2	10.6	-0.6	0.7	62.8	18.4
CO	59.4	69.8	62.4	31.1	4.4
CO_2	5.6	15.2	14.0	1.4	2.8
CH_4	12.2	1.4	2.2	2.1	16.2
C_2H_2	0.5	--	--	--	--
C_2H_4	5.3	5.0	8.2	1.6	5.3
C_2H_6	1.1	0.3	0.3	0.3	4.1
C_3H_6	1.9	6.0	6.7	0.7	5.3
C_3H_8	--	0.5	0.2	0.4	17.8
C_4H_8	1.1	1.2	1.2	--	2.3
iso-C_4H_{10}	--	0.2	0.1	0.1	14.4
n-C_4H_{10}	--	--	--	--	5.3
C_5+	2.4	--	3.9	--	3.6
Empirical Formula	$CH_{1.3}O_{0.65}$	$CH_{0.65}O_{0.82}$	$CH_{0.87}O_{0.61}$	$CH_{3.6}O_{0.8}$	$CH_{2.7}O_{0.05}$

*In this run, the methanol is thought to have thermally decomposed for the most part prior to reaching the catalyst.

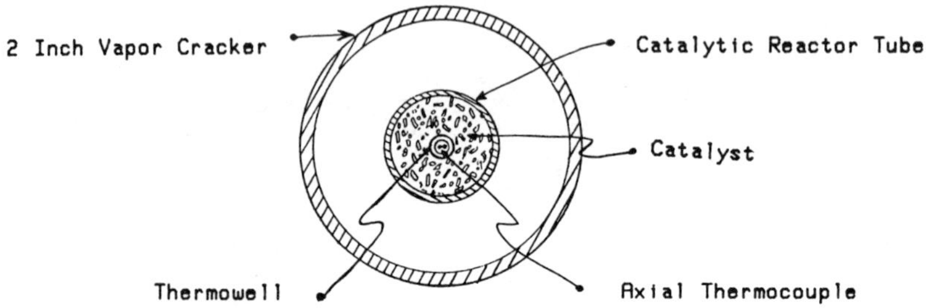

Figure 1. Cross-sectional view of catalytic reactor

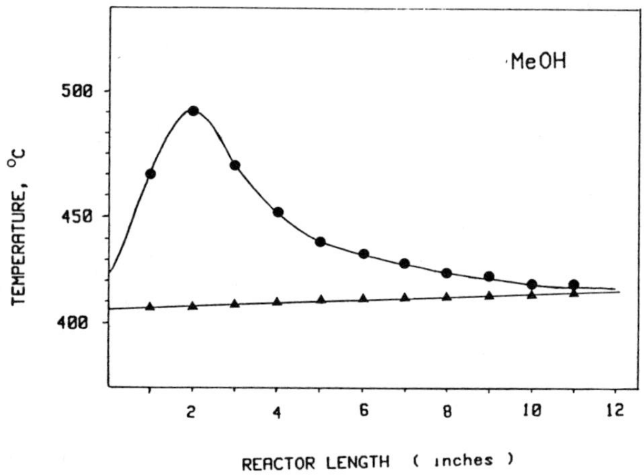

Figure 2. Temperature profile of reactor with methanol feed (lower curve prior to feeding)

in general agreement with data reported by Mobil for similar reaction conditions. (15)

Primary Pyrolysis Vapors. After a catalyst regeneration cycle to remove residue from the methanol experiments, a slipstream of the primary pyrolysis vapors was passed over the ZSM-5 catalyst. Steam was used as the carrier gas at a weight ratio of two parts of steam to one part of wood feed. The pyrolysis vapors were cooled from 510°C at the exit of the vortex reactor to 400°C at the entrance of the catalytic reactor. Figure 3 shows the temperature profile immediately prior to feeding the biomass, as well as 5-10 minutes later. The temperature profile with pyrolysis vapors as feed was not as large in magnitude as that seen with methanol and it had a very broad maximum. The heat of reaction of the pyrolysis vapors is less exothermic than for methanol and the steam carrier gas also has a moderating effect on the temperature rise. The broadness of the temperature profile reflects that the pyrolysis vapors are a complex mixture of compounds, which are probably reacting at different rates. The broader and lower temperature profile would make temperature control easier in a biomass-to-gasoline (BTG) reactor than in a methanol-to-gasoline (MTG) reactor. The location of the temperature maximum was monitored during the run, as shown in Figure 4. During the fairly short experiment, the temperature maximum was observed to move to the end of the reactor, indicating that a fairly rapid deactivation of the catalyst had occurred. During this time, the composition of the hydrocarbon products appeared to be relatively constant.

The gasoline fraction (eluting before naphthalene) made from wood is very similar by GC to that made from methanol, except for the presence of some phenolics in the former. The composition of the gas formed over the catalyst from the pyrolysis vapors is shown in Table II, as calculated from tracer gas concentrations before and after the reactor, along with the gas composition formed by the thermal decomposition of the pyrolysis vapors as determined previously (16); the catalytically formed gases had significantly less hydrogen and methane, but more carbon dioxide and propylene than the thermally formed gases. In comparing the composition of the gases formed by the catalytic conversion of the pyrolysis vapors to the catalytic conversion of the methanol, the relative hydrogen richness of the methanol becomes apparent in the relatively high hydrogen, methane, propane, and isobutane yields. This relative hydrogen richness is summarized by the empirical formulas for the two gas streams, in which the hydrogen-to-carbon ratio for the wood-derived gases is one-fourth that for the methanol-derived gases. This excessive amount of hydrogen in the methanol-derived gases suggests that methanol could be used as a hydrogen donor to hydrogen-poor feedstocks and this has been explored by several researchers. (7,9-11) The very low yield of isobutane from pyrolysis vapors would preclude the use of alkylation to incorporate the ethylene, propylene, and butenes into the gasoline product with a stand-alone process. Adsorption of the gaseous olefins from the by-product gases onto cold ZSM-5 or activated carbon at low pressures may be a viable method to recycle them into the catalytic reactor. Since gaseous olefins are known to be intermediate

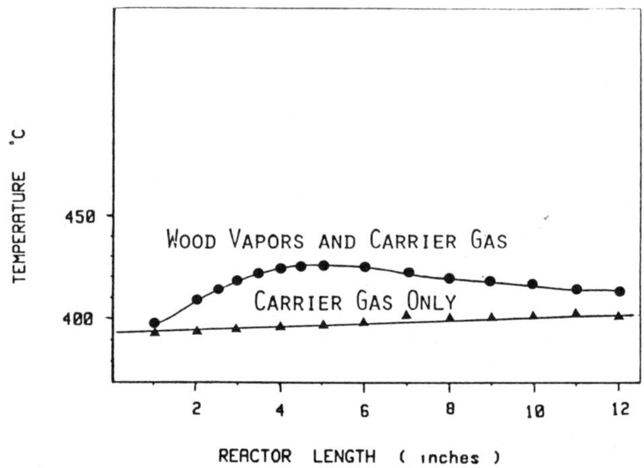

Figure 3. Temperature profile of reactor with pyrolysis vapors feed (lower curve prior to feeding)

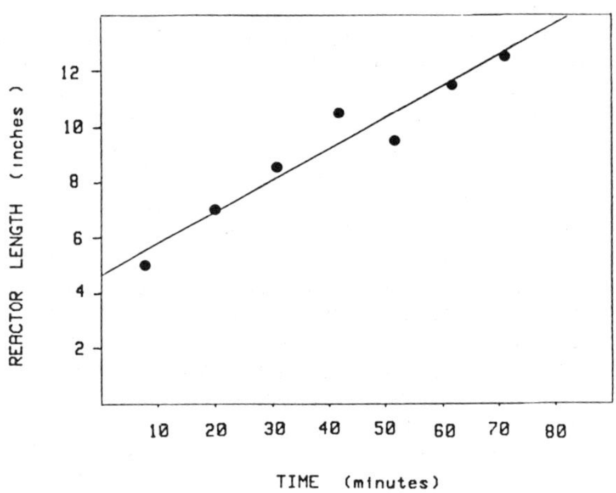

Figure 4. Location of temperature maximum with pyrolysis vapors feed

compounds in the methanol-to-gasoline process, recycling olefins would be expected to have a beneficial impact similar to cofeeding methanol with wood pyrolysis vapors.

A six-element in-line static mixer was installed at the entrance to the vapor cracker (transfer line) to ensure the complete mixing of the pyrolysis vapors, the tracer gas, and the carrier steam prior to reaching the slipstream catalytic reactor. Experimentation verified the elimination of diametrical gradients of tracer-gas concentration downstream of the static mixer. In Run 88-C, the steam-to-biomass mass ratio was 1.2 and the softwood sawdust feed rate was 18.5 kg/hr (41 lb/hr). The temperature of the pyrolysis product stream was 475°C at the exit of the vortex reactor. Nitrogen was used as a tracer gas at a flow rate of 0.224 kg/hr to result in a concentration of 10.1 wt % in the permanent pyrolysis gases and 3.6% in the permanent catalyst off-gases. Based on the nitrogen content and the flow measured with a gravimetrically calibrated orifice meter, the slipstream fed into the catalytic reactor was 1.4 wt % of the main stream. The weight hourly space velocity (WHSV) was 2.44 g dry biomass/g catalyst (MCSG-2) hr. By GC analysis, the gasoline hydrocarbon fraction of the organic condensate was about 75% and the naphthalenic/phenolic fraction was 25%. The gasoline fraction was qualitatively similar to that made previously and was heavily dominated by toluene, xylenes, and trimethyl benzenes. The olefins and C_5+ hydrocarbons in the catalyst off-gases were 4 and 2.4 wt % of the dry feed respectively. The total liquid hydrocarbon yield was 10.1 wt % which is equivalent to 28 gal/ton of dry feed, assuming the density of xylene. Assuming that the olefins can be recycled and converted to gasoline, the potential yield shown is 39 gal/ton. If the phenolics were reacted with methanol to form the methyl aryl ethers, the potential liquid fuel yields would then be increased to about 16% (16 wt % yield of the gasoline is one barrel per ton of dry wood). The gasoline boiling fraction of this fuel is projected to be about 85%; the other 15% is composed almost entirely of methylated naphthalenes which could be hydrocracked in a typical refinery to a gasoline fraction. The yield of methyl naphthalenes has been shown by the MBMS experiments (see chapter by Evans and Milne in this book) to be sensitive to the operating conditions in the zeolite reactor and it is expected that it can be reduced in the future.

The composition of the permanent gases formed over the catalyst was calculated by subtracting the pyrolysis gas yields from the total gas yields, as shown in Table II. These catalyst formed gases from Run 88 are similar to those previously shown in that the hydrogen yield was essentially zero and the yield of methane and other paraffins was very low compared to ethylene, propylene, and butenes. It is interesting to note that the amount of carbon dioxide and carbon monoxide calculated to have been formed by the catalyst in Run 88 corresponds to a potential yield of 16% gasoline (xylene) by the stoichiometry of global equations (12), which is in good agreement with the sum of the olefins and organic condensates produced. The overall yields from Run 88 are shown in Table III.

TABLE III. Run 88 Yields

Product	Wt % Dry Feed	Method of Determination
Pyrolysis		
Char	12	Gravimetric
Gases	11	Tracer gas/GC/ gravimetric
Catalysis		
Coke	8	Gravimetric
Fuel gases	16	Tracer gas/GC/orifice meter
Gaseous olefins	4	Tracer gas/GC/orifice meter
Gasoline	10	Gravimetric/GC
Phenolics/ napthalenics	2	Gravimetric/GC
Water (Total)	37	Difference
	100	

Conclusions

Pyrolysis vapors made by the fast pyrolysis of softwood are very reactive with ZSM-5 catalyst to form a liquid hydrocarbon product, which is similar to that formed from methanol. Although the catalyst is deactivated relatively quickly with pyrolysis vapors compared to methanol, experimentation indicates that the catalyst can be repeatedly regenerated by oxidation. This suggests that a catalytic reactor, which can maintain a high level of catalytic activity in spite of high coking rates, would be desired. This problem has been addressed and resolved by the petroleum refining industry which utilizes an entrained-bed reactor (riser-cracker), to crack heavy hydrocarbons to gasoline and about 5% to 15% coke, coupled with a fluidized-bed oxidative regenerator for the relatively slow, controlled oxidative coke combustion and catalyst regeneration. (17) For the conversion of biomass to gasoline, these preliminary data are quite encouraging that a nearly direct conversion of biomass to gasoline can be accomplished at atmospheric pressures in one very rapid thermal cycle without the cost of producing hydrogen and operating at high pressures. With the calculated thermal efficiencies, it appears that there will be sufficient energy in the by-products to operate the process, even including the drying of a rather wet biomass feedstock prior to pyrolysis.

The gasoline product is almost entirely alkylated benzenes with only a small amount of benzene. This gasoline would be expected to have octane ratings in excess of 100 and to have blending octane numbers between 115 and 135 based on reported blending octane values for the various alkylated benzenes. (18) Due to the expected continued demand for unleaded gasoline having higher octane numbers, the gasoline made from biomass by this process would be expected to command premium prices if sold to a petroleum refinery for blending purposes. The naphthalenic fraction of the organic products could easily be hydrocracked to increase the gasoline yields in a modern refinery. (19) At this time, the phenolic by-products appear to be present in minor amounts and could be reacted with methanol to form methyl aryl ethers which are proven octane enhancers. (20) Ongoing research is being directed toward olefin recycling and the improvement of product distribution and yields.

Acknowledgments

The financial support of the Biofuels and Municipal Waste Technology Division of the Department of Energy, FTP 619, with Mr. Simon Friedrich as DOE Program Manager and the technical direction of the Program Monitor, Dr. Don J. Stevens, of PNL, are gratefully acknowledged. Dr. M. Maholland of the Colorado School of Mines and Drs. T. A. Milne and R. J. Evans contributed product identification of many of the hydrocarbon products.

Literature Cited

1. Baker, E. G.; Elliott, D. C. Preprints of 193rd ACS Meeting, Div. Fuel Chemistry, Denver, CO, April 5-10, 1987, Vol. 32, No. 2, pp. 257-263.
2. Meisel, S. L.; McCullough, J. P.; Lechthaler, C. H.; Weiz, P. B. 1976. CHEMTECH February, 86-89.
3. Anon. Chem. Eng., March 3, 1986, p. 10.
4. Dejaive, P.; Auroux, A.; Gravelle, P. C.; Vedrine, J. C.; Gabelica, Z.; Derouane, E. G. J. Cat. 1981, 70, 123-136.
5. Weisz, P. B.; Haag, W. O.; Rodewald, P. G. Science 1979, 206, 5 October, 57-58.
6. Diebold, J. P.; Chum, H. L.; Evans, R. J.: Milne, T. A.; Reed, T. B.; Scahill, J. W. In Energy from Biomass and Wastes X, 1987, Klass, D., ed; Elsevier Applied Science Publishers: London, p. 801.
7. Chen, N. Y.; Degnan, T. F. Jr.; Koenig, L. R.; CHEMTECH 1986, August, 506-511.
8. Chang, C. D.; Silvestri, A. J.; J. Cat. 1977, 47, 249-259.
9. Hanniff, M. I.; Dao, L. H.; Energy from Biomass and Wastes X 1987, Klass, D., ed; Elsevier Applied Science Publishers: London, p. 831-843.
10. Chantal, P. D.; Kaliaguine, S.; Grandmaison, J. L. Appl. Cat. 1985, 18, 133-145.
11. Chang, C. D.; Chen, N. Y,; Koenig, L. R.; Walsh, D. E. Preprints of 185th ACS Meeting, Div. of Fuel Chem. 1983, Seattle, WA, March 20-25, 146-152.
12. Diebold, J. D.; Scahill, J. W. Annual AICHE Meeting, Houston, TX., March 29-April 2, 1987, Paper 39a. (to appear in Energy Progress).

13. Kam, A. Y.; Lee, W. Fluid Bed Process Studies on Selective Conversion of Methanol to High-Octane Gasoline, Mobil Research and Development Corp. 1978, U.S. DOE Contract No. EX-76-C-01-2490 FE-2490-15.
14. Chang, C. D. Hydrocarbons from Methanol, Marcel Dekker, NY, 1983.
15. Yurchak, S.; Voltz, S. E.; Warner, J. P. Ind. Eng. Chem. Process. Des. Dev. 1979, 18(3), 527-534.
16. Diebold, J. P. MS Thesis T-3007, Colorado School of Mines, Golden, CO 1985.
17. Decroocq, D. Catalytic Cracking of Heavy Petroleum Fractions, Gulf Publishing Company 1984, Houston, TX.
18. Lovell, W. G. Ind. Eng. Chem. 1948, 40(12), 2388-2438.
19. Gary, J. H.; Handwerk, G. E. Petroleum Refining, Technology, and Economics, Marcel Dekker, NY, 1975, pp. 121-141.
20. Singerman, G. M. Methyl Aryl Ethers from Coal Liquids as Gasoline Extenders and Octane Improvers. Gulf Research and Development Co., Pittsburgh, PA. 1980. DOE/CE/50022-1.

RECEIVED July 1, 1988

Chapter 24

Fluidized-Bed Upgrading of Wood Pyrolysis Liquids and Related Compounds

N. Y. Chen, D. E. Walsh, and L. R. Koenig

Mobil Research and Development Corporation, Central Research Laboratory, P.O. Box 1025, Princeton, NJ 08540

The effective hydrogen index (EHI) is a calculated indicator of the "net" hydrogen/carbon ratio of a pure or mixed heteroatom-containing feed, after debiting the feed's hydrogen content for complete conversion of heteroatoms to NH_3, H_2S, and H_2O. Compounds with EHI's less than about 1 are difficult to upgrade to premium products over ZSM-5 catalyst due to rapid catalyst aging in continuous fixed bed processing.

However, high conversions of such feeds (acetic acid, methyl acetate, and wood pyrolysis liquids) can be maintained in a fluidized bed system operating under methanol-to-gasoline conditions and employing frequent catalyst regenerations. Also, synergistic effects were observed for blends of low and high EHI model compounds, as well as for wood pyrolysis liquids coprocessed with methanol. Based on these results, a processing scheme is described in which the char from the wood pyrolysis step is gasified to make methanol which is cofed with the pyrolysis liquids over ZSM-5.

Zeolite ZSM-5 is particularly effective for the conversion of methanol to gasoline range hydrocarbons (1). In addition to methanol, other oxygenates, including their complex mixtures, can be converted as well (see, for example, references 2-8).

0097–6156/88/0376–0277$06.00/0

The effective hydrogen index is defined as:

$$(H/C)_{effective} \text{ or EHI} = \frac{H-2O-3N-2S}{C}$$

where H, C, O, N, and S are atoms per unit weight of sample of hydrogen, carbon, oxygen, nitrogen and sulfur, respectively.

Model oxygen containing compounds, or "oxygenates," having EHI's <1 produce excessive amounts of "coke", which leads to rapid catalyst deactivation (5)(9) using ZSM-5. Thus, as shown in Table I, initially complete acetic acid conversion declines to about 60% after only 3 hours on stream. At that point, the total hydrocarbon yield is less than 10 wt.%, and the gasoline yield (C_5+) is less than 7% (8).

Ligno-cellulosic biomass is a resource from which liquid hydrocarbon fuels potentially may be derived. Pyrolyzing the wood yields gas and liquid products, but a relatively large percentage of the original wood carbon can be lost to a low value char by-product. Furthermore, like the model oxygenates described above, the EHI of the pyrolysis liquid products is substantially less than 1.

The present study was undertaken to examine the potential of short contact time regenerative fluid bed processing to obviate the problems associated with rapid deactivation in fixed bed catalytic upgrading of such low EHI feeds.

In fixed bed operation, reduced catalyst aging rate and synergistic yields benefits have been reported for co-processing a low EHI (<1) model compound (e.g.acetic acid) with a sufficient amount of a high EHI (>1) compound (e.g. methanol) (8-9). In another fixed bed study, Chantal et al (10) coprocessed up to 10% methanol with an oil derived from supercritical extraction of wood chips. Although some benefits in yields from the addition of methanol were evident, based on the earlier work (8-9) the amounts of methanol used by Chantal were too small to reduce catalyst deactivation. It was of interest therefore, to further examine the potential benefits of methanol co-processing for both model compound feeds and wood pyrolysis liquids in short contact time regenerative fluid bed processing.

Table I Acetic Acid Conversion over ZSM-5
Fixed Bed Data
370°C, 1 atmosphere total pressure, 1 LHSV

Time on Stream, hr	0.3	1.3	3.0
Conversion, %	100	71	61
Product Selectivity, %			
CO		1	1
CO_2		16	23
H_2O		56	24
Oxygenates[+]		7	37
Hydrocarbons		20	15
Hydrocarbons, %			
$C_1 + C_2$		2	2
C_3		5	4
C_4		8	26
C_5+		85	68

[+] Mainly Acetone

Experimental

Catalyst. The catalyst used was a silica alumina bound HZSM-5.

Production of Wood Pyrolysis Liquids. Sawdust (primarily pine and fir) pyrolysis was carried out in a fixed bed reactor, which held about 200 ml of sample. The sawdust charge had the following elemental analysis (dry basis): C - 49.20%, H - 6.80%, O - 43.40%, ash - 0.60%. The reaction was conducted at atmospheric pressure by heating the sample at approximately 20°C/min. to 520°C in helium flowing at about 800 ml/min. The sample was then held at the final temperature for three hours with helium gas flowing.

Reactor Description and Run Procedures. All catalytic runs were performed in a computer controlled fluidized bed apparatus (11) (Figure 1) operating cyclically to effect successive and repeated reaction/regeneration intervals. Approximately 35 ml of the silica alumina bound ZSM-5, having a mesh size typical of fluid catalyst, was charged to the Vycor reactor along with 15 ml of 120 - 200 mesh Vycor. Helium fluidizing gas entered through a frit at the base of the tapered section of the reactor bottom. A small flow of helium also swept through the feed oil sidearm inlet line. The total helium flow (850 cc/min.) plus the vapor phase reactant and products maintained the bed in vigorous motion which, in turn, insured good temperature control. Runs were carried out at 1 WHSV based on the low EHI feed component, 410°C and atmospheric pressure. The catalyst was automatically oxidatively regenerated after each 10-20 minutes reaction interval.

When processing wood pyrolysis liquids, the two pyrolysis product liquid layers were homogenized (EHI of the blend was 0.32) by high speed mixing and fed immediately to the fluid bed catalytic reactor. When co-processed with methanol, the two pyrolysis liquid layers were dissolved in the methanol to provide a mixture having an apparent EHI of 1.2-1.3.

The aqueous phase product from the catalytic fluid bed was separated and analyzed by gas chromatography for oxygenates from which conversion could be

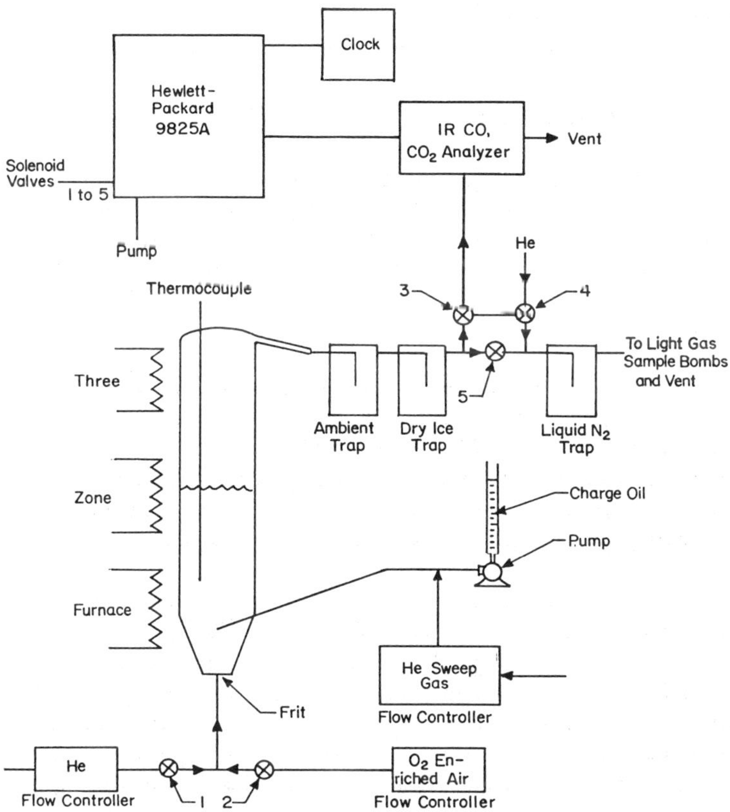

Figure 1. Cyclic Fixed Fluid Bed Apparatus

calculated. The hydrocarbon product layer contained only a very small amount of oxygenates (<1%).

The liquid hydrocarbon product was also analyzed by gas chromatography as were gaseous products, the latter being analyzed for both their hydrocarbon and CO_x contents. Coke was calculated by the computer from the data collected via the on-line CO and CO_2 infrared analyzer during each catalyst regeneration. Elemental and total material balances were generally >95%. Results presented were normalized to a no-loss basis.

Results and Discussion

Conversion of Model Compounds. Experimental data are presented in Table II. Two conversions are shown for each run. "Total conversion" represents the conversion to all products, while "conversion to non-oxygenates" represents conversion to all hydrocarbon, CO_x and H_2O products.

Methanol. The overall yields from the methanol experiment (EHI = 2.0) are in good agreement with the data obtained in the fluid bed MTG process (12). The hydrocarbon gas products, however, are higher in propene and lower in isobutane, probably due to the lower reaction pressure used in this study.

Acetic Acid and Methylacetate. The data obtained for acetic acid illustrate several interesting points which can be contrasted with the fixed bed operation cited above. First, total conversions greater than 90% may be maintained indefinitely provided periodic catalyst regeneration is employed. In spite of its having an EHI of 0, which assumes that oxygen is rejected as water, our experimental data show that decarboxylation takes place to a large extent. As a result, by rejecting oxygen as CO_x, substantial production of hydrocarbons is possible. Hydrocarbon liquid product yield is about 60% higher than that obtained at 2 hours processing time in the non-regenerative fixed bed operation and about 65% of the hydrocarbon product is C_5+ gasoline with a predominantly aromatic character. This high selectivity toward aromatics formation is consistent with the low effective hydrogen content of the "hydrocarbon" fraction of acetic acid.

Table II Model compound conversion data*
410°C, 1 atm., 1.0-1.1 WHSV, 20 min. reaction intervals
HZSM-5 in SiO_2/Al_2O_3

	Methanol CH_3OH	Acetic Acid CH_3COOH	Methyl Acetate $CH_3CO_2CH_3$	1.9/1 molar MeOH/ Acetic Acid	3.8/1 molar MeOH/ Acetic Acid
EHI of Charge	2.0	0.0	0.67	1.0 (1.0)	1.3
Total Conversion	98.6	91.2	89.4	>91.0 (94.9)	95
Conversion to Non-Oxygenates	98.6	79.8	86.1	90.4 (89.2)	--
Products (Wt.% of Charge)					
CO	0.0	3.7	6.2	2.1 (1.8)	1.1 (1.2)
CO_2	0.2	31.4	17.6	9.4 (15.8)	5.2 (10.6)
H_2O	55.8	28.4	21.5	45.3 (42.1)	48.8 (46.7)
Oxygenates	1.4	20.2	13.9	9.6 (10.8)	5.2 (7.7)
C_4- Hydrocarbon gas	19.0	3.8	6.0	7.9 (11.4)	9.9 (13.9)
C_5+ Liquid Hydrocarbon	23.3	10.6	32.1	24.9 (17.0)	28.7 (19.1)
Total Hydrocarbons	42.3	14.4	38.1	32.8 (28.4)	38.6 (33.0)
Coke	0.3	1.9	2.7	0.8 (1.1)	1.1 (0.8)
Wt.%'s of Hydrocarbon					
$C_1 + C_2$	5.4	1.5	5.6	7.2	7.6
$C_3°$	1.6	0.1	0.7	0.4	0.6
$C_3=$	25.9	5.2	6.7	13.8	14.2
$iC_4°$	5.5	0.5	0.3	0.4	0.5
$nC_4°$	0.4	0.3	0.0	0.1	0.1
$C_4=$	5.8	15.7	1.4	1.6	1.9
Total C_4-	44.6	23.3	14.7	23.5 (34.0)	24.9 (37.5)
C_5+ (gasoline)	54.7	65.0	78.7	74.1 (59.8)	72.3 (58.1)
Coke	0.7	11.7	6.6	2.4 (6.2)	2.8 (4.4)
H/C (effective)** of C_5+	1.7	1.3	1.3	1.4***	1.4***

* Numbers shown in parentheses are the calculated values which would be expected if the mixture's behavior was the arithmetic average of its two components.

** Small amounts of oxygen were observed in the C_5+ liquid. The use of the effective hydrogen index corrects for this.

*** Approximate numbers.

On a weight basis, acetic acid yields only 40% as much hydrocarbon as methanol. The lower yield in ZSM-5 processing results primarily from acetic acid's carbon loss due to oxygen rejection through decarboxylation to CO_2.

Methylacetate has an EHI of 0.67 and thus, ordinarily, would also be considered difficult to process. Its net hydrogen content, however, is substantially higher than that of acetic acid. Because of its higher carbon content (48.6% C), and despite decarboxylation and coking reactions, the observed hydrocarbon yield remains comparable to that of methanol. Moreover, hydrocarbon selectivity for direct conversion to C_5+ gasoline is higher than acetic acid or methanol. Thus, the direct yield of C_5+ gasoline is 32.1% on charge vs 23.3% for methanol.

From a hydrogen balance standpoint, both acetic acid and methyl acetate reject less H_2O and more CO_x than methanol, with resultant C_5+ liquids having effective H/C's of approximately 1.3 vs 1.7 - 2 for methanol processing.

<u>Mixtures of Acetic Acid and Methanol</u>. Processing a 1.9/1 or a 3.8/1 molar mixture of CH_3OH and acetic acid provided observations similar to those already disclosed by Chang et al (8), viz, an enhancement in C_5+ liquid yield at the expense of C_4- vs what might be expected if the mixture behaved as the average of its two components, the calculated values for which are shown in parentheses in Table II. The selectivities of the hydrocarbon products amplify the observed synergism with respect to C_5+ liquids. Furthermore, there is an enhancement in total hydrocarbon yield vs linear combination expectations.

As shown in Table II, the effect of increasing mole percent methanol in the MeOH/acetic acid charge is an attendant decrease in oxygen rejection as CO_2 and an increase in oxygen removal as H_2O. Thus, more carbon remains available to form hydrocarbon products, much of it becoming C_5+ liquids.

The above findings demonstrate that a short contact time fluid bed reactor operating in a cyclic mode can be used to process low EHI compounds to yield substantial amounts of C_5+ liquid hydrocarbon products.

By co-processing a low EHI material with a high EHI compound such as methanol, a shift in oxygen rejection from decarboxylation to dehydration takes place. The shift results in an increased yield of hydrocarbons.

The reaction of acetic acid may have potential application in converting fermentation products to hydrocarbons. Acetic acid is a major by-product in bacterial fermentation of biomass to ethanol (13). Mixtures of acetic acid and ethanol may also be processed to hydrocarbons (14).

Upgrading of Wood Pyrolysis Liquids. The products from sawdust pyrolysis at 520°C in flowing helium at atmospheric pressure produced the yields shown in Table III.

Table III. Wood Pyrolysis at 520°C and 1 atm.

Product	Wt.%
CH_4	1.4
CO	7.1
CO_2	8.5
Liquid Oxygenates	55.0
Char	28.0

Because the object of these experiments was to track the amount of wood carbon which could be converted to hydrocarbons by pyrolysis/ZSM-5 upgrading schemes, the amount of water produced by pyrolysis was not measured, and water in the pyrolysis liquids was fed along with the oxygenated products in subsequent ZSM-5 processing. Elemental analyses and the apparent EHI's (including any water) are presented in Table IV. Inspection of these data indicate that the liquid products contain about 31% of the original wood carbon. The char product accounts for another 49 wt% of the original wood carbon, and is available for indirect liquefaction via gasification and methanol synthesis. The remaining 20% of the wood carbon becomes CO, CO_2 and methane, about half of which (as CH_4 and CO) is also potentially available for conversion to methanol.

When processed in the presence of methanol, wood pyrolysis liquids exhibited synergisms and selectivity shifts similar to those discussed above for the model

Table IV. Elemental Analysis, wt%

	Sawdust	Aqueous Liquid Layer (51%)	Organic Liquid Layer (4%)	Char
C	49.2	25.9	55.0	87.3
H	6.8	8.8	7.5	3.9
O	43.4	65.3	37.5	8.0
Ash	0.6	-	-	-
EHI	0.3	0.3	0.6	-

compounds. The details of this behavior are presented below within the context of two potential processing arrangements shown in Figures 2 and 3. A major feature common to both is the use of the pyrolysis char as a cheap source of methanol.

Figure 2 shows the products obtained in a scheme in which direct upgrading of the wood pyrolysis liquids over ZSM-5 occurs in parallel with upgrading of methanol obtained from synthesis gas derived from gasification of the pyrolysis char. In Figure 3, the methanol is mixed with the pyrolysis liquids prior to co-processing over ZSM-5. Approximately 40 lbs. of methanol per 100 lbs. of dry wood feed is potentially available from the char and pyrolysis gas products. This amount would provide a weight ratio of methanol/pyrolysis liquids of 0.73.

In the parallel processing scheme, a total of 20.6 lbs. of hydrocarbon (approximately 85% C_5+ gasoline, including alkylate) per 95 lbs. of total feed (pyrolysis liquid and methanol) is obtained. In the co-processing mode, approximately 3 lbs. of additional hydrocarbon result concomitant with reduced oxygenates and coke.

Stated differently, without char gasification, only about 6% of the carbon in the wood could be upgraded to hydrocarbon products even if all the oxygenates produced by pyrolysis were recycled to extinction. Parallel upgrading of methanol derived from char gasification can increase this value to approximately 36%, i.e., about 6% from the pyrolysis liquids and 30% from methanol. Methanol co-processing boosts the percent of wood carbon transformed into

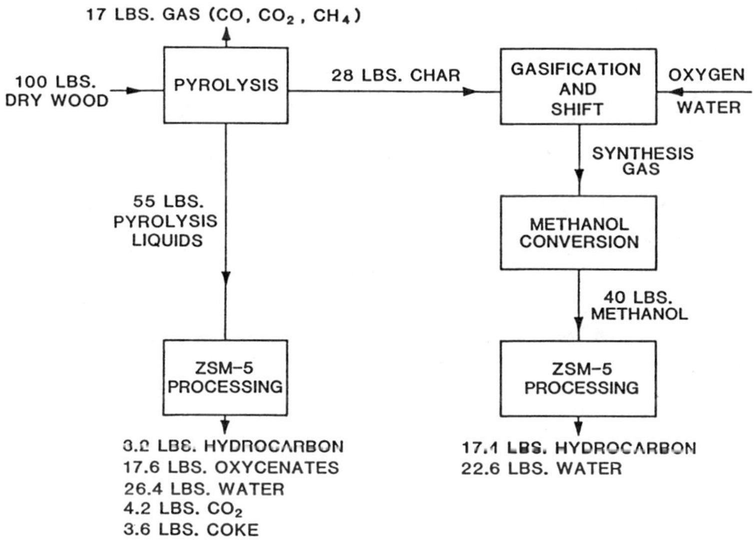

Figure 2. Parallel Processing Scheme

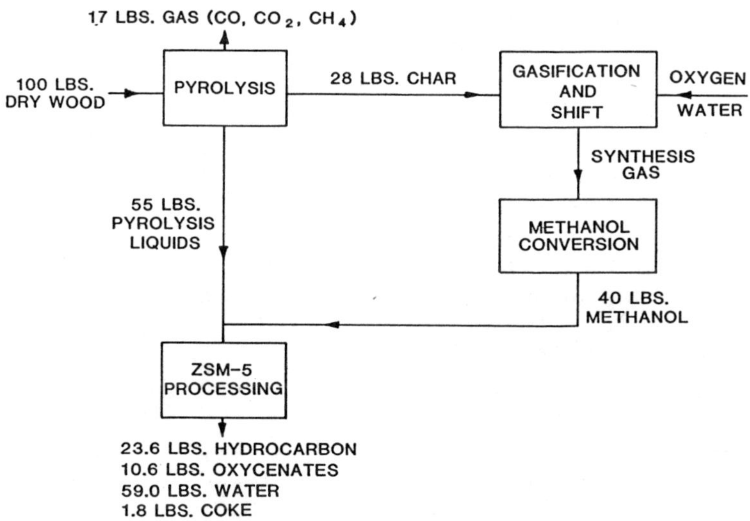

Figure 3. Coprocessing Scheme

hydrocarbon products to approximately 42%, i.e., about 12% from the pyrolysis liquids and 30% from methanol.

Conclusions

Hydrocarbon yields can be increased significantly when wood pyrolysis liquids are coprocessed with methanol over ZSM-5 catalyst vs separate processing of the two streams over the same catalyst. Thus, coprocessing pyrolysis liquids with methanol produced from char gasification is one means of producing hydrocarbon fuels from wood. Results obtained in this study provide a basis for comparison with other processing schemes such as wood gasification followed by either Fischer-Tropsch synthesis or methanol synthesis plus Mobil's MTG process.

Literature Cited

1. Chang, C. D. and Lang, W. H., U. S. Patent 3 894 103, 1975 .
2. Chang, C. D. and Silvestri, A. J., J. Catal. 1977, 47, 249.
3. Kuo, J. C., Prater, C. D. and Wise, J. J., U. S. Patent 4 041 094, 1977.
4. Ireland, H. R., Peters, A. W. and Stein, T. R., U.S. Patent 4 045 505, 1977.
5. Haag, W. O., Rodewald, P. G. and Weisz, P. B., U. S. Patent 4 300 009, 1981.
6. Brennan, J. A., Caesar, P. D., Ciric, J. and Garwood, W. E., U. S. Patent 4 304 871, 1981.
7. Weisz, P. B., Haag, W. O. and Rodewald, P. G., Science, 1979, 206, 57.
8. Chang, C. D., Chen, N. Y., Koenig, L. R. and Walsh, D. E., "Synergism in Acetic Acid/Methanol Reactions Over ZSM-5 Zeolites", Am. Chem. Soc. Div. Fuel Chem. Preprints, 1983, 28 (2) 146.
9. Chang, C. D., Lang, W. H. and Silvestri, A. J., U.S. Patent 3 998 898, 1976.
10. Chantal, P., Kaliaguine, S., Grandmaison, J. L. and Mahay, A., Appl. Catal., 1984, 10, 317.
11. Walsh, D. E., U. S. Patent 4 419 328, 1983.
12. Lee, W., Maziuk, J., Weekman, Jr., V. W. and Yurchak, S., "Mobil Methanol-to-Gasoline Process," Large Chemical Plants, Elsevier Scientific Publishing Co., Amsterdam, The Netherlands, 1979.

13. Cooney, C. L., Wang, D. I. C., Wang, S. D., Gordon, J. and Jiminez, M., Biotech. Bioeng. Symp., 1978, 8, 103.
14. Chen, N. Y., Degnan, T. F. and Koenig, L. R., Chemtech, 1986, 16, 506.

RECEIVED March 31, 1988

Chapter 25

Low-Pressure Upgrading of Vacuum-Pyrolysis Oils from Wood

M. Renaud, J. L. Grandmaison, Christian Roy, and S. Kaliaguine

Department of Chemical Engineering, Université Laval, Sainte–Foy, Quebec G1K 7P4, Canada

The oil fractions collected on the six hearthes of a wood vacuum pyrolysis process demonstration unit have been upgraded to hydrocarbons in the C_5-C_{10} range on a ZSM-5 catalyst. The use of a precoker at reaction temperature has been demonstrated and C_5-C_{10} hydrocarbons yields of the order of 30 wt% of oil fed, have been reached. GC/FTIR analyses of the heavier liquids collected after the precoker indicate that the acids in the oils play a dominant role in the gas phase reactions occurring in the precoker.

As biomaterials are structurally and chemically complex, biomass thermochemical conversion processes (1,2) produce complex fractions including a liquid fraction which, depending on the process, can be obtained in large (liquefaction, pyrolysis) or small yields (gasification). These liquids have found little utility because of their large contents in oxygen which implies low heat values, instability and corrosive properties. Two routes have been tested (3,4) in order to produce hydrocarbons from these liquids. The first one involves hydrotreatment with either H_2 or H_2 + CO over classical hydrotreatment catalysts. The second route is the simultaneous dehydration and decarboxylation over HZSM-5 zeolite catalyst in the absence of any reducing gas.

Soltes and Lin (5) achieved significant deoxygenation of various biomass tars from gasification/pyrolysis processes, in the presence of hydrogen-donor solvents (cyclohexane, tetralin, decalin) and various silica-alumina supported metals catalysts. Elliot and Baker (6) performed hydrodeoxygenation of the products of a process of liquefaction in the presence of CO at high pressure, using NiS, CoS and MoS_2 catalysts. Hydrodeoxygenation of a biomass tar can be achieved when phenolic compounds are in high concentrations. In wood pyrolysis however, the liquid products contain large amounts of low molecular weight organic acids, ketones, aldehydes and furans as well as phenolic compounds in the methoxy- or dimethoxy (6-8)

0097–6156/88/0376–0290$06.25/0

substituted forms. These mixtures are thermally instable in typical hydrotreating conditions (6).

The second route however seems more appropriate for the conversion of pyrolytic oils. It was indeed shown that a large variety of oxygenated compounds can be converted into hydrocarbons over H-ZSM+5 (9-20). In all cases hydrocarbons in the gasoline range are obtained due to the shape selective properties of this zeolite catalyst (21). Moreover a remarkable resistance to coking is observed due to the sterically restricted transition state selectivity effects, and these properties are also of considerable interest in the context of pyrolytic oils upgrading. Works along this line have been performed by Chantal et al. (22,25) working with oils derived from Aspen Poplar by dense gas extraction, Frankiewicz (23) who used a dual process combining pyrolysis of solid wastes with a catalytic converter and Mathews et al (24) who treated small fractions of the oils produced from wood by their thermochemical process.

In the present paper we report a study of the conversion over H-ZSM-5 catalysts, of the fractionated oil produced by vacuum pyrolysis of Populus Deltoides wood in a Process Demonstration Unit (PDU) (26).

Experimental

Pyrolytic Oil. The oil under study was produced by the vacuum pyrolysis process, currently under development at LAVAL University and CRIQ (26). The PDU is a six-hearth furnace, 2m high and 0.7 m in diameter. In this unit, the organic vapours and gaseous products are rapidly removed from the reactor through a series of outlet manifolds. Then the vapours are condensed in a primary cooling unit, and recovered as a liquid fraction for each hearth. As shown in Figure 1 four additional liquid fractions are collected in the secondary cooling unit. In this work only the fractions from the primary cooling unit have been studied and designated as oils #1 to #6 corresponding to the six hearthes from top to bottom.

In standard pyrolysis experiments using Populus Deltoides chips, the pressure in the reactor was lower than 80 Torr and the hearth temperatures were 215, 275, 325, 370, 415 and 465°C from top to bottom respectively. Table I gives some analytical results for these six oils. The low water contents are due to the fractional

Table I. Some Characteristics of the Pyrolytic Oils

Oil #	% Water*	Formic acid wt %	Acetic acid wt %	$\bar{M}w$** Average Molecular weight
1	7.8	2.64	3.63	342
2	5.4	2.34	4.13	528
3	4.4	3.73	5.18	572
4	3.9	4.40	3.26	393
5	7.0	7.25	3.16	233
6	3.8	2.60	2.26	123

* determined by Karl-Fisher
** determined by GPC

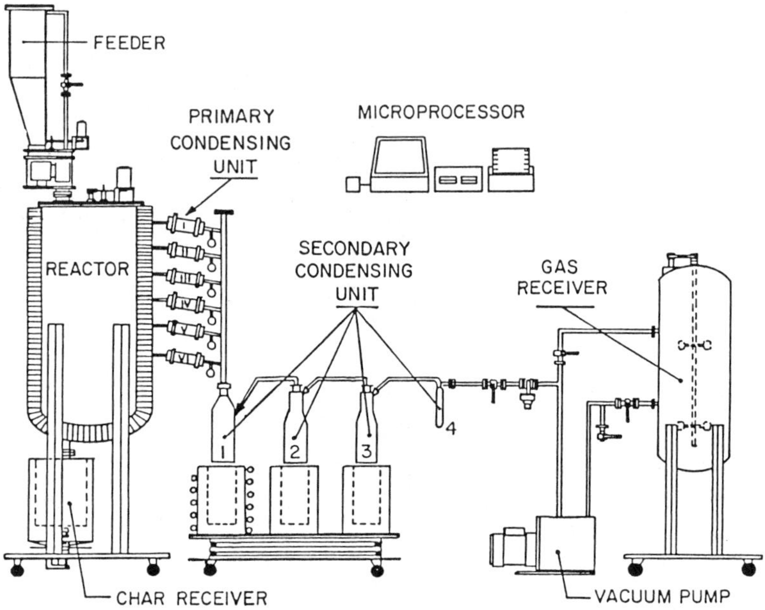

Figure 1. Scheme of the vacuum pyrolysis process development unit.

condensation of pyrolysis products in the primary cooling unit. Formic and acetic acids are considered the most abundant single organic compounds in these oils.

Conversion Reactor. The experimental set-up is a slightly modified version of a reactor described previously (22). As heating the pyrolytic oils to the reaction temperature is believed to induce thermal polymerisation/condensation reactions, a device was designed in order to preheat the vaporized feed and condense at an appropriate temperature the heavier products. This device is described in Figure 2. Using a syringe pump (SAGE 220) the liquid oil is injected in a controlled flow of helium (6.0 SCC/min) and first fed to an empty pyrex tube maintained at the same temperature than the tubular reactor [inlet oil partial pressure 50-300 Torr, residence time 5-15 s]. At the outlet of this tube the gas passes through a hot trap (150°C) where a heavy fraction is condensed. The gas flow then enters the microcatalytic pyrex tubular reactor and it is led to the heated (200°C) automatic sampling valve of a Gas Chromatograph (Perkin Elmer, Sigma 115) through a heated line (160°C). After the valve the products are collected as three fractions in three successive traps maintained at 4, -70 and 196°C respectively.

After a test the tubular reactors can be transported to another set-up where they can be regenerated in a controlled flow of dry air at 500°C. The CO_2 and H_2O produced are adsorbed over ascarite and drierite respectively and weighed. The sum of the masses of carbon in CO_2 and hydrogen in H_2O was determined and compared to the measured mass of the coke deposited in each of the two pyrex tubes.

Each experiment was performed using 1.0 ± 0.2 g of the catalyst designated as H-22, undiluted. This is the H form of a ZSM-5 sample prepared by the procedure described as method B' by GABELICA et al (28). It has Si/Al ratio of 22, and its sodium content is less than 220 ppm as determined by PIXGE (29).

The GC on line with the reactor was equipped with TC and FID detectors and a Porapak-Q column (6 ft, 1/8" OD, 80-100 mesh). In instances were peak identification was in doubt, the fraction in the 4°C trap was further analyzed using another GC (Perkin Elmer Sigma 3) equipped with a FID detector and a capillary column DB-1 (30 m).

In typical runs, oil was injected for about 3 000 s but chromatographic analysis was started at 2 000 s for all experiments.

Experimental Design. A statistical experimental design was employed to study the effect of the process parameters, namely, temperature (350-450°C), space velocity (LHSV = 0.5-2.5 h^{-1}) and oil fraction, on various response variables. In order to reduce the number of experiments, a Box-Behnken experimental design (30) was selected. This design is visualized as a cube in Figure 3. Since 6 oil fractions were studied two Box-Behnken cubes were used in order to minimize the number of experiments by studying three oils in each cube. Eventhough a discrete variable is employed to represent the oil, in reality the hearth number (1 to 6) follows the reaction temperature on the hearth which is continuous. The experiments were performed in a random manner.

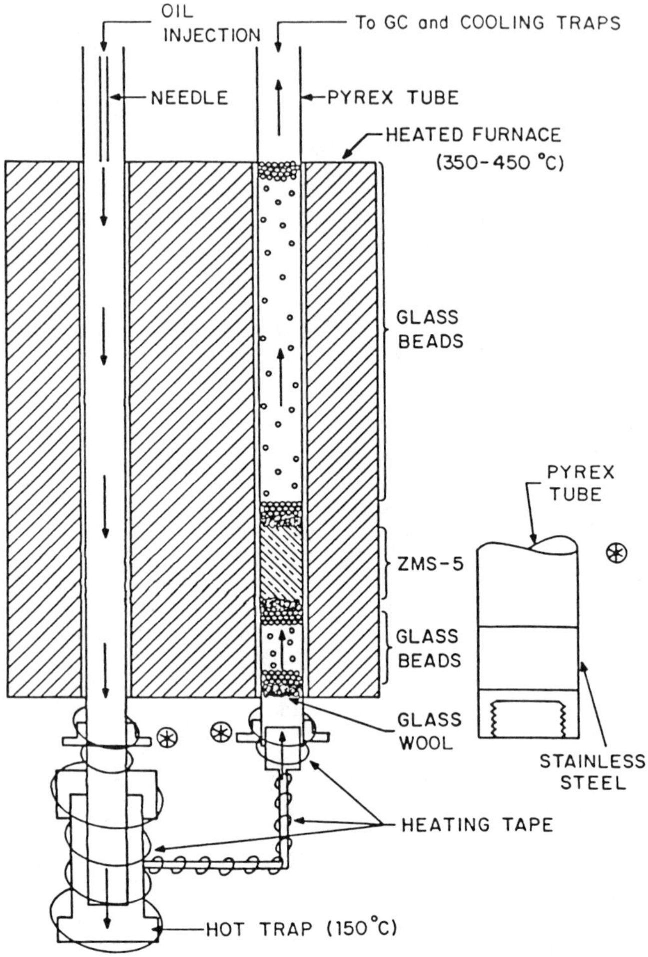

Figure 2. Scheme of the oil precoker and reactor.

GC/FTIR Analysis of Oils and Residua. Analyses of oils and residua (collected in the hot trap at the outlet of the empty Pyrex tube) were performed using a GC/FTIR Digilab FTS60 system equipped with a Hewlett-Packard 5890 gas chromatograph. The capillary column was a DB-WAX coated column of 0.25 mm diameter and 60 m length. The injection was in the on-column mode and the oven temperature program was 50°C (10 min) 5°C/min → 240°C. The spectral data acquisition rate was 1/1.1 s^{-1} (with co-adding of 4 spectra) and the spectral resolution was 8 cm^{-1}. The peak component identification was performed using an automatic library search against the vapour phase SADTLER library (9000 compounds).

Results and Discussion

Table II shows the experimental results and conditions for the catalytic upgrading tests of the six pyrolytic oils. It gives the weight percents of the various fractions of products collected. Coke #1 corresponds for example to the total weight of the material deposited on the wall of the empty pyrex tube whereas coke #2 is the one left in the tubular reactor (on the catalyst) at the end of a test. The tar collected in the hot trap is designated as the residuum, whereas the cumulative mass recovered in all three cold traps is indicated as "traps".

The composition of the "traps" fraction is the one of the stream leaving the reactor. It is given in Table II in a condensed manner as weight percents of various components including C_5-C_{10} hydrocarbons, C_{10}^+ hydrocarbons and oxygenates. These oxygenates are mostly comprising phenolic and furanic compounds. Figure 4 shows some of the characteristics of the six pyrolytic oils. The average molecular weight varies in a continuous manner with the hearth number, showing a maximum value for the oil produced on the 3rd hearth and a minimum value for the one from hearth No 6.

It must be noted that the sum of weight fractions of acetic and formic acids is following a similar pattern except for the oil No 5 which contains much formic acid and is therefore much more acidic. The low content in formic acid of oil No 6 is a clear indication that the solid residue leaving hearth No 5 is very different in nature from the one from hearth No 4 bearing much less intact or slightly degraded cellulose (the main source of formic acid) and much more recondensed material. Figure 5 shows for each oil the average values of the weight percents of coke #1, residuum and the yield in C_5-C_{10} hydrocarbons. It is interesting to note that at least for the first four oils these values are obviously correlated with both oil average molecular weight and acidity (Figure 4). As expected, coke #1 is higher when the oil is heavier, but the residuum is much smaller. This last result can only be understood if the residuum is seen as an intermediate product of gas phase depolymerization which would be faster for the most acidic mixtures. Indeed for example the low residuum value obtained with oil No 5 would then be the result of this oil being much more acidic than the others.

The C_5-C_{10} yield is also correlated with $\bar{M}_w$ (up to oil No 5) but this correlation is the result of a more intricate interaction of various factors. C_5-C_{10} yield does not depend only on molecular size or acidity of the oil but also on its chemical nature. Oil No 3 which contains more lignin fragments is liable to yield small phe-

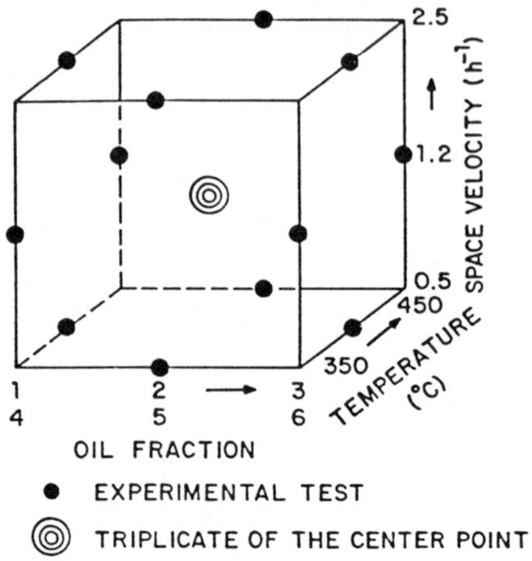

Figure 3. Box Behnken experimental design.

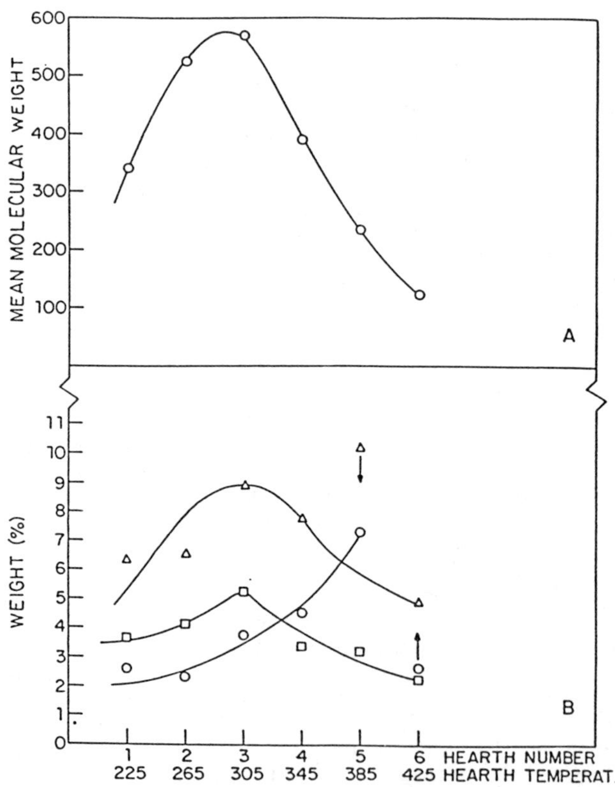

Figure 4. Characteristics of the pyrolysis oils. A: Average molecular weight; B: wt % of acetic acid (□), Formic acid (o) and acetic + formic acids (Δ).

Table II. Experimental Conditions and Results of Catalytic Upgrading of Pyrolytic Oils

Oil #	Temp. °C	LHSV h^{-1}	Coke wt % #1	Coke wt % #2	Residuum wt %	Traps wt %	Product Distribution, wt % CO	CO_2	H_2O	C_1-C_4	C_5-C_{10}	Oxyg.	BTX	C_{10}^+	Yield C_5-C_{10}
1	350	1.2	15.2	2.3	50.8	31.6	1.2	1.7	5.6	1.2	61.9	26.0	23.5	3.3	19.3
1	400	0.5	21.7	6.5	10.9	60.9	0.7	0.8	3.7	2.2	48.7	25.7	21.1	18.4	29.7
1	400	2.5	17.4	2.8	54.8	24.9	1.0	2.0	6.2	4.3	57.6	23.9	19.7	4.9	14.3
1	450	1.2	18.7	2.3	40.9	38.1	0.9	1.1	3.2	1.4	76.8	12.0	18.0	4.6	29.3
2	350	0.5	10.0	4.3	62.9	22.9	0.9	1.2	4.5	1.8	52.2	33.7	18.8	5.4	12.0
2	350	2.5	30.5	3.9	31.2	34.4	1.3	2.7	5.5	1.4	51.4	29.5	13.1	8.1	17.7
2	400	1.2	20.6	5.9	26.5	47.1	0.8	1.0	3.4	2.0	45.8	32.5	17.2	14.6	21.6
2	400	1.2	22.9	5.1	28.8	43.2	0.9	1.2	3.8	2.5	44.4	34.1	18.8	13.1	19.2
2	400	1.2	21.3	5.6	24.0	49.1	1.0	1.2	3.1	1.3	45.4	29.9	17.4	17.1	22.3
2	450	0.5	30.9	3.6	9.1	56.4	0.8	1.2	3.1	2.0	48.3	36.3	18.4	8.5	27.2
2	450	2.5	36.6	2.8	22.5	38.0	2.6	4.6	7.2	1.6	54.7	23.3	22.0	2.8	20.8
3	350	1.2	33.1	4.3	30.7	31.9	0.6	0.7	3.2	0.6	47.0	34.0	10.7	13.9	15.0
3	400	0.5	25.4	3.4	25.4	45.8	1.3	1.9	4.5	4.0	42.6	34.6	28.1	11.1	19.5
3	400	2.5	19.2	3.1	46.9	30.8	0.6	0.9	2.4	1.1	34.9	48.8	8.8	11.3	10.7
3	450	1.2	22.3	5.9	29.4	42.3	0.8	0.9	2.3	1.8	38.6	39.0	17.8	16.5	16.3
4	350	1.2	20.1	2.4	52.4	25.0	1.3	1.7	5.1	1.1	58.1	24.6	22.4	8.2	14.5
4	400	0.5	23.0	2.7	32.4	41.9	1.7	2.2	6.5	2.2	52.6	25.1	27.7	9.8	22.0
4	400	2.5	11.7	2.6	58.3	27.4	1.3	1.8	6.4	1.9	51.3	27.6	20.0	9.7	14.1
4	450	1.2	26.3	3.5	17.5	52.6	0.9	1.3	7.1	4.2	48.0	21.9	16.7	16.5	25.2

Continued on next page

Table II. Continued

Oil #	Temp. °C	LHSV h^{-1}	Coke wt % #1	Coke wt % #2	Residuum wt %	Traps wt %	Product Distribution, wt % CO	CO_2	H_2O	C_1-C_4	C_5-C_{10}	Oxyg.	BTX	C_{10}^+	Yield C_5-C_{10}
5	350	0.5	27.1	3.4	16.9	52.4	1.6	1.9	6.7	1.3	52.2	29.7	26.0	6.6	27.4
5	350	2.5	21.4	8.6	28.6	41.4	0.9	0.6	8.0	1.5	54.3	31.1	22.1	3.5	22.5
5	400	1.2	25.5	3.9	24.5	46.1	0.4	0.5	2.9	1.1	53.1	28.5	19.0	13.6	24.5
5	400	1.2	22.1	4.1	27.9	45.9	0.6	0.9	2.1	1.2	47.9	34.3	17.2	13.0	22.0
5	400	1.2	29.6	3.2	20.3	46.9	0.3	0.5	4.0	1.4	56.9	24.9	20.4	12.0	26.7
5	450	0.5	30.6	5.6	9.7	54.2	0.7	0.9	2.3	3.3	52.9	36.8	22.0	2.6	28.7
5	450	2.5	31.8	1.3	25.2	41.7	2.5	4.7	8.8	12.1	37.1	30.5	30.6	4.1	15.5
6	350	1.2	18.7	2.9	56.7	21.6	0.2	0.3	1.7	0.2	25.1	71.0	6.9	1.5	5.4
6	400	0.5	30.3	12.1	3.0	54.5	0.3	0.3	1.0	0.9	30.7	47.2	11.2	19.9	16.7
6	400	2.5	10.7	2.1	68.2	19.0	1.4	1.9	7.4	2.5	30.7	21.5	19.3	11.7	10.2
6	450	1.2	20.4	2.4	47.9	29.3	1.3	1.2	3.9	5.9	46.1	23.9	15.6	17.5	13.5

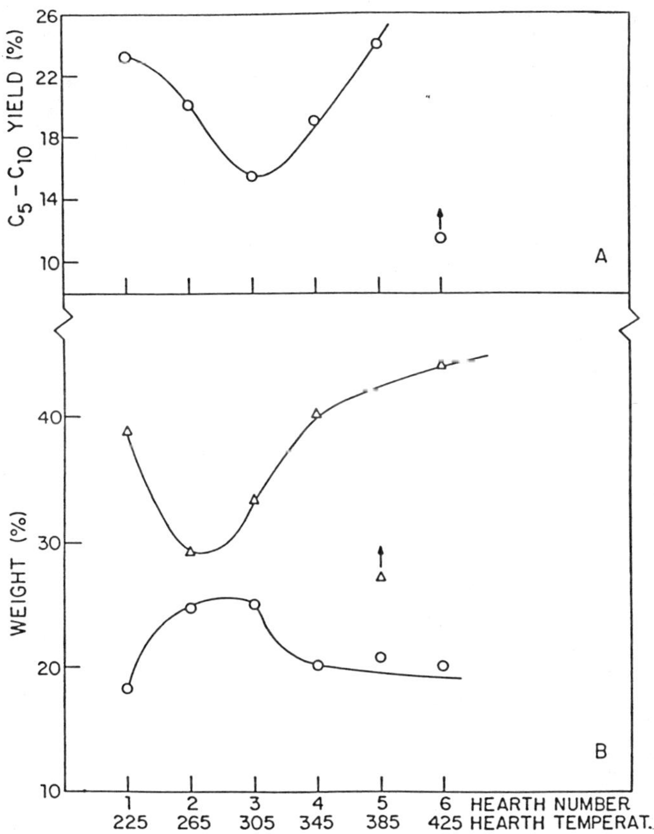

Figure 5. Average values for A: C_5–C_{10}; B: coke1 + coke2 (o), residuum (Δ).

nolics which are volatile enough to reach the catalyts but which are not converted to hydrocarbons. Similarily oil No 6 which is believed to contain less polysaccharides pyrolysis products shows a very low C_5-C_{10} yield.

The effect of the various parameters on the "traps" weight percent is reported in Figure 6. This term represents roughly the fraction of the oil reaching the catalyst and must therefore be considered as a yield for the preliminary thermal processing. The curves are drawn not taking into account the results for oil No 5. Comparing the curves obtained at 350 and 450°C (both at 1.2 LHSV) shows that increasing the temperature increases the depolymerisation of the oils components. The two curves at 400°C and 0.5 and 2.5 LHSV show that increasing the residence time in the empty tube and the hot trap increases also the depolymerisation yield.

These results show the interest of the experimental procedure adopted in this work, as this gas phase thermal depolymerisation which occurs at the same temperature than the catalytic process can be studied separately. This is not the case when the vaporized oil is directly injected onto the catalyst and both processes happen simultaneously.

Figure 7 gives the yield in C_5-C_{10} obtained for the various oils as a function of temperature and LHSV. Comparing the curves at 350 and 450°C shows that a rise in temperature increases this yield in proportions roughly similar to the increase of the "traps" fraction (see Figure 6) without affecting much the percent of unconverted oxygenates in the "traps" fraction (see Figure 8). This suggests that the most important effect of temperature is to increase the yield of thermal depolymerization in the precoker. This is confirmed by the fact that the oil No 5 which is the most acidic and is more thorougly depolymerized (yielding a higher "traps" fraction), is also showing the highest C_5-C_{10} yield. Similar observations can be made for the effect of contact time at 400°C.

It is interesting to note that for several oils (No 1, 2, 4 and 5) yields close to 30% have been reached. This is to be compared with the results of Chantal et al (22) who obtained maximum yields of 15-17% from SCE oils converted on ZSM-5 in a one step process. Figure 8 shows the variations of the percents of oxygenates in the traps. In line with previous discussion, maximum oxygenates are obtained from oils No 3 and 6 which contain respectively more lignin fragments and more product of high temperature decomposition of recondensed material and which therefore both contain more phenolic compounds.

Time on Stream Data. Figure 9 reports the results of a set of experiments (not shown in Table II) designed to assess the deactivation behaviour of the catalyst. In these experiments the feed was a mixture of the six oils. It was injected for a fixed time in the reactor assembly maintained at 400°C. Then the gases leaving the reactor were analyzed and the injection of oil was stopped until the chromatographic analysis was completed, at which time the injection was restarted for another fixed time and this operation was repeated three times.

Figure 9 shows that the composition of the product changes with time on stream, with deactivation appearing after 250-500 s. The decrease in deoxygenation activity is seen from the decrease in H_2O,

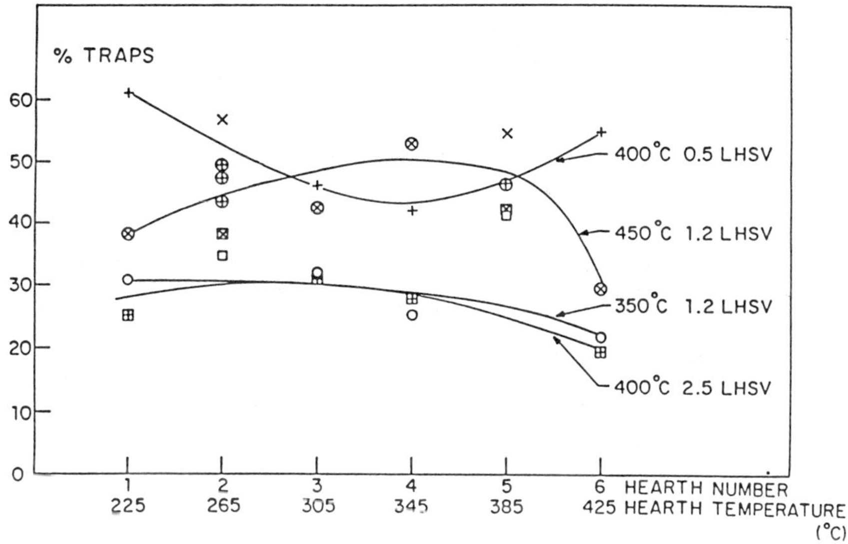

Figure 6. Weight % of the products collected at the reactor outlet.

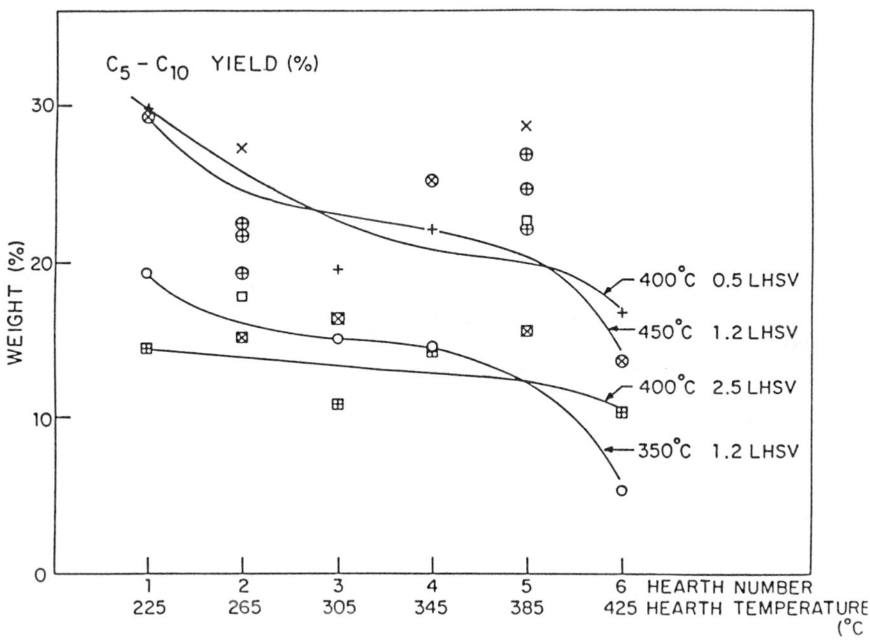

Figure 7. C_5–C_{10} hydrocarbon yields.

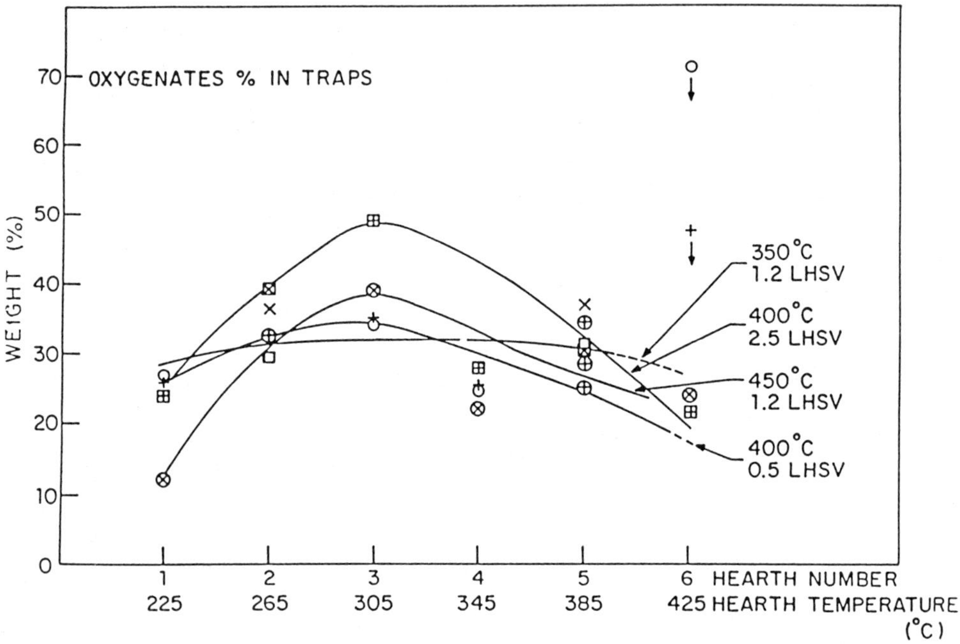

Figure 8. Weight % of oxygenates in the products at the reactor outlet.

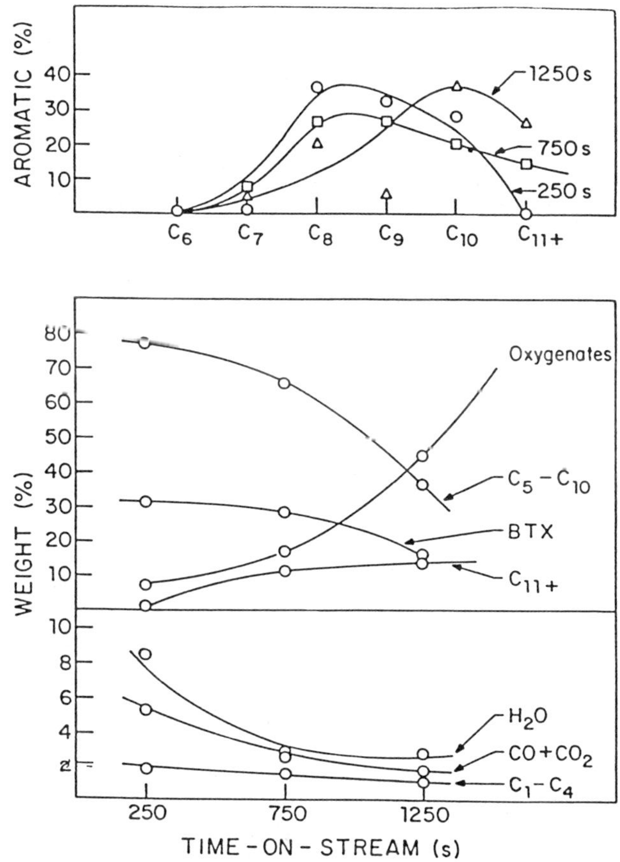

Figure 9. Product distribution as a function of time on stream.

CO, CO_2 and the increase in oxygenates. The appearance of C_{10}^+ aromatics in the products is of special interest. Since such large compounds cannot be generated within the pores of the ZSM-5 zeolite (21), they can only be generated either on the external acid sites or from the coke deposit as postulated recently (13). This would also explain the evolution of the C_6-C_{10} aromatics distribution with time on stream, showing a relative increase in C_{10} as well as C_{10}^+.

The time scale for catalyst deactivation is showing the great resistance of the ZSM-5 catalyst to coking. At similar temperatures for example a pellet of H-Y cracking catalyst is deactivated within 2-3 seconds in the FCC process.

GC/FTIR Data. Table III reports GC/FTIR analyses for oil #5 and three residua obtained in three experiments performed at 350, 400 and 450°C, using oil #5 as the feed. All compounds detected either as major (M), large (X) or trace (T) components have been characterized by their infrared spectra. The number following each identification in Table III is the fit quality index of the experimental spectrum to the one in the library. The lower this number, the better the fit.

Table IV. Compounds present in oil and not (or low) in residuum

Peak number	Abundance in oil		
13B	M	acids	Formic acid (0.05)
16	X	acids	Propanoic acid (0.03)
34	M	acids	Cycloalkane with a ketone and an acid functions
55	X	acids	Butyric acid; 2-hydroxy, 2 methyl
63	T	acids	Levulinic acid (0.08)
7	M	ketones	2-propanone-1-hydroxy (0.05)
8	T	ketones	1-cyclopentene-3 one (0.03)
62B	X	ethers	Ether, 2-methoxy ethyl vinyl (0.21)
72	T	ethers	Pyrocatechol, 3-methoxy (0.07)

Table IV lists all the compounds present in oil #5 and not found (or in very low concentration) in the residua. It includes five acids, two ketones and two ethers. Acetic acid is not included in this list since it is still present in residua with its concentration only decreasing at 450°C.

Table V gives all the compounds which according to the results of Table III, were present in the residua but were not found in oil #5. These compounds include eleven esters, some of which as major components, five lactones or diones, six phenolic compounds and only two acids found as traces.

From the comparison of Tables III, IV and V the following conclusions can be drawn on the chemical transformations occurring in the empty Pyrex tube used as a precoker:

1/ Acids in the oil are heavily esterified.

Table III. GC/FTIR Analysis

Peak number	Oil #5	Residuum 350°C 0.5 h^{-1}	Residuum 400°C 1.2 h^{-1}	Residuum 450°C 0.5 h^{-1}	Compounds (fitting)
1	-	T	M	X	Methyl formate (0.03)
2	-	M	M	X	Methyl acetate (0.03)
5	-	T	X	-	1,3-Cyclohexanedione (0.15)
7	M	T	-	-	2-Propanone-1-hydroxy (0.05)
9B	-	X	M	M	Glycolic acid, methyl ester (0.03)
10	-	-	X	-	Butyric acid, 4,4-dimethyl (0.15)
11	M	M	M	X	Acetic acid (0.03)
12	-	-	X	-	2-Propanone, hydroxy-acetate (0.08)
13B	M	-	-	-	Formic acid (0.05)
15	-	M	X	X	Furfuryl alcchol, tetrahydro, proprionate (0.08)
16	X	-	-	-	Propanoic acid (0.03)
20B	-	-	X	X	Propionic acid, ester
25	T	T	X	-	Butyric acid, 4-hydroxy, G-lactone (0.04)
30B	-	X	-	-	Adipic acid, monoethyl ester (0.10)
32	-	T	X	-	Butyrolactone, 2-carboxy, ethyl, ester (0.10)
33	-	M	M	M	Crotonic acid, 4-hydroxy, G-lactone (0.02)
34	M	-	-	-	Cyclic aliphatic + ketone + acid
35	X	M	M	X	Cyclopentane-1,2-dione, 3-methyl (0.03)
38	X	X	X	X	Guaiacol (0.04)
42	-	X	X	T	Cyclopentane-1,2-dione, 4-methyl
44	-	T	X	-	Lactone
45	T	-	T	X	4H-pyran-4-one, 3-hydroxy (0.06)
45B	T	T	X	X	4H-pyran-4-one, 3-hydroxy, 2-methyl (0.07)
46	X	X	X	X	p-Cresol, 2-methoxy (0.02)
48	-	M	M	-	Phenol (0.06)
51B	X	-	T	-	1-3-cyclohexanedione (0.16)
54	X	X	X	X	Unidentified aliphatic acid
55	X	-	T	-	Butyric acid; 2-hydroxy, 2-methyl

Continued on next page

Table III. Continued

Peak number	Oil #5	Residuum 350°C 0.5 h^{-1}	Residuum 400°C 1.2 h^{-1}	Residuum 450°C 0.5 h^{-1}	Compounds (fitting)
56	X	M	X	-	Acetic acid, methoxy or ethoxy
58	-	-	X	-	Lactic acid, methyl ester
61	M	M	M	M	Phenol, 2-6 dimethoxy (0.02)
62	-	M	M	M	Unidentified ester
62B	X	-	-	-	Ether, 2-methoxyethyl vinyl (0.21)
65	X	M	M	M	Phenol, 4-allyl-2,6-dimethoxy (0.04)
68	-	X	X	T	Phenol, 4-propenyl-2,6-dimethoxy
68B	X	-	-	X	Acrolein, diethyl acetal (0.16)
69	-	M	-	-	Benzoic acid (0.05)
71	-	X	-	-	Phenol 2-6-dimethoxy compound
77	-	X	T	-	Benzaldehyde, 4-hydroxy, 3-methoxy (0.03)
80B	X	-	X	-	Butyric acid, 4-hydroxy-G-lactone (0.10)
81	-	X	-	-	Acetophenone, 4PR-hydroxy, 3PR-methoxy (0.05)

Table V. Compounds not found in oil but present in residuum

Peak number	Abundance in residuum at 350	400	450° C		
1	T	M	X	esters	Formiate, methyl (0.03)
2	M	M	X		Acetate, methyl (0.03)
6	-	T	-		Lactic acid, methyl ester (0.08)
9B	X	M	M		Glycolic acid, methyl ester (0.03)
15	M	X	X		Furfuryl alcohol, tetrahydro proprionate (0.08)
20B	-	X	X		Propionic acid, ester
22	-	T	-		Pyruvic acid, 3 hydroxy-3 phenyl-methyl ester (0.15)
24	T	T	-		Acetic acid, 2 hydroxyethyl ester (0.15)
30B	X	-	-		Adipic acid, monoethyl ester (0.10)
58	-	X	-		Lactic acid, methyl ester
62	M	M	M		Unidentified ester
32	T	X	-	lactones and diones	Butyrolactone,-2 carboxy, ethyl ester (0.10)
25	T	X	-		Butyric acid, 4-hydroxy, G-lactone (0.04)
33	M	M	M		Crotonic acid, 4 hydroxy, G-lactone (0.02)
42	X	X	T		Cyclopentane-1,2-dione, 4-methyl
5	T	X	-		1,3-Cyclohexane dione (0.15)
48	M	M	-	phenolic compounds	Phenol (0.06)
68	X	X	T		Phenol, 4 propenyl-2,6-dimetoxy
69	M	-	-		Benzoic acid (0.05)
71	X	-	-		Phenol 2-6-dimethoxy derivative
77	X	T	-		Benzaldehyde, 4 hydroxy, 3 methoxy (0.03)
81	X	-	-		Aceto phenone, 4 Pr-hydroxy, 3 Pr-methoxy (0.05)
29	-	T	-	acids	Valeric acid, 4-hydroxy (0.11)
54B	-	-	T		Glutaric acid, 2 methyl

2/ Lactones and diones are obtained quite probably by dehydration and condensation reactions involving carbohydrate derivatives.

3/ Among the phenolic compounds, phenol which is absent in oil #5 is found as a major component in residua. Cracking reactions involving side chains like propenyl and methoxy groups on phenolic compounds have therefore occurred.

All three reaction types must be linked to the high concentrations of carboxylic acids in the oil. These acids obviously act as reactants in esterifications. Acids are also good catalysts for dehydration, condensation, cracking and hydrolysis of ether linkages in the range of temperature under consideration. It is therefore highly probable that the carboxylic acids in the oil act as homogeneous catalysts of these reactions. The high observed rates of cokefaction must also be associated with the high acidity of the gas phase.

These results underline the efficiency of the vacuum pyrolysis process in avoiding secondary reactions of acids since oil #5 was produced at 415°C in the vacuum pyrolysis demonstration unit and the reactions in the precoker were observed at 350°-450°C. It is therefore clear that the vacuum pyrolysis has an important asset due to its ability to preserve intermediate products of thermal degradation which may include highly priced fine chemicals.

Conclusions

The experimental set-up used in this work allowed to decouple the gas phase thermal conversion of the pyrolytic oils from the catalytic upgrading. These two sets of reactions happen in the same temperature range and are therefore simultaneous when the oil is directly fed to the catalyst. Preliminary thermal conversions include some coking and an important thermal depolymerization. Both reactions are accelerated in the presence of volatile acids in the gas phase. Moreover performing both reactions in a preheater is beneficial to the catalytic conversion first because the deactivation of the catalyst is less important and secondly because depolymerized fragments yield higher conversions to hydrocarbons. C_5-C_{10} hydrocarbon yields as high as 30 wt% have been consistently obtained for reaction times of 2 000 s'.

Literature Cited

1. SIMONS RESOURCE CONSULTANTS & B.H. Levelton, "A Comparative Assessment of Forest Biomass Conversion to Energy Forms" Phase I-Proven and Near-Proven Technology Vol I-IX - Phase II - Further development, ENFOR PROJECT C-258, Canada, 1983.
2. E. Chornet & R.P. Overend, "Biomass Liquefaction: An Overview" in "Fundamentals of Thermochemical Biomass Conversion" ed. R.P. Overend, T.A. Milne & L.K. Mudge, Elsevier 1985 p. 967-1002 - 2-B K.L. Tuttle, "Review of Biomass Gasification Technology" in "Progress in Biomass Conversion Vol. 5" ed. D.A. Tillman & E.C. John, Academic Press, 1984, p. 263-279.
3. S. Kaliaguine, "Upgrading pyrolytic oils from wood and other biomasses". Energy Project Office, NRCC, Ottawa, 1981.
4. S. Kaliaguine, Proc. Specialists Meeting on Biomass Liquefaction, Saskatoon, 1982, p. 75-125.

5. E.J. Soltes & S.C.K. Lin, "Hydroprocessing of Biomass Tars for Liquid Engine Fuels" in "Progress in Biomass Conversion" Vol. 5, Ed. D.A. Tillman & E.C. John, AC Press 1984, p. 1-68.

6. D.C. Elliot & E.G. Baker, "Upgrading Biomass Liquefaction Products through Hydrodeoxygenation", Biotechnology & Bioengineering Symp. 14, 159-174 (1984).

7. H. Ménard, C. Roy, A. Gaboury, D. Bélanger & G. Chauvette, "Analyse totale du pyroligneux provenant du Populus Tremuloides par HPLC", Compte-rendu du 4° Séminaire R & D Bioénergique, Winnipeg, Manitoba 29-31 mars 1982.

8. C. Roy, B. de Caumia, D. Brouillard & H. Ménard, "The Pyrolysis under Vacuum of Aspen Poplar" in "Fundamentals of Thermochemical Biomass Conversion", ed. R.P. Overend, T.A. Milne & L.K. Mudge Elsevier, N.Y. 1985, p. 237-256.

9. C.D. Chang & A.J. Silvestri, "The Conversion of Methanol and other O-compounds to Hydrocarbons over Zeolite Catalysts", J. Catal., 47, 249-259 (1977).

10. B.L. Maiorella, "Fermentation Alcohol: better to convert to fuel", Hydrocarbon Processing, August 1982, p. 95-97.

11. P.D. Chantal, S. Kaliaguine & J.L. Grandmaison, "Reactions of Phenolic Compounds on H-ZSM5", Catalysis on the Energy Scene, Elsevier, 93-100, 1984.

12. P.D. Chantal, S. Kaliaguine & J.L. Grandmaison, "Reactions of Phenolic Compounds over H-ZSM5", Appl. Catal., 18, 133-145 (1985).

13. M. Renaud, P.D. Chantal & S. Kaliaguine, "Anisole Production by Alkylation of Phenol over HSM-5", Can. J. Chem. Eng., 64, 787-791 (1986).

14. C.D. Chang, W.H. Lang, A.J. Silvestri, "Manufacture of Gasoline", U.S. Patent #3, 998-898 (1976).

15. W.O. Haag, D.G. Rodewald, P.B. Weisz, "Conversion of Biological Materials to liquid fuels", U.S. Patent #4,300,009, 1981.

16. P.B. Weisz, W.O. Haag, D.G. Rodewald, "Catalytic Production of High Grade Fuel (Gasoline) from Biomass Compounds by Shape-Selective Catalysis", Science, 206, 57-58 (1979).

17. Y.S. Prasad, N.N. Bakhshi, J.F. Mathews & R.L. Eager, "Catalytic Conversion of Canola Oil to Fuels and Chemical Feedstocks Part I Effect of Process Conditions on the Performance of HZSM-5 Catalyst", Can. J. Chem. Eng., 64, 278-293 (1986).

18. N.Y. Chen & L.R. Koenig, "Process for converting carbohydrates to Hydrocarbons", U.S. Patent #4,503,278 (1985).

19. P.A. Pesa, M.W. Blaskie & D.R. Fox, "Decarbonylation of N.-Butyraldehyde using Zeolite Catalysts. "U.S. Patents #4,517,400, (1985).

20. R.M. Gould & S.A. Tabak, "Multistage Process for Converting Oxygenates to Liquid Hydrocarbons with Ethene recycle", U.S. Patents #4,543,435, (1985).

21. E.G. Derouane, "Molecular Shape-Selective Catalysis in Zeolites - Selected Topics", Catalysis on the Energy Scene, Elsevier, p. 1-17 (1984).

22. P.D. Chantal, S. Kaliaguine, J.L. Grandmaison and A. Mahay, "Production of Hydrocarbons from Aspen Poplar Pyrolytic oils over H-ZSM5", Appl. Catal., 10, 317-332 (1984).

23. T.C. Frankiewicz, "Process for Converting Oxygenated Hydrocarbons into Hydrocarbons", U.S. Patents #4,308,411 (1981).

24. J.F. Mathews, M.G. Tepylo, R.L. Eager & J.M. Pepper, "Upgrading of Aspen Poplar Wood Oil Over HZSM5 Zeolite Catalyst", Can. J. Chem. Eng., 63, 686-689, (1985).

25. R. Labrecque, S. Kaliaguine & J.L. Grandmaison, "Supercritical Gas extraction of Wood with Methanol", Ind. Eng. Chem. Prod. Res. Dev., 23, (1), 177-182 (1984).

26(a). C. Roy, R. Lemieux, B. de Caumia & H. Pakdel, "Vacuum Pyrolysis of Biomass in a Multiple-Hearth Furnace", Biotechn. & Bioeng. Symp. 15, 107-113 (1985).

26(b). C. Roy, R. Lemieux, B. de Caumia and D. Blanchette, "Processing of Wood Chips in a Semi-Continuous Multiple Hearth Vacuum Pyrolysis Reactor", This symposium.

26(c). H. Pakdel and C. Roy, "Chemical Characterization of Wood Pyrolysis Oils Obtained in a Vacuum Pyrolysis Multiple Hearth Reactor", This symposium.

27. C. Roy, A. Lalancette, B. de Caumia, D. Blanchette et B. Côté, "Design et construction d'un réacteur de pyrolyse sous vide fonctionnant en mode semi-continu et basé sur le concept du four à soles multiples", in Third Biomass Liquefaction Specialists Meeting Sherbrooke 29-30 Sept 1983.

28. Z. Gabelica, N. Blom & E.G. Derouane, "Synthesis and Characterization of ZSM5 type zeolites III. A critical evaluation of the role of alkali and ammonium cations", Appl. Catal. 5, 227-248 (1983).

29. I.M. Szoghy, A. Mahay & S. Kaliaguine, "Elemental analysis of zeolites by PIXGE". Zeolites 6, 39-46 (1986).

30. G.P.E. Box & D.W. Behnken, "Some new three level designs for the study of quantitative variables", Technometrics, 2, (4), 455-475 (1960).

31. P.B. Weisz, "The remarkable active sites: Al in SiO_2", Ind. Eng. Chem. Fundam, 25, 58-62 (1986).

32. Anon. "SAS User's Guide", SAS Institute inc. Carry North Carolina, 1979.

RECEIVED March 31, 1988

Chapter 26

Molecular-Beam, Mass-Spectrometric Studies of Wood Vapor and Model Compounds over an HZSM-5 Catalyst

Robert J. Evans and Thomas Milne

Solar Energy Research Institute, 1617 Cole Boulevard, Golden, CO 80401

Molecular beam mass spectrometry and a quartz micro-reactor operating in the pulsed mode, have been used to characterize the direct conversion of biomass pyrolysis products over an HZSM-5 zeolite. This technique allows the observation of the reaction products in real time and is ideally suited to screen a variety of catalysts, operating conditions, and model compounds to determine relative reactivity and variation in product composition as a function of operating conditions. The relative advantages of shape-selective catalysts with strong acid sites are apparant for wood pyrolysis products. The effect of temperature and weight hourly space velocity on product composition was screened and shows differences in reaction pathways between biomass pyrolysis products and methanol. The synergistic effect of co-feeding methanol and biomass was observed, using this technique, and showed that more aromatic gasoline products were formed. Methods of calibration have been implemented so that yield estimates can be made. Replicate runs gave hydrocarbon yields from wood of 18 ±4 wt %, with equal amounts of light olefins and aromatics. Screening of model compounds has shown that the methoxyphenols, derived from lignins, give low yields of hydrocarbon products, while carbohydrate-derived ring structures, such as furfural and α-angelicalactones, are good sources of light aromatics.

The goal of the molecular-beam, mass-spectrometry (MBMS) studies of biomass pyrolysis product conversion over zeolite catalysts is to provide rapid characterization, in real time, of the fate of the complex reactants as a function of reaction parameters. This technique allows the qualitative observation of transient, reactive and high molecular weight reactants and products that might otherwise escape detection by conventional collection and analysis methods. The goal is to optimize gasoline yields from biomass, at this small scale, by evaluating a typical Mobil HZSM-5 zeolite in its ability

0097–6156/88/0376–0311$06.00/0

to deoxygenate wood vapors and to provide support and guidance for the bench-scale reactor studies described elsewhere in this volume. The screening results allow an assessment of several important aspects of the conversion process: the identification of promising catalysts; the determination of the critical variables in yield optimization; the effect of catalyst composition, structural properties, and physical form on product formation; the behavior of model compounds that represent the range of functionality found in biomass pyrolysis vapors; the determination of relative coking rates for various catalysts and feedstocks; and the regenerability of catalysts.

This screening approach is not a stand-alone technique, however, and bench-scale experimentation is needed to provide data for parameters which the screening approach does not adequately address: isolation and confirmation of isomers of products, product mass balance, long-term studies in accessing catalyst life, and the variation of certain reactor operating conditions which have limited ranges at this scale, such as feed partial pressure. The catalytic conversion of wood pyrolysis products will be discussed in a screening context in this paper.

Previous Work in Zeolite Conversion of Biomass Vapors. Applications of zeolites for the conversion of oxygenates in general and biomass pyrolysis products in particular has been previously reviewed (1,2). Over the last decade, Mobil researchers (e.g., 3-5) and others (6-17) have examined a variety of oxygenated compounds over HZSM-5 catalysts of various origins. Several investigators have converted wood pyrolysis products (10-14).

Recent work at SERI has focussed on the study of direct wood-vapor conversion without an oil condensation step (1). The MBMS experimental system (18-21) has been used to show that the biomass pyrolysis vapors were reactive on the HZSM-5 zeolite catalyst. The complete destruction of all primary products was observed, and a method of monitoring deactivation of the catalyst was demonstrated. Specifically, the carbohydrates were converted beyond the furans to light olefins and aromatics and the lignin products were also completely destroyed with the same classes of gaseous products. However, the coking tendency of each class of starting material was not quantified at that time.

In this paper, we report results for the effects of weight hourly space velocity (WHSV), temperature, the co-feeding of methanol, yield estimates, and the relative performance of selected model compounds.

Experimental

The experimental apparatus and approach have been previously described (18,19) and are summarized briefly here. The sampling mode is an extractive method in which a sampling probe is immersed in the system to be studied. A small orifice at the apex of the cone extracts gas and with a downstream pressure (stage one pressure = 10^{-2} torr) that is significantly lower than the source pressure, (one atmosphere in this case) fully developed sonic flow is established. Such an expansion leads to extensive cooling and a more or less abrupt transition to collisionless flow. A second conical orifice extracts or collimates a central core of gas leading to molecular beam formation. The molecular beam is introduced into the ion source of a quadrupole mass spectrometer. Low-energy electron ionization (EI) (approximately 25 eV) is used to minimize fragmentation within the ion source. The mass range of interest is repeatedly scanned and each scan is stored by the data system for later analysis.

In the case of the pyrolysis-vapor reactants, the mass spectrum is very complex and many ions have multiple origin (20,21). In addition, isomers can seldom be distinguished by mass spectrometry alone. Nevertheless, even in the worst case of primary oils composed of hundreds of compounds, a significant number can be identified with little ambiguity (20). In the case of the product slates observed so far over active ZSM-5, the mass spectrometry is quite simple and unambiguous, except for isomers. Thus pure compounds representing all the major products are added both to determine their fragmentation pattern, and possible interference, and to achieve semi-quantitative calibration. Only in the case of ethylene and carbon monoxide, is there an overlap of parent and fragment ion (28 amu). In this case, electron impact fragmentation is used beneficially (as is often the case) to permit the ion at 27 amu (C_2H_3+) to be used to moniter ethylene, while correcting the ion at 28 amu to measure CO. Proportions of isomers such as o-, p-, and m-toluene, will eventually be estimated by comparison with companion studies by Diebold et al. (14) using gas chromatography and GC-MS on collected gasoline-range products.

Batch pyrolysis experiments were carried out by inserting wood splints, or boats of wood powder or pure compounds into the preheated reactor with hot carrier gas flowing and simultaneously starting the data aquisition system. Under typical conditions, a 30 mg wood splint was pyrolyzed over approximately 30 s giving a weight of carrier gas/weight of wood vapor of approximately 4 (a vapor concentration of approximately 1% by volume in the helium assuming an average molecular weight of the primary vapors of about 100). This resulted in a range of WHSV's from 0.6 to 3, depending on the amount of catalyst present.

After initial screening, with pure HZSM-5 (1), the catalyst was replaced with one-sixteenth inch diameter extruded pellets of HZSM-5 in an unspecified binder. This catalyst was provided by Mobil (22) and was also used in the larger bench-scale testing (14). The catalyst was ground to a 25/45 mesh size (250/350 microns) and catalyst weight was varied to control the WHSV.

For initial screening, we have operated our micro-catalytic reactor in a pulsed mode. As is shown below (Results & Discussion, Figure 2), the catalyst appears to rather quickly achieve steady-state response, after the first few samples of wood (e.g., ≤100 mg wood per gram of catalyst). This is confirmed by the approximate mass closures (≥80 % typically) from summation of all gaseous species evolving from each wood-vapor pulse, plus measured char yields for each pulse, and overall coke yields of 5-10 weight percent (See Table II under Results & Discussion). There is some evidence of absorption of water and light aromatics when vapor first contacts the catalyst, so that the reported yields of liquid products are probably conservative. We plan to insert wood dowels in a continuous manner to verify the behavior under steady state of wood addition.

Results and Discussion

The goals of this catalyst screening project are to gain insight into the conversion of biomass pyrolysis products with a variety of catalysts and under a range of conditions. Three topics will be discussed here: catalyst and parameter screening, MBMS estimation of yield, and model compound conversion.

Methanol and Wood Conversion Product Classes. Methanol has been used in this screening work to ascertain catalyst activity. The methanol relative product distribution on an active, pure catalyst is shown in Figure 1A. (Table II gives the identification of the ions observed). No methanol (m/z 31 and 32) breakthrough was observed, and the first formed product, dimethyl ether (m/z 45 and 46), has been consumed to form a mixture of C_2 to C_6 olefins and toluene, xylene, and trimethyl-benzene. Note the lack of benzene and alkanes. With lower space velocities and higher methanol partial pressure, the alkenes are known to disproportionate to branched alkanes and to form more aromatics (11). The absence of products above m/z 120 indicates the well-known shape selectivity of the catalyst.

The conversion of the primary pyrolysis products of birch wood for 2 grams of fresh catalyst and for the catalyst subjected to the pyrolysis vapors from about 2.6 grams of wood are shown in Figure 1C and 1D, respectively. The primary pyrolysis products of birch are shown in Figure 1B, and the detailed interpretation of the ion peaks is given in Reference 20. All the primary products appear to be converted, regardless of size, even though most of the zeolite active sites are within 5.1-5.6 angstrom cavities. One possible explanation for this observation is that the larger molecules are initially cracked on acid sites on the macro surface of the catalyst or binder, or are converted to coke. The products on the active catalyst, however, do appear to be shape selective with most products below m/z 156. We also screened conversion behavior on other acid type catalyst including wide-bore molecular sieves and silica/alumnia. The formation of higher molecular weight products such as the methylated naphthalenes (m/z 142, 156, and 170) were much more pronounced than in the shape selective HZSM-5. The formation of water, CO, and CO_2 are the major paths for the rejection of oxygen and represent a substantial portion of the yield. They are under-represented by the intensities in the mass spectra because the instrument was tuned to favor heavy organics, so conversion of primary products could be monitored. See Table II and Figure 5 below, for more quantitative indications of the relative proportions of the various product species.

The products from the partially deactivated catalyst are shown in Figure 1D. They are a mixture of the peaks in the primary spectrum and the fully active catalyst spectrum. The lignin-derived peaks are more prevalent, such as m/z 154, 168, 180 and 194. The carbohydrate-derived products are still partially converted to furan derivatives (e.g., furan, m/z 68 and methylfuran, m/z 82). The lignin derived methoxyphenols are still partially converted to phenols: e.g., phenol, m/z 94, and cresol, m/z 108. These products are the only unique intermediate oxygen-containing products identified so far in the zeolite conversion of biomass primary vapors.

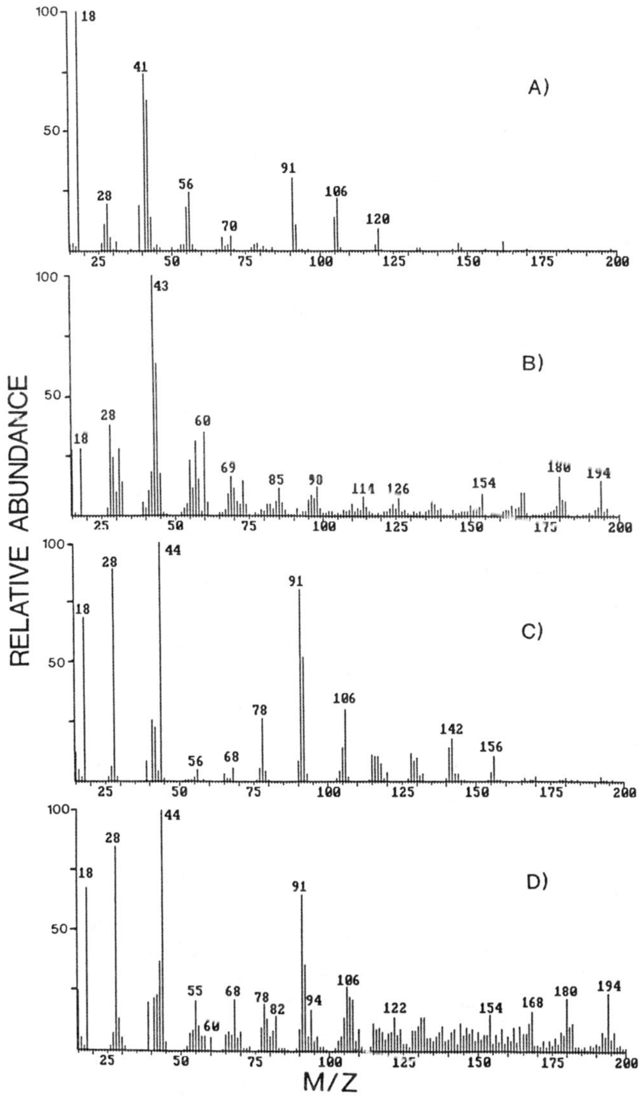

Figure 1. (A) The mass spectrum of methanol conversion products over HZSM-5 at 500°C and WHSV= 1.3; (B) The primary pyrolysis spectra of wood vapor; (C) The mass spectrum of wood pyrolysis products converted over active catalyst (512°C WHSV = 8.3; (D) The product spectrum over the deactivated catalyst after repeated uses for wood and methanol. The spectra do not contain corrections for sensitivity factors.

The higher proportions of aromatics relative to alkenes for wood vapors, vis-a-vis methanol, suggests a direct formation path not involving light olefins. The furans are only minor products on active catalysts, but are still formed on deactivated catalysts, even when the aromatics are reduced. This may indicate that the conversion of carbohydrates to furans occurs on weak acid sites, somewhat analogous to the formation of dimethylether from methanol, which also forms as an intermediate. Figure 2 shows the progressive evolution of a variety of products from catalyst conversions as 2.6 grams of wood were pyrolyzed and the vapors passed over 2 gms of catalyst. Also shown are the steady emergence of primary products.

The gradual and consistent changes in the different types of products shown in Figure 2 over the complete catalyst aging cycle are strong evidence against a significant contribution of transient processes to the pulse mode/mass spectrometric detection method of catalyst screening. The gradual decrease in toluene, an obvious product of internal pore catalytic sites, is consistent with the gradual increase of the intermediates, furan and phenol, and the primary pyrolysis products, represented by m/z 60 for carbohydrate-derived products and m/z 194 for lignin-derived products. The different rates of appearance of m/z 60 and 194 show the relative reactivity of these different classes of wood pyrolysis products on the catalyst. These changes in product ratios with catalyst life shown in Figure 2 support the early and late wood product spectra shown in Figure 1. The mass balances obtained in the quantitative studies that are described below provide further evidence that the experimental approach reflects complete wood product behavior on the catalyst and not transient phenomena dominated by macro surface chemisorption of single pulses of pyrolysis products.

Effect of Temperature and Weight Hourly Space Velocity. The proportions of organic products that form over HZSM-5 vary systematically with WHSV and temperature. The effects of these two parameters on the relative abundance of selected products from the conversion of pine pyrolysis vapors over HZSM-5 are shown in Figure 3. The temperature was varied from 300° to 550°C and the WHSV from 0.7 to 2. These are relative abundance plots and do not directly show the yield of these components since coke and inorganic gases will vary with these parameters and are not reflected in the product distribution. The plots of trimethylbenzene, toluene, and benzene show a trend of dealkylation with increased temperature. Trimethylbenzene decreased with temperature, while toluene increased to a maximum at 500°C and then decreased at higher temperatures. Xylene decreased steadily over this temperature range, but not as pronounced as trimethylbenzene. This indicates that the dealkylation of trimethylbenzene and xylene lead to increased abundances of toluene reaching a maximum at 500°C where the depletion of these heavier species and the dealkylation of toluene leads to a decrease in its relative abundance. Benzene increased throughout the temperature range studied. At higher temperatures, benzene's relative abundance also increased with lower WHSV's.

The intermediates are represented in Figure 3 by m/z 118 (probably benzofuran) which shows a maximum at 400°C for all three WHSV's although the relative abundance increased with WHSV. The

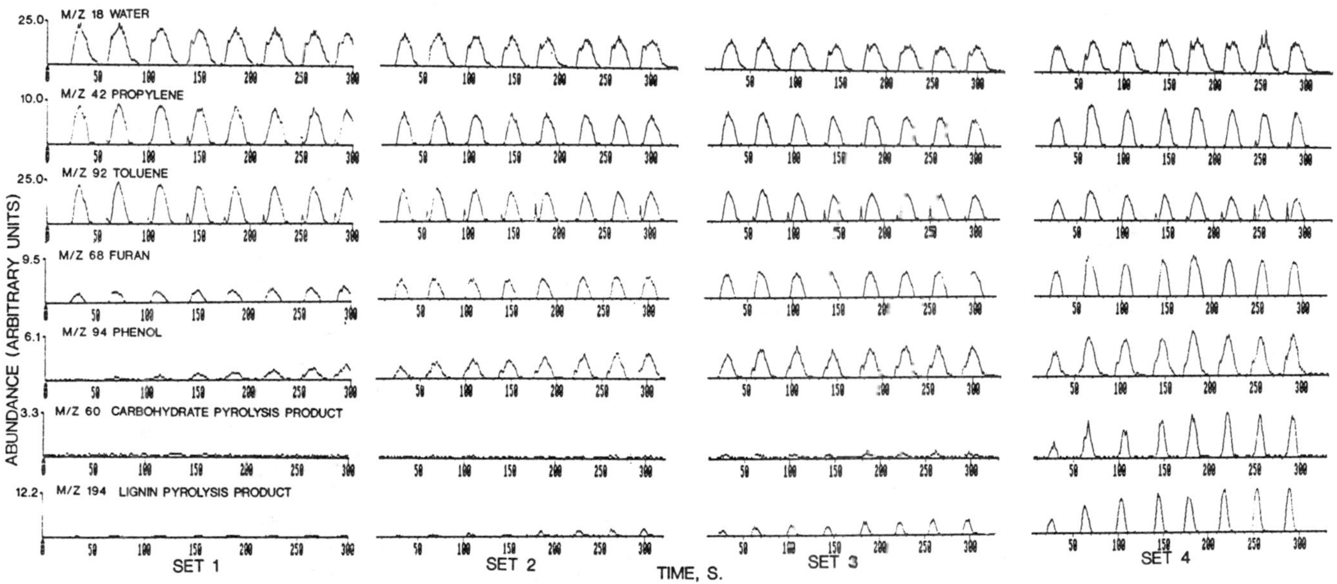

Figure 2. The product evolution curves for selected masses over the course of feeding 2.6 g of wood vapors over 2 g of catalyst. Each pulse represents approximately 65 mg of wood; 500°C; WHSV= 8.3. A total of ten additional wood samples were run at the beginning, between each set, and at the end for quantitative analysis as well as methanol to monitor catalyst activity. The catalyst was exposed to helium flow between sets when no wood was being fed.

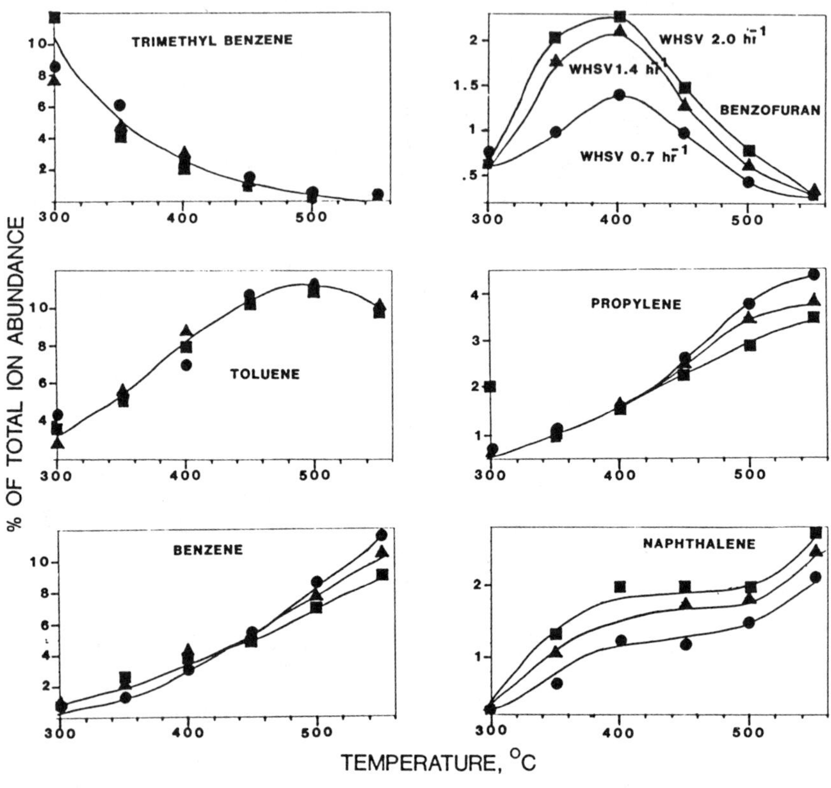

Figure 3. The effect of temperature and WHSV on the distribution of selected products from the conversion of pine wood pyrolysis vapors over HZSM-5 catalyst. (Note: % of total ion abundance should not be equated with % of neutral products since mass spectrometer tuning and cross sections favor aromatics.)

alkenes showed a steady increase with temperature and a dependence on WHSV at the higher temperatures as illustrated by propylene. The generation of alkenes under extreme reaction conditions indicate that the mechanism of aromatic formation from the alkenes in methanol conversion is not as dominant in wood pyrolysis product conversion. A large fraction of the aromatics formed from wood are apparently formed more directly by the deoxygenation and aromatization of the primary pyrolysis products rather than through alkenes, as in methanol conversion. Increased formation of the alkenes at higher temperatures may be associated with dealkylation of the aromatics. The formation of condensed aromatics is also enhanced at higher temperatures as demonstrated by naphthalene in Figure 3.

The Effect of Co-Feeding Wood and Methanol over HZSM-5. The relatively low levels of hydrogen in biomass, coupled with the high amount of oxygen to be rejected, are the reasons for the 34 wt % theoretical maximum yield of hydrocarbons from dry wood (where the hydrocarbon products are assumed to have the average stoichiometry of xylene) (1). The low hydrogen content of wood and many oxygenated materials that have been tested since the introduction of HZSM-5, has led several researchers to experimentation with the cofeeding of methanol to enrich the feedstock in hydrogen and increase the potential hydrocarbon yield.

Chang et al. (23) showed that hydrogen deficient oxygenates could be successfully converted on HZSM-5 if co-fed with an adequate amount of hydrogen rich material, such as methanol. Chantel et al. co-fed methanol with pyrolysis oil (9) and phenols (8,10), but only at 10 % methanol, which is below the optimum levels found by Chang et al. (23). Chen et al. (12) studied the conversion of wood pyrolysis oils with an equal weight of methanol and observed a net increase of 63% in hydrocarbons. We briefly explored the cofeeding of equal weights of methanol with wood pyrolysis vapors. The use of directly formed vapors distinguishes this experiment from the above studies. Selected changes in product ratios are shown in Figure 4 for wood, methanol, and wood plus methanol with the methanol-alone effect subtracted so that wood plus the synergistic effect is shown. The aromatic/inorganic ratio gives an indication of liquid fuel yield and the wood plus methanol enhancement is significantly higher than either the wood or methanol alone. This is in agreement with past results. As stated previously, the mode of oxygen rejection is an important consideration for biomass due to its hydrogen deficiency. This leads to the conclusion that the preferred mode of oxygen rejection from wood is as CO_2 or CO. Chen et al. (12) point out that when a hydrogen donor is present, the prefered route of oxygen rejection is as H_2O to preserve carbon. The CO_2/ H_2O ratio was increased by the methanol cofeedings in this experiment. This observation does not agree with past results (9,12) where increased water yields and decreased CO_2 yields were found when cofeeding methanol. Since only ratios have been calculated in this experiment, it is impossible to say whether the increase in the CO_2/H_2O ratio is due to changes in both CO_2 and H_2O or just one product. The presence in this experiment of the CO_2 and H_2O from the prompt gases and char formation in the initial pyrolysis step could contribute to the difference between these results and past results (9,12). The nature of the aromatic products is also improved by the methanol synergism. The trimethylbenzene/toluene is greatly increased due to the presence of methanol and moves the aromatic product distribution from wood closer to that observed for methanol. This means less benzene in the gasoline and more of the higher octane, methylated benzenes.

A hypothesis for the synergistic behavior is that methanol, in the presence of wood conversion products, does not have to undergo the dimethyl ether conversion step to alkenes, but rather can react directly with the reactive wood-derived products to alkylate aromatics. This increase in methylated benzenes would be expected to occur at the expense of the normal slate of products formed from pure methanol.

Chantal et al. (8,10) have studied the effect of methanol cofeeding on product distribution from phenolic compounds. They proposed an oxonium mechanism to explain alkylation where oxonium ions are generated from diphenylether and anisole, intermediates that were formed from phenolic starting materials. They used low levels of methanol in these experiments: 90/10 (phenol/methanol) vs 1/1 (wood pyrolzate/methanol) in the results reported in this paper and by Chen et al. (12). Under conditions of high methanol concentrations, and in the presence of carbohydrate-derived material, the formation of an oxonium ion from methanol ($CH_3OH_2^+$) as proposed by Aranson et al. (24) is also possible. Direct reaction with either wood-derived reactants or various products from zeolite catalysis could explain the observed synergistis effect.

Hydrocarbon Yields. Despite encouraging results that the wood pyrolysis products are all destroyed on the catalyst and the high quality composition of the organic products, the initial indirect estimates of yield (25) were approximately 11% hydrocarbon products or about a third of the theoretical yield based on oxygen rejection considerations (1). Therefore, it was important to include an estimate of yield in the MBMS catalyst screening work based on improved calibration procedures. The initial results are reported here.

Calibration factors for the products of interest were determined by injecting known amounts of liquid and gaseous products in the reactor prior to packing the catalyst. Before each run, the gaseous calibrants were also bled into the flowing helium above the catalyst bed and the calibration table was updated based on the observed response factors. The products calibrated were water, CO, CO_2, methane, ethylene, propylene, butene, butane, pentene, benzene, toluene, xylene, and methylnaphthalene. Products not calibrated were estimated by using calibration factors for similar materials (e.g., using the calibration factor for methylnaphthalene for naphthalene). The products not calibrated were present in low amounts, so error caused by this procedure was probably within the goals of this screening effort.

Estimates of yields for runs made in triplicate at 500°C and WHSV = 4, on two days, are shown in Figure 5. The error bars are the standard deviation (only positive deviation is shown) of three replicates on each day. Water was significantly different for the two days, but the CO and CO_2 estimates were quite close. The total

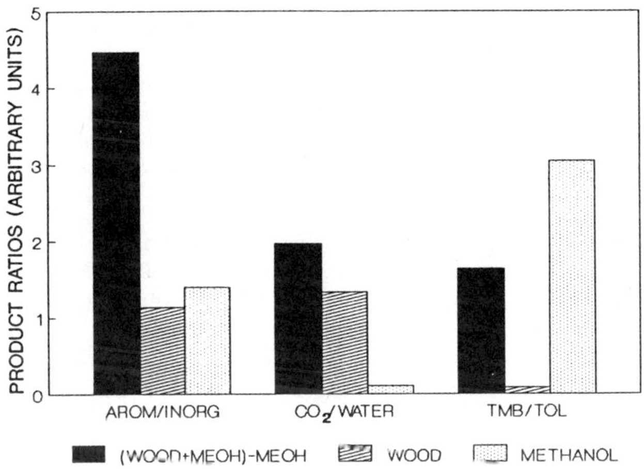

Figure 4. The effect of co-feeding methanol with wood. Selected product ratios for wood alone, methanol alone, and wood co-fed with methanol with the methanol products subtracted. The ratios are: light aromatics/ H_2O + CO + CO_2, (arom/inorg); CO_2/water; and trimethylbenzene/toluene (TMB/TOL). Temperature = 500°C; wood WHSV= 2.9; methanol WHSV= 2.8.

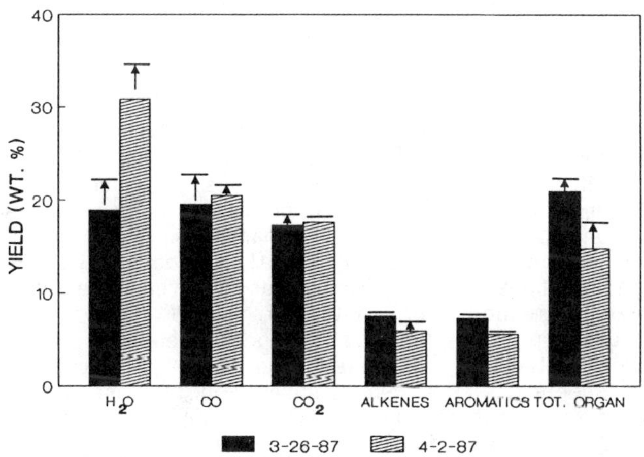

Figure 5. Yields on a dry-weight basis from wood pyrolysis products over HZSM-5 for two different runs. Error bars show positive standard deviation based on triplicate determinations. Total organics include alkenes, light aromatics, furnans, indenes and naphthalenes. Table II contains the breakdown of products.

organics yield was significantly less on the second day. The total organic yield contains the light aromatics, alkenes, furans, naphthalenes, and indenes.

The estimated product mass balances for these two runs are shown in Table I. These values assume 10% coke formation, which was determined in other experiments by weight gain of the catalyst. The 3-26-87 data are quite close to the calculated values, but the 4-2-87 data are significantly high in oxygen, which can be explained by the high water value for that day.

TABLE I. Product Mass Balances (dry weight basis) for the Average Product Composition for Runs on Two Days. Results Assume 10% Coke Which is All Carbon and a Char Composition of $CH_{.53}O_{.12}$

Element	Run 3-26-87	Run 4-2-87	Calculated
C	54	48	52
H	5.2	5.7	6.2
O	43	54	42
	102	108	100

The complete product slates for the yield estimates are given in Table II Summations ranged from 96 to 114, with an average of 106. The biggest source of error is probable the quantitation of coke and water (since H_2O is a major constituent detected with lower sensitivity than the other constituents).

Model Compound Products over ZSM-5 Pellets. The behavior of individual wood pyrolysis products on the catalyst gives insight into the behavior of the major compound classes. If certain compound classes are particularly good or bad performers, then perhaps pyrolysis conditions can be changed to enhance the good performers or minimize the bad ones. The experimental approach used here allows compounds to be tested that would not be feasible at a larger scale. However, methods for controlling WHSV and calibration are still under development. Results are presented here, for selected compounds over HZSM-5 that represent the major compound classes, as uncalibrated spectra. The behavior of model compounds has given some insight toward potential yield. The product spectra of four model compounds are shown in Figure 6. The best use of these model compound screening results is for product slate comparisons.

The spectra of hydroxyacetaldehyde (Figure 6A) is typical of the light oxygenates with high H_2O, CO, and CO_2 and moderate amounts of olefins and light aromatics. Even at the high WHSV of 6 all the starting material is destroyed. The 5-hydroxymethylfurfural (Figure 6B) products show higher relative amounts of organic products, including the furans and light aromatics, but also

TABLE II Yields of Total Products from Calibrated MBMS Runs With Birch Wood Dowels HZSM-5, 2 gms; WHSV≈ 4; T= 500°C; Values are in wt % yield (Moisture Free).

Assumed Species	Mass	3/26/87			4/2/87			Avg of
		#18	#26	#34	#13	#14	#15	6 Runs
Char		15.9	16.3	16.7	12.8	13.1	17.9	15.4
Coke*		10	10	10	10	10	10	10.0
H_2O	18	20.5	21.1	15.1	28.1	29.7	34.7	24.9
CO	28	21.0	21.9	15.9	20.2	20.3	21.0	20.1
CO_2	44	17.6	18.4	16.1	17.1	18.2	17.6	17.5
Methane	16	0.9	1.0	1.0	0.6	0.7	0.2	0.7
Butane	58	1.2	1.3	2.5	0.1	1.5	0.3	1.2
Ethylene	28	2.9	3.0	2.8	2.5	2.6	2.0	2.6
Propylene	42	2.0	2.1	1.8	1.9	2.0	1.8	1.9
Butenes	56	1.6	1.9	1.9	1.3	1.2	1.0	1.5
Pentenes	70	0.4	0.6	1.8	1.2	0.5	-	0.8
Benzene	78	1.3	1.3	1.1	0.8	0.8	0.7	1.0
Toluene	92	3.9	4.1	3.1	2.9	3.1	2.7	3.3
Xylene	106	2.5	2.8	2.2	2.1	2.1	1.6	2.2
Naphthelene	128	0.6	0.5	0.5	0.4	0.5	0.3	0.5
Me-Naphthelene	142	0.9	0.7	0.7	0.6	0.7	0.4	0.7
Indene	116	0.6	0.6	0.6	0.3	0.4	0.3	0.5
Me Indene	130	0.5	0.6	-	0.3	0.3	0.2	0.3
Furan	68	0.5	0.6	0.7	0.4	0.4	0.4	0.5
Me Furan	82	0.2	0.2	0.6	0.1	0.1	0.1	0.2
Benzofuran	118	0.5	0.5	-	0.4	0.3	0.2	0.3
Me Benzofuran	132	0.2	0.3	0.5	NM**	NM**	NM**	0.3
∑ Aromatics (monocyclic)		7.7	8.2	6.4	5.8	6.0	5.1	6.5
∑ alkenes		6.9	7.6	8.3	6.9	6.3	4.8	6.8
∑ Furans		1.4	1.6	1.8	0.9	0.8	0.7	1.2
∑ Organics		20.5	21.9	22.1	16.3	17.4	12.3	18.4
Total		106	110	96	104	109	114	106

* Estimate ** Not measured

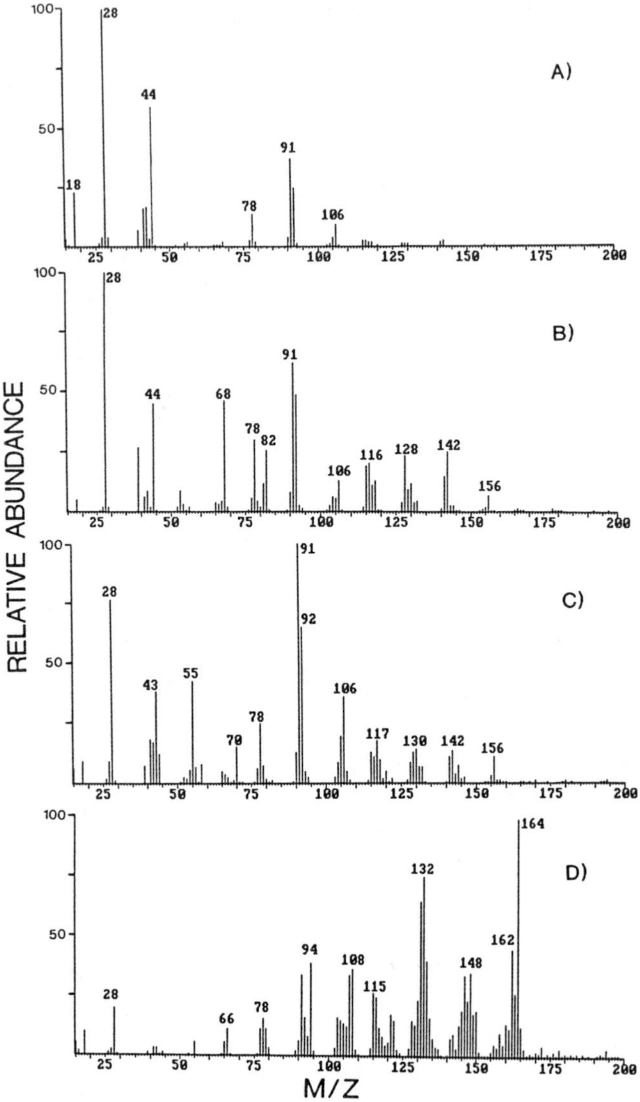

Figure 6. Mass spectra of the products of the conversion of Model compounds over HZSM-5: (A) hydroxyacetaldehyde, 500°C WHSV ~ 6; (B) 5-hydroxymethylfurfural, 500°C WHSV ~ 4; (C) α-angelicalactone, 500°C WHSV ~ 4; (D) isoeugenol, 500°C, WHSV ~ 2. Spectra do not contain corrections for sensitivity factors.

high amounts of the naphthalenes. The α-angelicalactone results (Figure 6C) show the highest abundances of aromatics of the model compounds that were studied. The products include more m/z 70 (likely pentene) than the spectra for other feedstocks. The isoeugenol results (Figure 6D) show larger amounts of unconverted feedstock (m/z 164) and phenols (phenol, m/z 94; cresol, m/z 108) than observed for whole wood at this WHSV. This low reactivity is confirmed by the work of Chantal et al. (8,10). The species at m/z 132 could possible be explained by the cyclization of the side chain to form hydroxyindene with the loss of the methoxy group. It may be that at the effectively lower space velocities for lignin vapors in whole wood, the complete disappearance of lignin products is due to coking.

Several trends are summarized here for the model compounds studied to date: (1) The light oxygenates, such as hydroxyacetaldehyde, gave high yields of water, CO and CO_2 with relatively low yields of hydrocarbon products; (2) The carbohydrate-derived ring compounds, such as α-angelicalactone, gave the greatest yields of aromatics; (3) The methoxyphenols showed lower conversion rates at the relatively high space velocities used. It is thought that these materials have a high coking potential, based on tests performed at lower space velocities. The presence of unconverted materials and phenols in the product stream may be more desireable than increased coking rates.

Summary

The use of MBMS for real time screening of wood pyrolysis product conversion over Mobil's HZSM-5 allows the rapid, qualitative evaluation of performance of different catalysts, reaction conditions and feedstocks. This information can help guide more extensive testing at the bench-scale level and provide insight into the chemistry of conversion. The following conclusions can be drawn from the work to date:

(1) No significant breakthrough of primary wood vapor organics occurs until the weight of wood/weight of catalyst is about 1 (at a WHSV of 4-5);
(2) There were few intermediates from wood (furans from the carbohydrates and phenols from the methoxyphenols) on the way to alkenes, light aromatics, and naphthalenes;
(3) The light aromatics/alkenes ratio is 1 for the nonoptimized and low-partial-pressure conditions studied to date;
(4) Much of the oxygen is favorably rejected as CO and CO_2;
(5) The coke yield is ~10% in these experiments at around 500°C in He;
(6) The hydrocarbon yield was ~18±4 wt % based on two sets of preliminary calibrated experiments;
(7) Methanol is a poor model for wood because the mechanism of hydrocarbon formation is apparently different. The co-feeding of methanol with wood appears to provide synergistic benefits.

Future work will focus on the optimization of parameters, such as the partial pressure of the pyrolysis vapors, for maximizing the yield of C_5-C_{10} hydrocarbons, and the effect of HZSM-5 structural parameters on biomass conversion.

Acknowledgments

The authors are pleased to acknowledge the technical assistance of Mike Maholland and Kristal Seder and stimulating discussions with Jim Diebold and Helena Chum. The support of the Biofuels and Municipal Waste Technology Division of DOE is appreciated, as is the technical monitorship of Don Stevens, Gary Schiefelbein, and Si Friedrich. Mobil Research and Development Corporation supplied the samples of HZSM-5.

References

1. Diebold, J. P.; Chum, H. L.; Evans, R. J.; Milne, T. A.; Reed, T. B.; Scahill, J. W. In Energy from Biomass and Wastes X; Klass, D. L., Ed.; Institute of Gas Technology, Chicago, and Elsevier Applied Sciences Publishers: London, 1987; p 801.
2. Kaliaquine, S. In Wood Liguefaction Specialists Meeting; Univ. of Saskatchewan; Feb. 16-17, 1982; p 75.
3. Chang, C. D.; Silvestri, A. J. J. Catalysis 1977, 47, 249.
4. Chang, C. D. Hydrocarbons from Methanol; Marcel Dekker, Inc: New York, 1983; p 129.
5. Chen, N. Y.; Degnan, T. F.; Koenig, L. R. Chemtech 1986, 506.
6. Frankiewicz, T. C. In Proceedings of Specialists' Workshop on Fast Pyrolysis of Biomass; Diebold, J., Ed.; Copper Mountain, Colorado, Oct. 19-22, SERI/CP-622-1096; 1980; p 123.
7. Hanniff, M. I.; Dao, L. H. In Energy from Biomass and Wastes X; Klass, D. L., Ed.; Institute of Gas Technology, Chicago, and Elsevier Applied Sciences Publishers: London, 1987; p 831.
8. Chantal, P. D.; Kaliaguine, S.; Grandmaison, J. L. In Catalysis on the Energy Scene, Elsevier Science Publ.: Amsterdam, 1984; p 93.
9. Chantal, P. D.; Kaliaguine, S.; Grandmaison, J. L.; Mahay, S. Applied Catalysts 1984, 10, 317.
10. Chantal, P. D.; Kaliaguine, S.; Grandmaison, J. L. Applied Catalysis 1985, 18, 133.
11. Matthew, J. F.; Tepylo, M. G.; Eager, R. L.; Pepper, J. M. Can. J. Chem. Eng 1985, 63, 686.
12. Chen, N. Y.; Walsh, D. E.; Koenig, L. R. Prepr. of 193rd ACS Mtg., Div. Fuel Chem., Denver, CO, April 5-10, 1987, 32, No. 2, p 264 and this volume.
13. Renaud, M.; Grandmaison, J. L.; Roy, C.; Kaliaguine, S. Prepr. of 193rd ACS Mtg., Div. Fuel Chem., Denver, CO, April 5-10, 1987, 32, No. 2, p 276 and this volume.
14. Diebold, J. P.; Scahill, J. W. Prepr. of 193rd ACS Mtg., Div. Fuel Chem., Denver, CO, April 5-10, 1987, 32, No. 2, p 297 and this volume.
15. Costa, E.; Uguina, A.; Aguadok, J.; Herinandez, P. J. Ind. Eng. Chem. Proc. Des. Dev. 1985, 24, 239.
16. Prasad, Y. S.; Bakhshi, N. M. Applied Catalysis 1985, 8, 71.
17. Prasad, Y. S.; Bakhshi, N. N.; Matthews, J. F.; Eager, R. L. Can J. Chem. Eng. 1986, 64, 278.
18. Milne, T. A.; Soltys, M. N.; J. Anal. Appl. Pyrol. 1983, 5, 93.
19. Milne, T. A.; Soltys, M. N.; J. Anal. Appl. Pyrol. 1983, 5, 111.

20. Evans, R. J.; Milne, T. A.; Energy and Fuels 1987, 1, 123.
21. Evans, R. J.; Milne, T. A.; Energy and Fuels 1987, 1, 311.
22. Mobil Research and Development Corporation, courtesy of N. Y. Chen.
23. Chang, C. D.; Lang, W. H.; Silvestri, A. J. U.S. Patent 3 998 898, 1976.
24. Aronson, M. T.; Gorte, R. J.; Farneth, W. E. J. Catalysis 1986, 98, 434.
25. Evans, R. J.; Milne, T. A. Preprints of 193rd ACS Meeting, Div. Fuel Chem., Denver, CO, April 5-10, 1987, 32, No. 2, p 287.

RECEIVED March 31, 1988

Chapter 27

Reactions of Model Compounds of Biomass-Pyrolysis Oils over ZSM–5 Zeolite Catalysts

Le H. Dao, Mohammed Haniff, André Houle, and Denis Lamothe

Laboratoire de Recherche Sur les Matériaux Avancés, INRS–Energie, Institut National de la Recherche Scientifique, 1650, Montée Sainte–Julie, Varennes, Quebec J0L 2P0, Canada

The catalytic conversion of model compounds (e.g. cyclopentanone, cyclopentenone, furfural, glycerol, glucose, fructose and their derivatives usually found in the pyrolysis oils of biomass materials) to hydrocarbon products over H-ZSM-5, Zn-ZSM-5 and Mn-ZSM-5 zeolite catalysts have been studied in a fixed bed microreactor at temperatures ranging from 350°C to 500°C. Although the deoxygenation of cyclopentanone was completed at 400°C, low yields of hydrocarbons were obtained for furfural, glycerol and carbohydrate derivatives. This is due possibly to thermal polymerization at reaction temperatures of 400°C and higher. Increasing the $(H/C)_{eff}$ ratio of the feed with methanol increased significantly the hydrocarbon yield of furfural but not that of carbohydrates. Reaction mechanisms for deoxygenation and tar formation are proposed.

It has been shown that synthetic zeolites such as ZSM-5 can be used to convert oxygenated compounds derived from biomass materials into hydrocarbons which can be used as fuels or chemicals feedstocks (1,2,3,4). However, the pyrolysis oils obtained from biomass materials by different thermal and thermochemical processes (5,6) showed poor hydrocarbon yields and high tar content when contacted over ZSM-5 zeolite catalysts at high temperatures (7,8). Since the pyrolysis oils are composed of a wide variety of oxygenated compounds such as cyclopentanone, cyclopentenone, furfural, phenol, carbohydrate and carboxylic acid derivatives (9,10); it is difficult to point out exactly which family of compounds is contributing more to the observed tar and to the rapid deactivation of the catalysts. Catalytic studies on model compounds which are usually found in the biomass pyrolysis oils are therefore primordial in order to determine the best catalytic system for the up-grading of pyrolysis oils to useful hydrocarbon products. The reactions of some phenolic, carbonyl and carboxylic acid derivatives over ZSM-5 catalysts are already

0097–6156/88/0376–0328$06.00/0

reported (7,11). The present paper thus reports the results for the conversion of cyclopentanone, cyclopentenone, furfural, glycerol, glucose and fructose, and their isopropylene derivatives over H-ZSM-5 and Zn and Mn-exchanged ZSM-5 zeolite catalysts at temperatures ranging from 350°C to 500°C. Some reactions are supplemented with methanol in their feeds, so as to determine the effect of increased $(H/C)_{eff}$ ratio (see Table 1 for definition) on the deoxygenation yields.

Experimental

Preparation and characterization of the catalysts. The synthesis of the sodium form of ZSM-5 was done according to procedure #2 given in the patent of Derouane and Valyocsik (12). The only difference was the use of silica fiber (Strem Chemicals) instead of silica gel powder. The Na-ZSM-5 zeolite was exchanged several times with 10% aqueous solution of NH_4NO_3 (10 ml per g of zeolite) at 80°C. The zeolite was then dried at 100°C and calcinated at 500°C so as to obtain the H-ZSM-5 catalyst. Mn-ZSM-5 catalyst was prepared by heating the ammonium exchanged zeolite with a 10% solution of manganese nitrate at 80°C for 3 hours (10 ml per g zeolite). After washing and drying, the catalyst was calcinated at 500°C. Zn-ZSM-5 catalyst was prepared in a similar fashion to that of the manganese form using a 10% solution of zinc nitrate. The X-ray diffraction pattern of the zeolite is similar to those reported in the literature. The chemical composition is shown in Table 1.

Preparation of the carbohydrate isopropylene derivatives (13). 20 g of the carbohydrate was mixed with 250 ml of acetone and 10 ml of conc. H_2SO_4. The mixture was stirred at room temperature for about 24 hours and then filtered to remove any unreacted solid. The filtrate was neutralized with solid $NaHCO_3$ dried over $MgSO_4$, filtered and the excess acetone was removed on a rotary evaporator to yield the yellow solid derivatives.

Apparatus. The catalytic conversion was studied in a continuous flow quartz microreactor with a fixed-bed of diluted catalysts. The reaction conditions are reported in Table 1. After an experimental run (about 3 to 4 hours), the tar on the catalytic bed was determined by taking the difference in weight of the reactor before and after placing it in a furnace set at 500°C in the presence of air. The reaction products were analyzed by GC and GC/MS (Table 1).

Results

Figure 1 shows the yields of conversion and the products distribution (C_1-C_6 hydrocarbons, aromatic, polyaromatics and tar) as a function of reactor temperature for pure cyclopentanone over H-ZSM-5/bentonite (80/20 Wt.%) catalyst. The conversion is completed at 350°C. The main reaction is a thermal decarbonylation of cyclopentanone, giving C_4 hydrocarbon fragment that reacts further on the catalytic bed to produce aliphatic, aromatic and polyaromatic hydrocarbons. Cyclopentenone which is partially deoxygenated (32%) over H-ZSM-5/bentonite (80/20 Wt.%) at 450°C, can be completely converted

TABLE 1

Chemical composition of ZSM-5 samples

Component (WT%)	H-ZSM-5	Mn-ZSM-5	Zn-ZSM-5
Na_2O	0.55	0.69	0.49
Al_2O_3	2.25	2.33	2.10
SiO_2	86.86	91.05	87.62
MnO	---	0.78	---
ZnO	---	---	0.26
TiO_2	0.59	0.64	0.38
L.O.I.*	7.62	3.26	6.68
Molar Ratio			
SiO_2/Al_2O_3	65.47	66.32	70.80
Na_2O/Al_2O_3	0.40	0.49	0.38

* L.O.I. means loss on ignition of sample weight.

REACTION CONDITIONS

Catalyst weight	:	10 g (80% ZSM-5 + 20% bentonite)
Temperature	:	350-560°C
Pressure	:	atmospheric pressure
Inert gas	:	helium (~ 3 mℓ/min)
WHSV **	:	variable
Reaction time	:	3 hours

** The weight hourly space velocity (WHSV) is defined as:

$$WHSV = \frac{\text{g of injected feed per hour}}{\text{g of catalyst}}$$

ANALYTICAL CONDITIONS

- Gas chromatography: HP 5890 GC with DB-5 (SE-54) column (30m x 0.25mm, 1.0 μ)

  For liquid : 70°C (4 min), then 4°C/min to 160°C then 20 min at 160°C

  Gas : 33°C (isothermal)
- GC/MS : HP 5890 GC and MS detector Pona column and DB-5 column

HYDROGEN/CARBON EFFECTIVE RATIO (11a)

The effective hydrogen index (EHI) is defined as:

$$EHI = (H/C)_{eff} = \frac{H - 2O - 3N - 2S}{C}$$

where H, C, O, N and S are atoms per unit weight of sample of hydrogen, carbon, oxygen, nitrogen and sulfur, respectively.

to hydrocarbons by the addition of methanol to the feed (cyclopentenone/methanol 70/30 Wt.%).

Table 2 shows the reaction of pure furfural ($(H/C)_{eff}$ = 0) over various cation exchanged ZSM-5 catalysts at 400°C. The hydrocarbon yields range from 6.6 to 9.4% and the oxygenated compounds yields vary from 25.6 to 50%. These values reflect the poor performance of all the three catalysts in deoxygenating furfural. The large quantities of furan and CO observed are mainly due to the thermal decarbonylation of furfural. Higher tar contents were obtained for the metal exchanged catalysts (21.2 and 25.9%) than the H-ZSM-5 form (14.2%). The low water contents (ranging from 2.7 to 7.0%) produced from these different reactions indicate poor catalytic deoxygenation of furfural. CO_2 which is produced from pyrolytic reaction is obtained in low yields. The yields for the aliphatic (9.0%) and olefin (8.5%) are relatively smaller than those for the aromatic (47.7%) and polyaromatic.

Figure 2 shows the product distributions for the reaction of various furfural/methanol mixtures over H-ZSM-5/bentonite (80/20 Wt.%) at 450°C; only the major components in the products are shown. The abscissa in Figure 2 is given in both increasing percentage of methanol and increasing $(H/C)_{eff}$ ratio for the feed. As the content of methanol increases, the yields for hydrocarbons and water increase while those for tar, furan and CO decrease. The drastic augmentation of hydrocarbons and water yields conjugated with the diminution of furan with increasing methanol concentration indicate that significant catalytic deoxygenation is taking place. For a mixture of 55/45 Wt.% methanol/furfural ($(H/C)_{eff}$ = 0.85), furan is completely removed from the reaction products and only small quantities of other oxygenated compounds were present (≤ 0.3%). The yields for the other products were similar to those observed previously for furfural. Only at 70/30 Wt.% methanol/furfural mixture there was a significant reduction in the tar content (6.7 or 14.1% on carbon basis). The average CO_2 present was 0.3%.

Table 3 shows the product distribution for reactions of furfural/methanol (30/70 Wt.%) over various concentration of H-ZSM-5/support at 450°C. By diluting the catalyst with bentonite from 80 to 18 Wt.%, the hydrocarbons yields raised from 30.6 to 41.4%. Changing the support material from bentonite to SiO_2-Al_2O_3 caused a small reduction in the hydrocarbon yield (36.3%); however, there was less tar formed when compared to the other cases. The products selectivities were similar to those of previous cases.

Table 4 shows the products from the reaction of glycerol [$(H/C)_{eff}$ of 0.67] over various cation exchanged ZSM-5 zeolites at 400°C. The reaction with Zn-ZSM-5 gave the best yield of hydrocarbons (14.4%) and the lowest yield of oxygenated compounds (11.5%). The tar and CO contents for the three catalytic systems were high (the tar ranged from 14.0 to 19.3% and the carbon monoxide from 5.1 to 7.8%). It should be noted that the major component in the oxygenated products is 2-propenal.

Table 5 shows the products distribution for the reaction of a mixture glycerol/methanol feed (55/44 Wt.%) with $(H/C)_{eff}$ = 1.25 over H-ZSM-5 at 400°C. Compared to the results in Table 4, an increase in hydrocarbons yields and a decrease in tar and oxygenated compounds yields were observed. By diluting H-ZSM-5 with bentonite

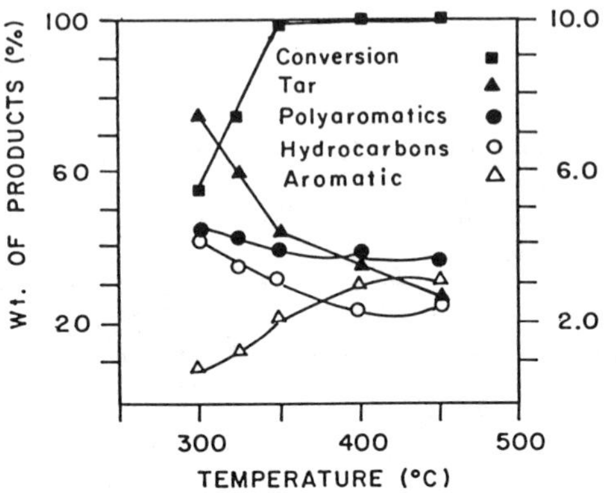

Figure 1. Reaction of cyclopentanone over H-ZSM-5/bentonite (80/20) at a different reactor temperature.

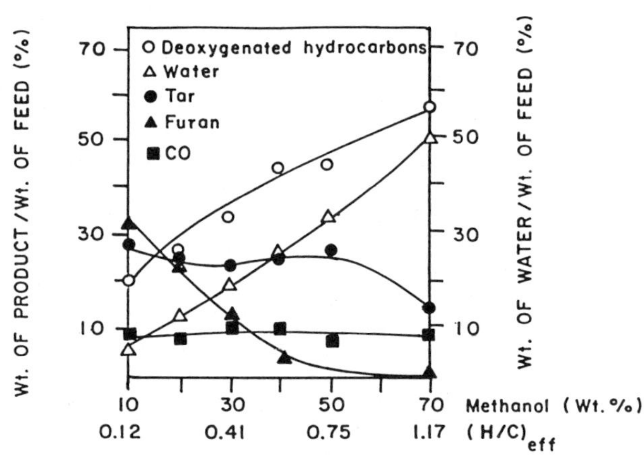

Figure 2. Reaction of furfural/methanol mixtures over H-ZSM-5/-bentonite (80/20g) at a reactor temperature of 400 ^{o}C and a WHSV of 0.238 ± 0.018 hr^{-1}.

TABLE 2
Reaction of furfural over cation exchanged ZSM-5 at 400°C and WHSV of 0.281 hr^{-1}

Experimental conditions:			
Catalyst composition:	80% H-ZSM-5, 20% bentonite,	80% Zn-ZSM-5, 20% bentonite,	80% Mn-ZSM-5 20% bentonite
Total product distribution (Wt.%)			
Furan	47.0	23.8	27.8
$\sum$ Oxygenated hydrocarbons	50.0	25.6	40.7
Tar	14.2	21.2	25.9
CO	22.8	36.3	17.9
CO_2	0.6	4.8	1.9
H_2O	4.4	2.7	7.0
Hydrocarbons	8.00	9.42	6.6
Product selectivity (Wt.%)			
Aliphatics, $C_1 - C_8$	8.8	7.6	10.5
Olefins, $C_2 - C_6$	8.5	10.6	6.5
Aromatics	50.6	48.8	43.9
Polyaromatics	32.1	32.9	39.1

TABLE 3
Reaction of furfural/methanol (30/70) feed $\{(H/C)_{eff} = 1.17\}$ over H-ZSM-5 at 450°C

Experimental conditions:			
Catalyst composition:	80% H-ZSM-5, 20% bentonite,	18% H-ZSM-5, 82% bentonite,	18% H-ZSM-5 82% $SiO_2-Al_2O_3$
WHSV (hr^{-1})	0.029	1.26	1.26
Total product distribution (Wt.%)			
Oxygenated hydrocarbons*	---	1.4	0.8
Tar	6.7	7.6	4.9
CO	9.4	7.8	7.2
CO_2	0.5	0.6	0.3
H_2O	52.7	41.6	50.4
Deoxygenated hydrocarbons	30.6	41.1	36.3
Product selectivity (Wt.%)			
Aliphatics, $C_1 - C_8$	1.8	11.6	20.7
Olefins, $C_2 - C_6$	2.9	3.0	3.6
Aromatics	83.5	72.8	60.8
Polyaromatics	11.9	12.6	14.8

* Mainly furan and benzofuran, derivatives and dimethylether.

TABLE 4
Reaction of glycerol ($(H/C)_{eff}$ = 0.67) over various cation exchanged ZSM-5 at 400°C and a WHSV of 0.228 hr^{-1}

Experimental conditions:			
Catalyst composition:	80% H-ZSM-5, 20% bentonite,	80% Zn-ZSM-5, 20% bentonite,	80% Mn-ZSM-5 20% Bentonite
Total product distribution (Wt.%)			
Oxygenated hydrocarbons*	32.7	11,5	16,7
Tar	14.0	15.8	19.3
CO	7.8	6.1	5.1
CO_2	0.3	3.7	1.4
H_2O	40.2	48.2	49.6
Hydrocarbons	5.0	14.4	7.8
Product selectivity (Wt.%)			
Aliphatics, C_1 - C_8	15.8	5.5	9.4
Olefins, C_2 - C_6	5.9	9.6	8.4
Aromatics	60.1	63.5	44.1
Polyaromatics	18.5	21.4	38.1

* Mainly 2-propenal and traces of acetone

TABLE 5
Reaction of glycerol/methanol (55/45) feed {$(H/C)_{eff}$ = 1.25} over H-ZSM-5 at 400°C and WHSV of 1.44 hr^{-1}

Experimental conditions:		
Catalyst composition:	80% H-ZSM-5 20% bentonite	18% H-ZSM-5 82% bentonite
Total product distribution (Wt.%)		
Oxygenated hydrocarbons*	18.0	25.6
Tar	11.8	5.9
CO	4.9	1.1
CO_2	0.9	0.6
H_2O	45.4	50.6
Hydrocarbons	19.1	16.2
Product selectivity (Wt.%)		
Aliphatics, C_1 - C_8	15.2	16.0
Olefins, C_2 - C_6	8.9	11.7
Aromatics	56.8	59.2
Polyaromatic	19.0	13.0

* Mainly 2-propenal, and traces of methanol and dimethylether

from 80 to 18%, less tar was formed (5.9%) but the yield of hydrocarbons also diminished.

Table 6 shows results for reactions of glucose and glucose derivative done over H-ZSM-5 at 450°C. With the addition of methanol, thus an increase of $(H/C)_{eff}$, there was an increase in the hydrocarbon yields for the glucose/water/methanol (20/50/30) and glucose derivative/water/methanol (27.6/12.2/60.2) cases compared to that of the glucose/water (20.3/79.7) case. Also, there was a simultaneous decrease in tar content with an increase in $(H/C)_{eff}$ of the feed. However in all cases the hydrocarbon yields are low and the tar contents too high for a viable catalytic process. The high water content observed in all experiments (29.2 to 66.1%) is not only due to catalytic deoxygenation through loss of water but also due to polycondensation reactions of glucose and its derivative. These condensation reactions produce polymeric oxygenated compounds which are responsible for the high tar content observed. Oxygenated compounds, CO and CO_2 are minor products which are normally obtained from thermal decomposition of biomass materials (14). The product selectivity indicates a high percentage of aliphatic (23.6 to 50.5%) and aromatic (34.2 to 53.2%). Except for the reaction with glucose/water, the olefin contents are low for the other two reactions. The production of polyaromatics (mainly indene and naphthalene derivatives) is rather high.

To test the effect of support and the catalyst efficiency, experiments were done with 18% H-ZSM-5 (instead of 80%) dispersed in 82% bentonite. The catalytic bed would have less acidic sites which are known to promote polymerization of the carbohydrates and hence to reduce the hydrocarbon yields. Table 7 shows the results for experiments done on the diluted catalytic bed. Only in the methanol added feeds there was a significant increase in the hydrocarbon yields and a decrease in the tar content. Thus, reducing the number of acidic sites does not seem to reduce the extent of the polymerization of carbohydrates. When the catalytic support material was changed from bentonite to SiO_2-Al_2O_3, the yields of hydrocarbons increased for both cases as shown in Table 8. Deoxygenation of glucose/water feed (20.3/79.7 Wt.%) with Mn and Zn exchanged ZSM-5 are reported in Table 9. There was a reduction in the hydrocarbons yields when compared to similar reactions with H-ZSM-5. The tar contents were as high as those reactions reported before.

The deoxygenation of fructose and its derivative over ZSM-5 catalysts are shown in Table 10. The results obtained are similar to those reported before for glucose and its derivative. As the $(H/C)_{eff}$ ratio increased, the hydrocarbons yields increased and the tar contents decreased.

Discussion

Cyclopentanone can be deoxygenated in high yield to hydrocarbons over H-ZSM-5 above 350°C. The main reaction is a thermal decarbonylation of cyclopentanone to give CO and C_4H_8 fragment that can react further on the catalytic bed to produce aliphatic, aromatic and polyaromatic hydrocarbons. Cyclopentenone with $(H/C)_{eff} = 0.8$ is more difficult to deoxygenated. The addition of methanol raises the $(H/C)_{eff}$ ratio and permits the complete deoxygenation over H-ZSM-5.

TABLE 6

Reaction of glucose and glucose derivative over 80% H-ZSM-5 and 20% bentonite at 450°C

Experimental conditions:			
Reactant composition:	20.3% glucose, 79.7% water,	20.0% glucose 30.0% methanol, 50.0% water	27.6% glucose derivative 60.2% methanol 12.2% water
WHSV (hr^{-1})	0.062	0.195	0.313
$(H/C)_{eff}$	0.0	1.59	1.36
Total product distribution (Wt.%)			
Oxygenated hydrocarbons	---	0.7 *	3.9 **
Tar	65.1	33.3	14.1
CO	2.1	0.5	0.2
CO_2	1.5	1.4	0.9
H_2O	29.1	59.7	66.1
Hydrocarbons	2.2	4.4	14.8
Product selectivity (Wt.%)			
Aliphatics, $C_1 - C_8$	23.6	35.7	50.5
Olefins, $C_2 - C_6$	41.6	3.8	2.7
Aromatics	34.8	53.3	43.2
Polyaromatics***	---	7.2	3.6

TABLE 7

Reaction of glucose and glucose derivative over 18% H-ZSM-5 and 82% bentonite at 450°C

Experimental conditions:			
Reactant composition:	20.3% glucose 79.7% water	20.0% glucose 30.0% methanol 50.0% water	27.6% glucose derivative 60.2% methanol 12.2% water
WHSV (hr^{-1})	0.862	1.195	1.12
$(H/C)_{eff}$	0.0	1.59	1.46
Total product distribution (Wt.%)			
Oxygenated hydrocarbons	0.6 *	1.0 *	4.2 **
Tar	51.3	21.0	11.1
CO	3.8	1.8	0.8
CO_2	2.8	0.3	0.8
H_2O	39.4	57.9	64.3
Hydrocarbons	2.1	18.1	18.8
Product selectivity (Wt.%)			
Aliphatics, $C_1 - C_8$	29.0	43.6	17.5
Olefins, $C_2 - C_6$	7.3	1.2	14.0
Aromatics	45.2	45.3	55.8
Polyaromatics***	18.3	10.0	12.7

* Mainly furan and benzofuran derivatives; ** Mainly acetone and furan derivatives; ***Mainly indene and naphthalene derivatives

TABLE 8

Reaction of glucose and glucose derivative over 18% H-ZSM-5 and 82% $SiO_2-Al_2O_3$

Experimental conditions:		
Reactant composition:	20.0% glucose 30.0% methanol 50.0% water	27.6% glucose derivative 60.2% methanol 12.2% water
WHSV (hr^{-1})	1.18	1.11
$(H/C)_{eff}$	1.59	1.46
Total product distribution (Wt.%)		
Oxygenated hydrocarbons	0.1	0.2
Tar	14.3	6.9
CO	1.6	1.8
CO_2	0.4	0.5
H_2O	62.4	53.5
Hydrocarbons	21.2	37.2
Product selectivity (Wt.%)		
Aliphatics, $C_1 - C_8$	19.0	21.0
Olefins, $C_2 - C_6$	12.0	8.3
Aromatics	65.6	62.0
Polyaromatics	3.4	8.7

TABLE 9

Reaction of 20.3% glucose and 79.7% water over manganes and zinc exchanged ZSM-5

Experimental conditions:		
Catalyst composition	80% Mn-ZSM-5 20% bentonite	80% Zn-ZSM-5 20% bentonite
WHSV (hr^{-1})	0.046	0.055
$(H/C)_{eff}$	0.0	0.0
Total product distribution (Wt.%)		
Oxygenated hydrocarbons	---	---
Tar	72.3	53.1
CO	2.6	2.2
CO_2	1.9	4.2
H_2O	22.3	39.8
Hydrocarbons	0.9	0.7
Product selectivity (Wt.%)		
Aliphatics, $C_1 - C_8$	26.0	35.4
Olefins, $C_2 - C_6$	42.8	2.6
Aromatics	31.0	62.0
Polyaromatics	---	---

TABLE 10

Reaction of fructose and fructose derivative over various ZSM-5 at 450°C

Experimental conditions:					
Reactant composition:	19.8% fructose 80.2% water	19.8% fructose 29.8% methanol 50.4% water	27.6% fructose derivative 60.2% methanol 12.2% water		
Catalyst composition:	80% H-ZSM-5 20% bentonite	80% H-ZSM-5 20% bentonite	80% H-ZSM-5 20% bentonite	80% Zn-ZSM-5 20% bentonite	80% Mn-ZSM-5 20% bentonite
WHSV (hr^{-1})	0.06	0.14	0.22	0.19	0.157
$(H/C)_{eff}$	0.0	1.17	1.49	1.49	1.49
Total product distribution (Wt.%)					
Oxygenated hydrocarbons	0.0	0.1*	3.6**	0.6**	0.3**
Tar	37.3	28.9	14.91	10.7	13.2
CO	2.6	2.8	0.8	1.2	1.3
CO_2	2.2	2.1	1.2	1.6	1.0
H_2O	60.6	58.5	46.2	61.4	58.7
Hydrocarbons	0.7	7.7	33.4	24.9	25.5
Product selectivity (Wt.%)					
Aliphatics C_1-C_8	17.4	21.5	33.2	9.8	8.6
Olefins, C_2-C_6	18.8	14.1	5.6	8.1	9.7
Aromatics	48.5	52.1	54.8	70.7	71.4
Polyaromatic***	15.3	12.3	6.3	11.4	10.2

* Mainly furan and benzofuran derivatives and dimethylether

** Mainly acetone and furan derivatives and benzofurane derivatives and dimethylether

*** Mainly indene and naphthalene derivatives

Both furfural and glycerol can undergo pyrolytic reactions at the temperature studied. Both the pyrolytic products and the feeds can take part in the catalytic deoxygenation over ZSM-5 to produce hydrocarbons. The high tar content observed in many of the reactions reported can be explained by the pyrolytic reactions. At high temperatures, furfural can undergo many thermal reactions. It has been shown (15) that furfural, under acidic conditions, can polymerized by an aldol-type reaction to produce high molecular resin referred as furfural black. The major pyrolytic reaction of furfural is decarbonylation to produce furan and CO as reported in the literature (16) and also observed in most of the reactions of furfural (Tables 2 and 3). Under high temperature and acidic conditions, furan can polymerized to humin (16). Glycerol can undergo dehydration at high temperature to produce acrolein (2-propenal) (17) which can react further to form polymers (polyacrolein) (18) (Tables 4 and 5). Once all these polymers are formed, they remain on the catalytic bed contributing then to the tar contents and the deactivation of the catalysts.

Tables 6 to 10 show that the yields of hydrocarbons obtained for most of the reaction of carbohydrates over ZSM-5 catalysts were low when compared to the yields of tar formed on the reactor bed. The results also indicate that the zeolite catalysts are not responsible for the high tar contents observed, since most of the tars were formed on the top of the catalytic bed. Two possible reasons for the high tar content are the low $(H/C)_{eff}$ ratio in the feed as also reported by other authors (11) and the polymerization of the carbohydrates and their derivatives at temperatures above 150°C. Firstly, if the $(H/C)_{eff}$ is lower than 1, the conversion to hydrocarbon will be small because the main deoxygenation reaction is the elimination of water which is dependant of the availability of the hydrogen in the organic components of the feed (2,4,11). Glucose and fructose and their derivatives with $(H/C)_{eff}$ ratio ≤ 0.7 are therefore expected to give poor hydrocarbons yields (Tables 6-10). However, supplementing these carbohydrates with compounds with high $(H/C)_{eff}$ (e.g. methanol with $(H/C)_{eff} = 2.0$), it is possible to increase the $(H/C)_{eff}$ of the feed. Tables 6, 7 and 10 show the effect of adding methanol to the carbohydrates feeds; there was a simultaneous increase in hydrocarbons yields with increase $(H/C)_{eff}$. Nevertheless, as the increase is small and the tar content still too high, the catalytic upgrading would be difficult in a fixed-bed reactor. Therefore more costly processes such as a fluidized bed system (11a) are necessary. The high tar content can be produced by the decomposition and polymerization of carbohydrates. It is known that glucose can undergo thermal polymerization and decomposition at temperatures higher than 150°C (19). Polycondensation of glucose catalyzed by acid produced polymers (polyglucose) with a wide range of molecular weights. Other non-volatile decomposition products which contribute to the tar content are levoglucosan and 1,6-anhydroglucofuranose (20). One consequence of the non-volatile products on the catalytic bed is that they can block the pores of ZSM-5 and, hence, preventing deoxygenation of the volatile compounds. Thermal decomposition of glucose can also produce volatile products (5-hydroxymethyl furfural, furfural, furyl hydroxymethyl ketone) which can undergo deoxygenation over ZSM-5 catalysts to yield the small amount of hydrocarbons observed. The experimental results suggest that the rate of production of volatile products is much slower than the rate of production of non-volatile products in a fixed-bed catalytic system. Hence, the poor hydrocarbon yield and the high tar content. The derivatives of glucose and fructose give better hydrocarbon yields, probably because of their higher H/C_{eff} ratio and also because most of the hydroxy groups responsible for the formation of polyglucose are blocked. However, the tar contents for these derivatives reactions are still too high.

Conclusion

Cyclopentanone can be deoxygenated with high yield to hydrocarbons over ZSM-5 at 400°C. The addition of methanol to cyclopentenone permits to raise the $(H/C)_{eff}$ ratio of the feed and, hence, its complete deoxygenation. The reaction of furfural and glycerol over ZSM-5 catalysts at temperatures between 400°C and 500°C produces pyrolytic products of different degree of volatility. The volatile fraction is deoxygenated by the catalysts to produce hydrocarbons while the non-volatile fraction remained on the catalytic bed causing the deactivation of the zeolite and the enhancement of the tar content. Glucose and fructose undergo thermal reactions which produce a significant amount of tar and a small amount of volatile products. The volatile fraction is deoxygenatd by ZSM-5 catalysts to produce hydrocarbons.

Acknowledgments

This work was supported by grants from the Natural Sciences and Engineering Research Council of Canada and from the Quebec Ministry of Sciences and Technology.

Literature cited

1. Weiz, P.B., Hagg, W.O., Rodewald, P.G., Science, 1979, 206, 57.
2. Frankiewicz, T.C., U.S. Patent 4 308 411, 1981.
3. Dao, L.H., Haniff, M., Preprint tenth Canadian Symposium on Catalysis, 1986, p. 278.
4. Chen, N.Y., Koenig, L.R., U.S. Patent 4 503 278, 1985.
5. Hasnain, S., Editor Fifth Canadian Bioenergy R&D Seminar; Elsevier: London, 1984.
6. Chornet, E., Overend, R., Editor Compte-rendu de l'atelier de travail sur la liquefaction de la biomasse, 1983.
7. Chantal, P.D., Kaliaguine, S., Grandmaison, J.L., Applied Catalysis, 1985, 8, 133.
8. Dao, L.H., Canadian Patent 1.201.080, 1985.
9. Schirmer, R.E., Pahl, T.R., Eliot Fuel, 1984, 368.
10. Dao, L.H., Hebert, P., Houle, A., Haniff, M., Proceeding of the Ninth Biennal Congress of the International solar Energy Society, 1986, Vol. 3, 1812.
11. a. Chen, N.Y., Walsh, D.E., Koenig, L.R., Preprint Amer. Chem. Soc. Div. Fuel Chem, 1987, 32(2), p. 264.; b. Chang, C.D., Silvestri, A.J., J. Catalysis, 1977, 47, 249; c. Chang, C.D., Silvestri, A.J., U.S. Patent 3,998,898, 1976.

12. Derouane, E.G., Valyacsik, E.W., European Patent 157521, 1985.
13. Shafizadek, F., Fu, Y.J., Carbohydr. Res., 1973, 29, 113.
14. Neuth, F.H., Adv. Carbohydr. Chem, 1951, 6, 83.
15. Kraushaar, B., Kompa, H., Schrolliens, H., Schulz-Eskloff, G., Acta Phys. et Chim. (Szeged), 1985, 31, 581.
16. Acheson, R.M., An Introduction to the Chemistry of Heterocyclic Compounds, Johns Wiley and Sons, NY, 1976, p. 126.
17. Segur, J.B., In Glycerol, Miner C.S., Dalton N.M., Editor Reinhold: NY, 1953, p. 335.
18. Derouane, E.G., J. Catalysis, 1981, 70, 123.
19. Smith, P.C., Guehtlein, H.E., Converse, A.O., Solar Energy, 1982, 28, 41.
20. Houminer, Y., Patai, S., Israel J. Chem, 1969, 7, 513.

RECEIVED March 31, 1988

INDEXES

Author Index

Affiliation Index

Subject Index

Production by Barbara J. Libengood
Indexing by Deborah H. Steiner and Colleen P. Stamm

Elements typeset by Hot Type Ltd., Washington, DC
Printed and bound by Maple Press, York, PA

Recent ACS Books

Biotechnology and Materials Science: Chemistry for the Future
Edited by Mary L. Good
160 pp; clothbound; ISBN 0–8412–1472–7

Chemical Demonstrations: A Sourcebook for Teachers
By Lee R. Summerlin and James L. Ealy, Jr.
Volume 1, Second Edition, 192 pp; spiral bound; ISBN 0–8412–1481–6

Chemical Demonstrations: A Sourcebook for Teachers
By Lee R. Summerlin, Christie L. Borgford, and Julie B. Ealy
Volume 2, Second Edition, 229 pp; spiral bound; ISBN 0–8412–1535–9

Practical Statistics for the Physical Sciences
By Larry L. Havlicek
ACS Professional Reference Book; 198 pp; clothbound; ISBN 0–8412–1453–0

The Basics of Technical Communicating
By B. Edward Cain
ACS Professional Reference Book; 198 pp; clothbound; ISBN 0–8412–1451–4

The ACS Style Guide: A Manual for Authors and Editors
Edited by Janet S. Dodd
264 pp; clothbound; ISBN 0–8412–0917–0

Personal Computers for Scientists: A Byte at a Time
By Glenn I. Ouchi
276 pp; clothbound; ISBN 0–8412–1000–4

Writing the Laboratory Notebook
By Howard M. Kanare
146 pp; clothbound; ISBN 0–8412–0906–5

Principles of Environmental Sampling
Edited by Lawrence H. Keith
458 pp; clothbound; ISBN 0–8412–1173–6

Phosphorus Chemistry in Everyday Living
By Arthur D. F. Toy and Edward N. Walsh
362 pp; clothbound; ISBN 0–8412–1002–0

Chemistry and Crime: From Sherlock Holmes to Today's Courtroom
Edited by Samuel M. Gerber
135 pp; clothbound; ISBN 0–8412–0784–4
